T. Thorell

On European Spiders

Salzwasser

T. Thorell

On European Spiders

1. Auflage | ISBN: 978-3-84605-368-3

Erscheinungsort: Frankfurt, Deutschland

Erscheinungsjahr: 2020

Salzwasser Verlag GmbH

Reprint of the original, first published in 1869.

ON EUROPEAN SPIDERS

BY

T. THORELL.

P. I.

REVIEW OF THE EUROPEAN GENERA OF SPIDERS,

PRECEDED BY SOME OBSERVATIONS ON

ZOOLOGICAL NOMENCLATURE.

WITH ONE PLATE.

(Presented to the Royal Society of Upsala, the 13 Febr. 1869).

UPSALA,
Printed by Ed. Berling.
1869.
S

I.

Among the Faunistic works, that treat of European Arachnida, BLACKWALL'S *Spiders of Great Britain and Ireland* (1861—1864) and WESTRING'S *Araneæ Suecicæ* (1861) undoubtedly occupy the first place, both on account of the high degree of perfection, to which these two works have carried our knowledge of the Spider-fauna of these two countries, and for the light they spread upon a large number of previously unknown or only imperfectly known species of the group of animals whereon they treat.[1]) A comparative examination of these two almost contemporaneous works is in more than one respect a matter of the greatest interest; indeed a comparison of the synonymous denominations of the various species described in them is absolutely necessary, for, as each of these authors appears to have been ignorant of the other's works — even those, which had been published previously to the works above mentioned — it has happened, that a large number of species common to both have been described in each with a totally different nomenclature. The results, to which I have been led by such a comparison, have been the primary cause of my making, and also constitute the principal subject of, the following annotations, in which I have first and principally endeavoured to fix the nomenclature of the spiders known to me, that are described in the works of WESTRING and BLACKWALL, adding such remarks as I supposed to possess any synonymical importance or utility. To these notes on WESTRING'S and BLACKWALL'S works I have added some

1) A faunistic work now in progress, and which, when complete, may worthily take its place beside the works of BLACKWALL and WESTRING, is MENGE'S *Preussische Spinnen*, Danzig 1866—68. Another less extensive but highly meritorious performance is OHLERT'S *Die Araneiden oder echten Spinnen der Provinz Preussen*, Leipzig 1867. To both these works we shall in the following pages often have occasion to refer.

remarks on a third almost contemporaneous performance, SIMON'S *Histoire Naturelle des Araignées* (1864), or, more properly speaking, on the "*Catalogue Synonymique des Aranéides Européennes*" which follows it, for that catalogue appears to me in many points to require a thorough revision, to which I was desirous of offering some sporadic contributions.

As the value of remarks on species of animals and plants is often somewhat equivocal, unless they be accompanied by descriptions or some other guarantee that the species are rightly identified, I consider myself bound to inform my readers, that I can with perfect confidence refer to the descriptions in WESTRING'S *Araneæ Suecicæ*, as really belonging to the spiders declared by me in the succeeding pages to be identical with species described by him. During many years' residence in Göteborg and constant intimacy with this gentleman, my respected teacher and friend, I have had the opportunity of becoming accurately acquainted with by far the greater number of the species described by him in that work, and all Swedish spiders, that I have since collected, I have sent to be examined by him, wherever there was the least doubt about their classification. Moreover the species described by WESTRING, which are wanting in my own collection, I have, with very few exceptions, had the opportunity of examining; some of them have been sent me for examination from the Zoological department of the National Museum in Stockholm by the kindness of Prof. C. STÅL, and others I have received from WESTRING himself.[1]) As I have, in identifying the spiders described by SUNDEVALL, followed WESTRING, who had SUNDEVALL'S own collection at his disposition, and whose determinations of the species found in that author are accordingly perfectly trustworthy, and as I have moreover myself examined a collection of spiders made by CLERCK, and have consecrated a great part of the last twenty years to arachnological researches in just that province (Uppland) of Sweden, where CLERCK, LINNÉ and DE GEER lived and laboured,[2]) I may reasonably make pretensions

1) I take this opportunity of openly expressing my thankfulness not only to Mr. WESTRING, to whom my thanks are more particularly due on account of the numerous and valuable communications that I have received from him concerning our Spider-Fauna, but also to Prof. STÅL, to Dr. HAGLUND and other friends who have sent me the Swedish spiders they had collected. I also beg to express my most sincere gratitude to Prof. LOVÉN and to Mr. AHLSTRAND, Librarian to the R. Acad. of Sciences in Stockholm, for the indefatigable kindness and attention, with which they have procured me the loan of several important works of arachnological litterature, to which I could not otherwise have had access.

2) On the Swedish species of spiders described by the older Swedish arachnologists I have already published the two following works: *Recensio critica Aranearum*

to intimacy with the distinguishing features of most of the species described by both the older and younger *Swedish* arachnologists, and which are taken up in the following work.

As regards *non*-Swedish species, I cannot, it is true, lay claim to the same degree of certainty. By means of the collections of Arachnoidea, which I have formed during several journeys and visits of considerable length to different countries of Europe, (as e. g. many parts of Germany, Switzerland, France and Italy), as well as through presents of various European spiders and other valuable communications from several Zoologists (among whom I may with thankfulness mention the late Prof. AL. V. NORDMANN, Dr L. KOCH, Count E. KEYSERLING, Dr E. OHLERT and Director L. REDTENBACHER), I have however acquired a tolerably good view of the European spider-fauna and have arrived at certainty in several complicated questions of synonymity. My remarks upon non-Swedish species are however confined to such forms as are either generally known or easily determinable, and I have moreover, both a regards Swedish and other spiders, specially noted, by placing an asterisk before the name, all the cases, in which I have not learned by actual inspection to know the species or genus I treat of.

The rules, which I consider ought to be observed in deciding controverted questions of zoological nomenclature, and which I have alluded to and endeavoured to apply in my *Recensio critica Aran.*, are generally in accordance with those laid down in *Annals and Magazine of Natural History*, Ser. 1, Vol. XI, p. 259 et seq. under the title: *Series of Propositions for rendering the Nomenclature of Zoology uniform and permanent, being the Report of a Committee for the consideration of the subject appointed by the British Association for the Advancement of Science.*[1]) These propositions are for the most part merely a repetition or development of the principles already laid down by LINNÉ in his *Philosophia Botanica*, and which FABRICIUS afterwards in his *Philosophia Entomologica* applied to Entomological Nomenclature. Since however my views differ on a few points from those of the British Committee, and since moreover its above cited work is far less generally known than it deserves to be, I think it best here to give a brief account of the rules I have in the following pages applied.

Suecicarum, quas descripserunt CLERCKIUS, LINNÆUS, DE GEERUS (in *Acta Reg. Societatis Scientiarum Upsal.* 1856; and *"Om Clercks Original-Spindel-samling"* [On Clerck'sOriginal Collection of Spiders] (in *Öfvers. af K. Vet. Akad. Förhandl. 1858).*

2) Compare also O. A. L. MÖRCH, Observations on Conchological Nomenclature, *ibid.* 3 Ser., Vol. II, p. 133; ASA GRAY, On Scientific Nomenclature, *ibid.* 3 Ser. Vol. XIII, p. 517.

As fundamental principle we of course go out from the so called *Lex Prioritatis*, which ordains that *each genus preserve the generic name, and each species the specific name, by which it was first made known;* the name of the person, who *first* described or figured such genus or species under the aforesaid name, being added as "*authority*". The reasonableness of this law is so clear and evident, that I should not have thought it necessary to mention it, were it not that there are persons (even among arachnologists), who seem to live in total ignorance of its existence and of every one's duty to conform to it.[1]) Simple as this rule is, some difficulties present themselves in its application, which may give rise to differences of opinion. One might e. g. ask *in what manner* a name ought to be given in order to have the right of being preserved. We conceive that the name ought to be followed by a *definition or characterization of the object named*, i. e. either (which is preferable) a *description* (diagnosis), or a *figure*, or at least a *reference* to some previously existing description or figure; moreover that such name and accompanying characterization ought to be *printed* and *published;* that accordingly no other denominations than those, which have been given in the above named manner, can, in fixing the scientific nomenclature of animals (and plants), be taken into consideration. Hence it follows that no one needs pay any attention either to names published in print *unaccompanied* by descriptions,[2]) nor to denominations given to natural

1) That the scientific names of animals and plants must be *Latin* (i. e. have a Latin form) would seem to be selfevident. Any person then, who describes a new species by e. g. a *French* name only, cannot expect that a such denomination should be respected on the ground of priority. If such names, for example, as *Athèlgue cladophore*, *Prostèthe cannelé* (Vid. HESSE, Mém. sur deux nouv. genres de l'ordre des Crust. sédentaires etc. *in* Ann. des Sciences Nat., 4 Ser., Zool., Tom. 18) are to be generally used, they must first be provided with a Latin form, and the right of *priority* (and *authority*) must be assigned to the work, where these animals are first entered with their Latin denominations. For this reason we consider that e. g. the genus called by LATREILLE in his Cours d'Entomologie, 1831, *Gastéracanthe* and which SUNDEVALL in his Conspect. Arachn. 1833 calls *Gasteracantha (Gastracanthus* WESTW. 1835), must be properly designated as *Gasteracantha* (SUND.) 1833.

2) If, when a genus has been once set up and characterized, there should be given, as a type of it, some species, named indeed, but not farther described, but *concerning which no uncertainty is possible*, I think that also such *specific name* ought to be retained Not only the genus but the species is in this case sufficiently distinguished by the characteristics of the genus. An example of this is afforded by *Filistata testacea* LATR. 1810 (= *F. bicolor* WALCK. 1820—25).

Neither does it appear reasonable, when a species has been described under a new generic name, to reject such *generic name* simply because the characteristics,

objects in *manuscripts* ("in litteris") or in *private* or *public collections* and *museums.*[1]) It follows further, that the date which ought to be accepted as the epoch of a denomination, is *the time when it was in the above manner made public*, but not that, at which it was written down or announced in a verbal lecture, or that at which it was delivered to the editor of a periodical or to some learned Society to be published under their auspices.[2])

---

which distinguish the genus, have not been *separately* set forth in the description. Nevertheless that now very common method of forming new genera is by no means so deserving of commendation and imitation as it is easy and convenient.

To reject a name, as some have proposed to do, on account of *defectiveness* in the definition, would seem not to be right, as leaving room for much arbitrariness. What seems to one good enough may to another appear insufficient or faulty. When one only knows with certainty what is to be understood by such a name, every one can either alter or improve the characterization for himself. A new genus on the contrary, that has been distinguished merely by referring to some particular species of an older genus as it's type, without in any way indicating, which of the characteristics of the species is to be considered as the mark of the new genus, no one can indeed be looked upon as bounden to acknowledge; nevertheless it appears to me advisable to do so, especially if the species referred to deviate in any generally known manner from the typical species of the old genus, and always if the new genus has been once received and acknowledged by a subsequent investigator; the right of priority ought also then to be assigned to him who first proposed the name.

1) One is of course no more at liberty to take an *authority* from such sources than to take a name from them. It is for this reason that e. g. for the names of the spiders described in REUSS' *Zool. Miscell.* (Mus. Senck. I) I always cite REUSS as the authority, although he has in most cases appended "WIDER" to the names: I do not in fact consider myself at liberty to doubt, that REUSS is the author of these *Zoologische Miscellen* and of the descriptions that occur in them, and have nowhere seen it stated that they were written by WIDER. Probably in WIDER's collection and his notes thereupon these spiders have borne the names, under which REUSS has published them.

Again when it is certainly known, that the person, who has published for example a description, is not really the author of the same, then it is evident that the name, that ought to be cited as authorithy for the described genus or species, is that of the *real author.*

2) I am perfectly aware that, especially on this last mentioned point, considerable differences of opinion exist, and that many consider that the priority of a work ought to be reckoned from the day, when it was delivered to the academy or society. Others are of opinion that every separate printed sheet ought to bear the date of its delivery to the press and that from such date priority ought to be reckoned. Against the first of the opinions here urged it may be objected, that it would frequently mislead a person who endeavoured to ascertain the true date, when an observation was first made or a species first described, because an author has frequently the opportunity of making extensive corrections and alterations in his manucript and proofsheets,

This last case it is especially important to take notice of, for a considerable time frequently elapses between the day, when a paper is thus delivered, and that, on which it is made accessible in print to the public; neither ought it to be forgotten that printed works often bear upon their title-page a date different from that, at which they really appeared, and which accordingly ought to stand there. — If a name has been published *without* caracterization, and this latter be supplied in a subsequent work, the name should be considered as originating at the epoch of this latter and not at that of the former work.[1])

Another question requiring an answer is the following: *How far backward in time ought the application of the law of priority to be extended?* — Here it would seem that a difference ought to be made according as the question regards the name of a genus, or that of a species, and the priority of *generic* and *specific* names to be decided independently of each other. Firstly and principally as regards the *names of species*, it will probably without difficulty be admitted, that, since the *Linnæan binominal nomenclature* for all species both in the vegetable and animal kingdoms is that which is universally received, *the introduction of that nomenclature into science* ought to constitute the epoch, from which priority should be reckoned, at least in the case of specific names. The præ-Linnæan authors, as is well known, distinguished the different species of a genus, not by a "*nomen triviale*", as LINNÉ calls the specific names consisting of a single word, which he introduced, but by a brief diagnosis, "*nomen specificum*" or "*differentia specifica*", which generally consisted of several words, though occasionally it might be comprised in but one, and in this latter case assumes to the eye the appearance of a *nomen triviale.* Some modern writers occasionally go back to these præ-Linnæan denominations, and receive the *differentiæ specificæ* that consist of a single word, or even the first word

even until the last sheet leaves the press. We object to both alternatives, that no one is bound to know of a work as long as it, either as manuscript or even as printed, lies concealed in the author's, publisher's, or any learned society's stores. When it *has been made accessible to the public in general*, then, and not previously, can it be said to have been *published.* Many disagreeable controversies concerning the right of priority might be avoided, if it were the general custom to register *conscientiously* upon every work the day *on which it was offered to the public in the booksellers' shops*, or, in cases where no exposure for sale takes place, when the distribution of the work was effected by some other process.

1) According to this rule some of the species-names used by WALCKENAER in his *Tableau des Aranéides* (1805) will have to give place to others, published at a subsequent period.

of those that consist of several, as names of species. This custom we look upon as one, that should altogether be rejected: it is easily perceived that it opens wide the door to unlimited arbitrariness, and that it is incompatible with the fixing of any determinate limit to the application of the law of priority. We assume then as a rule, *that in determining the priority of a specific name notice should be taken only of those works* (or independent portions of works), *in which the now received Linnæan nomenclature is exclusively and consistently employed.* We therefore leave unnoticed; 1°. all works published previously to the year 1751, when LINNÉ'S *Philosophia Botanica* appeared, in which his new system of nomenclature was first fully and distinctly propounded;[1]) 2°. all writings published *subsequently* to that epoch, *in which that nomenclature has either not at all, or not consistently and constantly been employed.*[2]) Such names as *Tarentula Apuliae* ALDROVANDUS (instead of *T. Apuliæ* WALCK.), *Textrix fuliginea* LISTER (instead of *T. denticulata* OLIV.) cannot therefore be received, because both ALDROVANDUS and LISTER lived long before LINNÉ'S time; neither can GEOFFROY, LEPECHIN or GOEZE be cited as authorities for the specific names of spiders, for, although they were acquainted with LINNÉ'S system of nomenclature, the first named author has never used it,[3]) whereas the other two use *in the same work*

1) LINNÉ had, it is true, already in his Academical Dissertation *Pan Suecus* (Amœnitates Acad., II, p. 225 - 262) for the sake of brevity ("ut brevitati studeam" says he) reduced the differentia specifica to a single word: it was however in the *Philosophia Botanica* (§ 257) that he for the first time proposed the laws of his new system of nomenclature: the term *nomen triviale* is here introduced, and it is stated that this *nomen triviale*, or specific name, shall consist of

"Vocabulo uno" and
"Vocabulo libere undequaque desumto",

whereby it's essential difference from the old diagnosis or *differentia specifica* is indicated. — LINNÉ in that work still continues to use the expression "nomen specificum" as synonymous with differentia specifica: and it is in the *Species Plantarum* (1753) that "nomen specificum" first occurs in its now generally accepted signification, i. e. as identical with nomen triviale or species-name.

2) It is however to be remarked (Conf. *Recensio crit. Aran.*, p. 4.) that some authors, and among them LINNÉ himself, have, in works, in which they must still be considered as having consistently employed the binominal nomenclature, sometimes used trivial names compounded of two, usually closely connected words, (e. g. *Carabus crux major* LINN., *Araneus* x *notatus* CLERCK, *Aranea resupina domestica* DE GEER, *Micryphantes ferrum equinum* GRUBE), a custom by no means deserving of imitation. If the two words, of which such a specific name consist, be *not* closely connected, so as to express a single idea (as is the case with "*resupina domestica*" DE GEER), the name ought in all instances to be rejected.

3) Except in the supplement to the 2nd Edit. of his Hist. Abrégée des Insectes.

sometimes nomina trivialia, sometimes verbose differentiæ specificæ to distinguish the species they treat of.[1])

It appears furthermore from this, that we ought not, as in some quarters has been proposed, to fix upon either the *Xth*, or still less the *XIIth Edition of Linné's Systema Naturæ* as the starting point from which priority in specific names is to be reckoned.[2]) For *most* Classes of animals nomina trivialia have been first employed by Linné himself, and that in the *Xth* Ed. of the Systema Naturæ; but this is not the case with all, and as regards *Spiders* in particular, Clerck has already 1757, in his famous work *"Svenska Spindlar, Aranei Suecici"*, applied Linné's nomenclature with perfect consistency, and accordingly the denominations given by him in that work have right of priority in preference to the Linnæan, as I have more fully shown in my *Rec. crit. Aran.* p. 4 et seq.

As regards *generic names* the above named Committee seems to assume, that for them, as for specific names, priority ought not to be reckoned farther back than to the date of *Systema Naturæ Edit. XII* (1767): Sundevall on the contrary considers *Edit. I* of that work (1735) as the limit that ought to be chosen. The most reasonable and consistent method would perhaps seem to be, either to reckon the priority of generic names also *from the epoch of the introduction of the binominal system into the science*, with the same limits, that we have indicated in the case of specific names, i. e. from 1751; or else to take into account, in determining that priority, *all works in which species have been consistently grouped in "genera"*, in the modern meaning of that word, quite as, in determining the priority of specific names, account should be taken of all works, in which nomina trivialia are consistently used. Against the first alternative the important objection may be made, that since in *Botany* a large number of far older generic names has been generally accepted — botanists in fact rec-

1) Gœze has (in "Listers Naturgeschichte d. Spinnen") undertaken to give names to a number of spiders described or figured by some older authors, as Albin and Schæffer; but as among these names some occur of such a form as for example *"Aranea tetra abdomine mucronato"*. *"A. atro-alboque lineata"*, there is surely no reason to burden the lists of synonyms with these names, nor to make any account of them in determining questions of priority. For the names of the spiders figured by Schæffer, priority should be reckoned from Panzer's *Syst. Nomencl.* to Schæffer's Icones Insect. Ratisbon. (1804).

2) The above named British Committee proposes Ed. XII (1767), Sundevall (*in* Årsberättelse öfver Zoologiens framsteg 1840—42) with more reason Ed. X (1758) — as being that in which the binominal system was first applied to both kingdoms of organic nature — as the starting point for specific names.

kon the priority of these names from TOURNEFORT'S[1]) *Institutiones Rei Herbariæ* (1719) — the admission of that alternative would cause too great a difference between the rules of zoological and botanical nomenclature. As for the second alternative, it cannot be thought of for the simple reason, that it would certainly be impossible to determine, when and by whom the term genera, in the sense in which it is now usually understood, was first applied. Now there being in Zoology contemporaneously with TOURNEFORT'S *Institutiones Rei Herbariæ* no such epoch-constituting work to go out from — for it must be admitted that, with respect to nomenclature, that is not the case with the famous works of RAY, — it would seem to be the best course and that which requires the least change in the existing nomenclature, to commence, as SUNDEVALL has proposed, reckoning the claims of priority for *generic* names from LINNÉ'S *Syst. Nat. Edit. I* (1735), the first in a systematic respect epoch-constituting zoological work, subsequent to the time of RAY and TOURNEFORT, and that in which for the first time real genera are arranged and defined consistently throughout the animal kingdom. — Some few zoologists indeed remove the limit of priority to a much earlier period: WILLOUGHBY, RONDELET, ALDROVANDUS, and even ARISTOTELES (who did not write in Latin!) have been cited as "authority" after generic names, although for several af these authors genera, in the modern meaning of the word, had no existence. Moreover it would be a matter of no small difficulty for those, who go back to so remote times, to discover who first employed such generic names as e. g. *Canis*, *Perca*, *Musca*, *Aranea!* — In *Arachnology* the manner in which this question may be determined is fortunately of no consequence, as all the genera comprehended in the classification of Spiders have been formed subsequently to the commencement of the present century.

It follows immediately from the law of priority, that if *the same* name should have been given to two *different genera of animals*, it belongs to the genus *first* described under that name; the other genus receives the next oldest of the names under which it has been made known, or in the absence of such, receives a new name[2]). The same rule of course holds

---

1) "TOUNEFORTIUS primus characteres genericos ex lege artis condidit": LINN., Philos. botan., § 209.

2) If a genus has been described by two different names, and has resumed (or ought to resume) the elder of them, the younger name, or synonym, ought not to be considered as free and unappropriated, and should not be employed as name of any other new genus than one formed by dividing the genus to which it was originally applied.

if two different *species* belonging to the same genus have obtained the same "nomen triviale". If several genera be united in one, that one ought to be distinguished by the name of one of them (preferably the *oldest)*, and on no account be called by a new denomination. And again if one genus be resolved in several, that genus which contains the *typical species* [1]) of the old genus ought to retain the old generic name; the other new genera either receive new names, or (as is preferable) are distinguished by synonyms, if such exist, of the genus, at the expense of which they have been formed. [2]) Entirely to reject the old generic name and form new names for all the new genera that result from the division, is in general a reprehensible course. An exception may be made of the cases in which the old name is an *ordinary nomen appellativum*, which is equally applicable to all the species included under the old name, and is or might be used as the denomination of a whole *Order* or *Class*, as is the case e. g. with the name *Aranea:* [3]) an exception may also be permitted, when the genus divided *does not constitute any natural unity*, i. e. when there is no species that can be considered as typifying it. We cannot therefore complain that such generic names as e. g. *Monoculus* LINN. and *Binoculus* GEOFFR. have been rejected by later naturalists, though we do not mean to maintain that such a step was either necessary or deserving of imitation. [4])

---

1) LINNÉ and FABRICIUS say the *commonest*, "vulgatissima": Phil. bot., § 246; Phil. entom., § 30. As however opinions may be divided as to whether a species be most common in, or typical of a genus, it seems to me desirable, when a genus is divided, and the person, who made the division, has determined for what species he would preserve the ancient name, not to make any alteration in it. Thus for example, although the spider called by SUNDEVALL *Salticus formicarius* is neither the commonest species within the old genus *Salticus* LATR., nor yet typical of that genus, still we retain with SUNDEVALL, who was the first who divided the genus, LATREILLES generic name for just that very species.

2) "Nomina generica, quamdiu synonyma digna in promptu sunt, nova non fingenda": LINN., Phil. bot., § 244. "Antiquum si disjungitur genus, nova nomina effingenda non sunt, quamdiu antiqua adsunt:" FABR., Phil. entom., p. 113, § 28.

3) It will hardly be questioned that it is better with SUNDEVALL to call the order of Spiders *Araneæ* than for instance *Araneides*, an ill-concocted word, that sounds no better than for example *Avides* instead of *Aves* or *Serpentides* instead of *Serpentes!*

LINNÉ even lays down as a general rule, that "Nomina generica, Classium et Ordinum Naturalium nomenclaturis communia, omittenda sunt". (Phil. bot., § 233). This rule however must be considered as bearing with a little modification: at least a generic name cannot (except in the above mentioned cases) be rejected because some Class or Order has *subsequently* received the same appellation.

4) If, on the division of a genus, the nomen triviale of one of the species

Besides the cases here mentioned, in which deviation from the law of priority is necessary or allowable, one more deserves to be noticed. When a word taken from the Latin or any of the more modern languages, and the signification of which is unquestionable, is applied as the scientific name of a genus, which, according to all ordinary rules of etymology, it can by no means indicate, it cannot be other than fitting to reject such generic name and replace it with another. Thus the name *Tarantula* FABR. (1793) e. g. has very properly been generally discarded in favour of the newer name *Phrynus* OLIV.; and the former name is now rightly applied to that genus of Lycosoidæ, which includes the *Tarantula* so often spoken of both by ancient and modern authors.

The names of different genera are often not indeed absolutely identical, but so similar, that it may be doubted whether they can be allowed to remain together or not. It is however only when the names are properly speaking identical, and the difference confined to the spelling, that I have thought it necessary to reject the later name or names: thus for example two such names as *Ariadne* and *Ariadna*, *Galene* and *Galena*, *Sphodros* and *Spodrus* cannot of course be allowed to exist beside one another. Many names differ only in *gender* and in having *different terminations*: and, though one ought of course in future to avoid forming names distinguished only in this manner from others already accepted, it appears to me that, when they have once come into general use, they may be retained; for the opposite course would be attended by too great changes in the existing nomenclature. I do not therefore consider that in the names *Atta* and *Attus*, *Aulonia* and *Aulonium*, *Euryopis* and *Euryopa* and such like, the use of the one name excludes that of the other.[1])

I cannot agree with the British Committee in considering that a known and received zoological generic name ought to be rejected, if it should previously have been used to denote a *botanical genus*, or *vice versa*, as it is scarcely possible that any misunderstanding or other inconveniences can arise from the retaining of such names. The consistent carying out of such a

---

belonging to it be taken as the generic name of that species, it ought no longer to be at the same time retained as specific name, but the species should receive the next oldest specific name, by which it has been described, or, in the absence of of such other name, a new one. Such names for example as *Tarentula Tarentula*, *Trutta Trutta* ought accordingly to be rejected.

1) In some instances persons have taken upon themselves to change whole series of generic names, so as to give them all the *same termination*. Such changes I do not think it worth while to notice.

wholesale doom of cassation would lead to much confusion both in zoological and botanical nomenclature.[1])

It is rightly observed by the British Committee, that a name once published is the property of the science, and cannot therefore be revoked or altered, not even by the person who has imposed it. Exceptions however exist, and we have already (pag. 10, 11) mentioned a couple: the Committee also admits, that there are names which ought unquestionably to be *discarded*, those namely, which in their signification are *absurd* or *false*. It would have been desirable that this sentence of reprobation had been extended also to certain classes of those names which the Committee only considers that naturalists ought in future *to abstain from forming* ("objectionable names"). Such are for instance *mongrel* names (compounded of two or more different languages) — e. g. *Cirrhifera* from κιρρός and *fero* — and names manufactured by mutilating and mangling other names, e. g. *Cypsnagra* from *Cypselus* and *Tanagra.*[2]) To this class belong also the equally barbarous denominations that have arisen from the ridiculous practice of composing unmeaning generic names of arbitrarily combined letters, usually in the form of an anagram: e. g. *Rocinela*, *Conilera*, *Cirolana*, *Anilocra*, formed from the letters in *Carolina*. We hope the time will come when also such names as those just mentioned will be *rejected*,[3]) though this is not yet the case. But certainly

1) LINNÉ is even more severe than the British Committee in this matter: not only will he not permit the same generic names to be used in botany and zoology or mineralogy (Phil. bot., § 230), but he even adds: "nomina Generica cum Anatomicorum, Pathologorum, Therapeuticorum, vel Artificum nomenclaturis communia omittenda sunt": ibid., § 231. FABRICIUS lays down the same rule (Phil. entom., § 21, p. 108); but it would be vain now to attempt to get it applied. — Some modern authors have gone into the opposite extreme, and maintain that two or more genera of animals ought to be allowed to have the same name, if only they do not belong to the same *Order*. This assumption is in direct opposition to the hitherto universally received praxis in most branches of zoology. In arachnology e. g. the names *Lycæna*, *Hecaërge*, *Macaria* have been discarded, because these generic names had been previously given to animals of another *Class*.

2) Some other equally ill compounded names have very properly been discarded by more modern zoologists. Thus for example SUNDEVALL has rechristened the bird-genus *Malherbipicus* (from MALHERBE, the ornithologist, and *picus*) *Pediopipo* [Conspectus Avium Picinarum p. 77 (1866)]; and GÜNTHER [Catalogue of the Fishes in British Museum, Vol. V, p. 387 (1864)] has changed into *Coccia* the crazy name *Ichthyococcus*, given to a genus of fishes in honour of an Italian ichthyologist, and compounded of his name, COCCO, and *ἰχθύς!*

3) In confirmation of this view I beg to adduce the following citations:

one ought to be at liberty to *amend* such in other respects appropriate denominations as are in a less serious degree *erroneously formed.* This right — which is far from being universally acknowledged, although defended and used by several good zoologists — ought to belong not only to the person who first published the name, but also to every one who observes and *can* correct the error. That e. g. LATREILLE changed his *Micromata* to *Micrommata*, and the absurd name *Clubiona lapidosa* WALCK. to *C. lapidicola*, and that these latter appellations have been generally received, ought therefore to be approved; and in virtue of the same right we alter e. g. WALCKENAER'S *Drassus rubrens* to *D. rubens* (as in fact MENGE and OHLERT have already done), his *Epeira myabora* to *E. myiobora*, *Latrodectus* to *Lathrodectus*, *Lipistius* to *Liphistius* (λείπω, ἱστός), *Deinopis* to *Dinopis*, as also it is now usual to write *Loxia pityopsittacus*, *Hyponomeuta*, *Histiophorus*, *Chiromys* etc. instead of *L. pytiopsittacus*, *Yponomeuta*, *Istiophorus*, *Cheiromys.* The right of making such improvements must be granted, in order to prevent the nomenclature of zoology from gradually assuming an appearance absolutely disgusting to a person possessing even the slenderest classical attainments. As long as the scientific names of animals and plants are to be *Latin*, we have a right to require that they do not sin against the simplest laws of that language. One is not, it is true, obliged to learn Latin and Greek in order to occupy oneself with Natural History: we are fully aware that a man may be a very distinguished naturalist without having had a classical education; but he who does not know sufficient Greek and Latin as to be able of himself to compound a scientific name for an animal or plant, might surely obtain the assistance of some more competent individual, if he find himself under the necessity of imposing a name. As most *generic* denominations are derived from the *Greek*, it follows, that it is principally words drawn from that language, that, in the process of composition and reduction to the Latin form, are most frequently subjected to barbarous misusage. Without

"Nomina generica ex vocabulo græco et latino similibusque hybrida, non agnoscenda sunt." LINN., Phil. bot., § 223. Conf. FABR., Phil. entom., § 18, p. 107.

"Nomina generica ex uno vocabulo plantarum generico fracto, alio integro composita, Botanicis indigna sunt." LINN., Phil. bot., § 224. — "Per anagramma orta non placent." SPRENGEL, in LINN., Phil. bot., Ed. 4, § 229.

"Nomina barbara, quæ quidam in Entomologia in novissimis temporibus introduxerunt, omnino rejicienda, quum nullo modo intelligantur et difficile pronuntientur." FABR., Phil. Ent., p. 109, § 24. — LINNÉ even says that all generic names should be rejected, "quæ a lingua græca vel latina radicem non habent" (Phil. bot., § 229), but he has not himself strictly adhered to this rule, and it would now be impossible to get it acknowledged and consistently carried out.

exposing ourselves to the charge of pedantry, we may at least venture to urge, with regard to them, the observance of the two following simple rules: 1.° *If a name be formed of two or more Greek words, these ought to be put together according to the simple rules for the formation of Greek compounds*; and 2.° *When the Greek word is transmuted into a Latin form, it ought to receive a Latin termination, and the Greek letters ought to be replaced by such Latin letters, as correspond to them in the works of Roman authors.*[1])

Now since a great number of names are in this respect most erroneously formed, I consider it not only as a right, but as a duty to correct them, e. g. to correct *Uptiotes* (from ὕπτιος) to *Hyptiotes*, *Megamyrmækion* (μεγαμυρμηκίον) to *Megamyrmecium*, *Ozyptila* to *Oxyptila*, *Arkys* to *Arcys*, *Deinopis* to *Dinopis*, and so forth. Such corrected names ought not to be considered as new, but to preserve their original rights, and be followed by the name of the individual, as authority, who first formed the name. Even names (at least generic names) formed of words taken from *other languages*, or of *proper names*, ought to be furnished with a Latin termination and, as far as is possible, with a Latin orthography. Generic names of unknown or uncertain etymology, but generally known and accepted, (e. g. *Epeira, Filistata, Clubiona)* must not be altered; and in general more freedom may be allowed in the formation of generic than of (adjective) specific names,[2]) which latter ought always to be in full conformity with the rules of etymology.

With the help of the rules above stated I have here endeavoured to fix the original *specific names* of number of spiders admitted into the works of WESTRING, BLACKWALL, and SIMON, as well as those of some other European spiders known to me, and to correct such errors of synonymism as have crept into the works of these authors. Ar regards the *genera*, I have endeavoured to restore also to them their original denominations, where they have been dispossessed by subsequent ones; as regards however the *bounds* and *compass* of the various genera, such difference of views prevails, that I cannot of course hope to gain for my own opinions on this subject more than a partial assent. In the case of the *families*, the law of priority is, as

1) "Nomina generica latinis literis pingenda sunt". LINN., Phil. bot., § 247.
"Sonus nominum, quantum fieri potest, facilitandus, ideoque nec græca nec barbara admittimus; et terminationem græcam in latinam mutamus". FABR., Phil. ent., p. 114, § 31.

2) We have accordingly, for example, preserved unaltered the *termination* in *Lathrodectus* (from λάθρᾳ and δήκτης) and *Epesinus* (ἐπισινής), though unquestionably *Lathrodectes* and *Episines* is the proper orthography; neither have we adopted corrections, which would greatly alter the appearance of the word (e. g. *Oops* instead of *Oonops*).

is generally known, not applied, and I have accordingly, in conformity with the practice of most modern arachnologists, adopted SUNDEVALL'S denominations, derived from the most prominent genus within the family, employing however the termination *-oïdæ*, as being more etymologically correct than *-ides* or *-idæ*. — I have no doubt in many points been guilty of real mistakes, but I venture nevertheless to hope for a mild judgement from persons acquainted with the subject, who are aware of the difficulties to be encountered in a work like the present. I ought especially to remark, that I have been unable to determine with certainty the exact date of the publication of some of the arachnological works here cited; this has been especially the case with a couple of works published in numbers without date, as also with some papers published in periodicals. As regards these latter, I have in dubious cases assumed the year for which the periodical is published, as the date of the articles it contains, though in many instances this may not be right, because the latter numbers of a journal commonly appear the year following. When the year of a work's printing is expressed, I have of course accepted that as the date of publication, whenever I did not know with certainty that such date was incorrectly given [1]).

In restoring the first or original specific names I have endeavoured to observe all the cautiousness so necessary in such a process. The species of the older writers are, as is well known, often difficult, sometimes impossible to determine with certainty: with respect to them I have, in applicable cases, laid it down as a rule to *preserve the determinations accepted by modern arachnologists who have lived in the country where the species described by the author in question have been collected.* It is evident that a *French* naturalist has the best opportunities for studying the *French* spiders described by FOURCROY, DE VILLERS, LATREILLE etc., a *German* the German species of SCOPOLI, FABRICIUS and PANZER, and so forth, as also we *Swedes* ought to be best acquainted with the *Swedish* forms described by CLERCK, LINNÉ and DE GEER. *Tradition* has here a significancy that must not be undervalued. It is only in cases in which I have supposed myself able to show that an evident mistake has been made, that I have deviated from this rule [2]).

1) This is for instance the case with WALCKENAER'S Hist. Nat. d. Ins. Apt., Tome II, which bears on its titlepage the date 1837, but did not come out till 1841.

2) Regarding the rules, which, in determining the species of the older authors, ought in doubtful cases to be applied, I beg to cite the following from Rec. Crit. Aran.: "... maximi nobis esse momenti crediderim penitus cognovisse, quæ formæ in iis regionibus gignantur, ubi vixerit et animalia collegerit scriptor, cujus species sint

As a complete registration of synonymous generic and specific names does not enter into the plan of these remarks, I have in general taken up in the *lists of synonyms* only such citations, as were necessary to show the origin and date of the various denominations given to each species and genus: I have however frequently, in the case of specific names, also referred to some work where the species in question is fully and unmistakably described or figured, as also invariably to WESTRING'S and BLACKWALL'S great works, and, for genera, to SIMON'S. Names from mere *lists* of species I have only in a few cases admitted among the synonyms, for in most instances we are destitute of all guarantee that such names really correspond to the species, to which the names properly belong. The common synonym for a number of generic names, *Araneus* CLERCK, *Aranea* LINN. (and other authors), I have not considered it necessary to include, neither have I in the synonyms for the genera formed at the expense of the old genera of LATREILLE and WALCKENAER, admitted these, unless it, for some especial reason, appeared to me desirable. When I have admitted into the lists a synonym, which I look upon as uncertain, I have placed before it a?

---

definiendæ; exclusis enim omnibus, quæ ibi non reperiuntur, ita sæpe minuitur et circumscribitur numerus formarum, in quibus dubitetur, ut nullo interdum negotio vere judicare possimus . . . . At si qua descriptio, licet hoc modo intra terminos quosdam coërcita, tamen in duas vel plures species æque quadrat, nec scriptoris verba vel figuræ ullam ansam ad unam earum, rejectis aliis, eligendam præbeat; nec denique ab omnibus receptum est, nomine veteri speciem quandam *ex iis, in quibus dubitatur*, significare; tum ita equidem censeo, quæ ex iis in provincia vel in patria illius scriptoris maxime sit vulgaris, eam nomine, quo ille usus sit, esse appellandam. Eodem quoque modo judicandum est, quum evidenter apparet, scriptorem aliquem duas vel plures diversas species confudisse: nisi si *figuram* addiderit, quæ unam earum manifeste repræsentet; tum enim nomine, quod ille adhibuerit, hanc speciem vocare, satius mihi videtur. — — — Quum autem in uno eodemque opere varietates ejusdem speciei ut diversæ species descriptæ et nominatæ sunt, difficile interdum videri potest judicare, ex nominibus datis quod retinendum sit et speciei imponendum. . . . . Definiendum est, quæ sit *forma principalis* sive *primitiva*, cujus nomen sibi adsciscat species necesse est, et cujus varietates igitur reliquæ sunt habendæ. *Forma* vero *principalis* ea existimanda est, quæ frequentissime invenitur in patria ejus, qui primus nomina, de quibus agitur, dedit. Si id dijudicari non potest, vel si apparet, scriptorem illum veram formam principalem non cognovisse: tum primum ad alias rationes est confugiendum, et ex nominibus, *quæ dederit*, id eligendum, quod exempli gratia magis quam reliqua *in hac specie* tritum atque usitatum sit, vel quod magis aptum et idoneum videatur — et id genus alia."

When a species, concerning which there is a difference of opinion, is not found in the country where the describer resided, it is evident that what has here been said of that country, must be considered as holding good for the locality where the species in question was taken.

In conformity with an alternative proposed by the British Committee, I have, in this as well as in my previous works, in giving the authority for a *generic name*, placed the author's name *within parentheses* whenever the limits of the genus received by me are different from those of that author, but *without parentheses* when the genus is considered as possessing its original compass. If I wish to indicate that a genus ought to be taken in the meaning proposed by some *other* particular author, I have usually added the name of that author after that of the original describer. Thus *Epeira* WALCK. signifies the genus *Epeira* as limited by WALCKENAER, who first set up that genus. *Epeira* (WALCK.) is the same genus, but with *different* limits; *Epeira* (WALCK.) WESTR. (sometimes, but only for the sake of brevity, *Epeira* WESTR.) means the same genus with the limits assigned to it by WESTRING.

After a *complete* name (including both the generic and specific names) the authority has been placed *without* parentheses, when the species occurs under the same both generic and trivial name in the author cited, but *within* parentheses, when the generic name used by him is different. I write, for example, *Epeira angulata* (CLERCK) with, but *E. adianta* WALCK. without parentheses. If a specific name appear to be not fully certain, I have generally placed after the authority cited for it the name of some other author, in whose works it indicates the species I refer to. *Erigone rufipes* (LINN.) SUND. thus indicates the spider, which SUNDEVALL describes as the *Aranea rufipes* of LINNÉ [1]).

---

1) The ordinary custom in botanical works, of appending as authority to the complete name of a species the name of the author, who first employed the whole name (both generic and specific), and of passing in silence over the writer, who first made known the species, if he should have used another generic name, has not been much followed by zoologists. What advantages that custom can offer, I am unable to discover. By the opposite method of notation, adopted by me and by most zoologists for indicating authorities, one obtains reference to the epoch, when the species was first made known, and from which the priority of the name is to be reckoned, and that is, I suppose, in most cases the main advantage gained by appending an authority. This method does not of course prevent the citation of a later author after a complete name, if the occasion be such as to require the making of a distinction between his description or figures of the species and those of others. We may accordingly very well speak e. g. of *Epeira quadrata* WESTR. and *Epeira quadrata* KOCH, though at the same time we assume, that the *Epeira quadrata* of both these writers ought in strict propriety to be called *Epeira quadrata* (CLERCK).

A † placed before a generic name indicates that the name, as being previously engaged or found unsuitable for some other reason, has been abandoned in favour of some younger name; this mark, when placed before the complete name of a species, has the same signification with respect to the specific name.

A date placed after a generic name indicates the year, in which that genus was made known and defined; after a *complete* or *specific* name it has the same signification with respect to the *specific* name. The addition of these dates to the names I look upon as of the greatest utility for preserving the proper denominations of the various species.

Instead of setting forth the observations I have thought fit to make on the *genera* (and families) recognized by our three authors, in the order in which they appear in their works here referred to, and mixed with disquisitions respecting the species, I have preferred to treat these larger groups separatety. I have therefore first made up a systematical list or review of the sub-orders, families, sub-families and genera of European Spiders recognized by me. Each generic name is accompanied by the name of the author, who first published it, and the year when this took place; moreover by its etymological derivation, its synonyms, and the name of the species that typifies the genus; and lastly are subjoined such synonymical and critical remarks as I have thought appropriate. In almost all the genera which I have had the opportunity of examining, I have subjoined a short description of the form and armature of the tarsal and palpal *claws*, which organs have not yet attracted all the notice they appear to deserve [1]). — To this list, when in going through our authors, I have come upon the different genera, I have always subsequently referred.

Under the head of each family I have introduced a short account of the characteristics of the sub-families and genera it comprises. These characteristics I have endeavoured as far as possible to derive from the number and position of the eyes and the form of the organs of the mouth, partly because such distinctive features are easily verified, partly because they are most generally (often too exclusively) used, at least in determining the limits of the *generic* groups. But I have also endeavoured to make use of the different form and number of the spinners, of differences in the conformation of the cephalothorax and abdomen, in the relative lengths and

1) As regards the claws of spiders I may refer to Savigny's admirable figures in "Déscription de l'Égypte", as also to Ohlert's important treatise: "Beiträge zu einer auf die Klauenbildung gegründeten Diagn. u. Anordn. der Preuss. Spinnen."

armature of the legs, the number of claws on the tarsi, etc. Genera, which rest exclusively on such characteristics as belong *only to one sex*, leaving the other undetermined, I have not adopted, but consider that they ought to be unreservedly rejected. I ought to call especial attention to the circumstance, that *exotic* forms have not been taken into consideration in the formation of these schematic reviews, which accordingly can be used as a clew in classifying such species only, as belong to the *European* fauna. The characteristics of the *sub-orders*, as they cannot be expressed in few words, and indeed may be considered as generally known, I have not thought it necessary to repeat, but refer for them to e. g. LATREILLE'S, SUNDEVALL'S, WESTRING'S and OHLERT'S works.

In the catalogue of *arachnological litterature*, with which I have opened this treatise, I have included all the works known to me on *now existing* European spiders, of a *descriptive*, *systematical* and *zoo-geographical* character, with the exception however of such writings as belong to the *præ-Linnæan* period, of which only a small number of works, referred to in the following pages, have been admitted. Works belonging to that period, among which I also reckon writings of later date, in which LINNÉ'S binominal system is not fully adopted, are in the list marked with a †. Moreover for reasons, that are easily understood, zoological handbooks and compendia, in which no new facts relating to our subject are communicated, have been excluded. Of works which as regards spiders contain only notices of their anatomy, economy etc., I have taken into my list only those, which I have occasion in this work to cite, and they are distinguished from others by their titles being included in brackets []. Of the litterature that treats exclusively of *exotic* spiders, I have similarly admitted only such works, as I had occasion to cite. Their titles are printed in smaller types. Some of the works in the catalogue I have not myself had the opportunity of consulting; these are marked with an asterisk, and whenever I have been obliged to quote such a work, I have always indicated the source from which I have derived the citation [1]).

Before proceeding to the special examination of the three works before us, I ought perhaps to give a short general account of each of them.

1. WESTRING'S *Araneæ Suecicæ* contains complete and accurate descriptions of 308 species of Spiders found in Sweden and Norway, of which 34 belong to the family *Epeiridæ*, 115 to *Therididæ*, 63 to *Drassidæ*,

1) For rectifications or additions to this catalogue, either privately or publicly communicated, the author will feel very thankful.

30 to *Thomisidæ*, 35 to *Lycosidæ* and 31 to *Attidæ*. Not only the species, but also the genera and families are in this work characterized in detail: by this the author has been enabled to avoid the error of taking up in the description of the species a number of distinctive marks common to whole series of species, an error, which makes the descriptions given by sundry other writers so deficient in characteristics, in spite of their often wearisome prolixity. WESTRING has succeeded in finding sharp and certain distinctive marks for the species he describes: we would especially call attention to the excellent characters he has discovered in the different number and distribution of the *spines* on the extremities. Equally important are the distinguishing features pointed out by WESTRING in his detailed descriptions of the males' palpi: nor has he quite overlooked the circumstance, that similar sharp characteristics may be found by studying the external sexual organs of the females. What immediately strikes a reader on looking into WESTRING's book, is the singular diligence and conscientiousness that it evinces: his descriptions have not been made independently of each other, they have not been written down once for all as the different species came under the author's eye, but they are the result of most accurate and many times repeated comparisons of the various species [1]). They have thus become

1) WESTRING's rigorous comparative treatment of the Swedish spiders has been considerably facilitated by the method in which his collection of spiders is preserved. He does not preserve his spiders in spirits, but impaled upon pins, after having first been dried by a process invented by himself and described with full details in his paper: "Anvisning att ändamålsenligt insamla och conservera Arachnider, förnämligast med afseende å spindlarne." We shall very briefly describe this method, which it is true at first seems difficult and tedious, but which one, after a little practice, finds as easy as it is appropriate. It is a characteristic of the method that the *spider's abdomen*, and that part *only* of ist body, is *hardened by heat*. The following simple instruments are required for the operation: 1:o, a glass cylinder of about 1^{in} or $1\,{}^{1}/_{2}{}^{in}$ diam. and about 4^{in} long, one end of which is closed with a cork: in this cylinder the spiders abdomen is hardened over the flame of a candle; 2:o, a small and very fine pair of scissors, as also a stronger and coarser pair: with the former the abdomen is cut off, with the latter the pin, which is used as a spit; 3:o, a little cylindrical shaft encircled at the one end by a cylindric metal ring filled with a cork, in which cork the spit is fastened during the operation; 4:o, a fine pair of tweezers, and a few small slices of cork about 2 lines thick, insect-pins, blotting-paper, and a lighted candle. When the spider has been in a proper manner killed (e. g. by vapour of ether or by heat) it is to be impaled on an appropriate insect-pin passed through *the right side* of the cephalothorax; the abdomen is then cut off (the animal being holden in the half-closed left hand, in which the abdomen, on being separated, falls) close to the cephalothorax, and the incision is dried with blot-

strictly *comparative*, a quality we do not often meet with in the descriptions of this group of animals; and we probably do not say too much when we assert, that WESTRING surpasses all his predecessors in the accuracy and sharpness of his descriptions, and that his work, in its descriptive character — if we overlook the occasionally somewhat lengthey diagnoses [1]) — may be considered as a model for those who come after him.

---

ting-paper. The head of another insect-pin of about the same substance having been cut off, the blunt end is introduced into the severed abdomen (through the opening caused by the abscission) up to the spinners, and is fastened by its point into the above-named shaft. By holding the pin a moment in the flame of the light, the abdomen is easily made to sit fast upon the little spit. The glass cylinder is then taken in the left hand and holden horizontally over the flame; with the right hand the spider's abdomen is introduced into the open end of the cylinder, and holden there immediately over the flame. In consequence of the heating of the air in the cylinder, the abdomen is gradually hardened, under which process it must be turned on all sides and brought nearer to or removed farther from the heated glass as occasion may require; but care must be taken not to employ too great a heat, as the abdomen would then be burned or crack, nor too small a heat, as the skin would in that case wrinkle and collapse. One must every now and then try with a fine needle whether the abdomen be everywhere firm so as not to yield to pressure: and the hardening process must be continued till this is the case. The pin (spit) is now cut off obliquely (so as not to be too blunt), at such distance, that a portion of about $^3/_4$ the length of the cephalothorax is left standing out from the abdomen. By means of the tweezers this portion of the pin fixed in the abdomen is introduced into the cephalothorax through the opening made by clipping the petiolum. When the abdomen and cephalothorax have been thus reunited, and placed in their natural position, the pin for mounting the spider is stuck perpendicularly into a slice of cork, so that the spider remains at a short distance from the cork; the legs are extended and fastened by means of pins in their natural position (as in the specimens in my collection), or else somewhat bent under the body (as in WESTRING's collection, in which case they are not so easily broken off); in this condition the animal must remain in a dry place, until the cephalothorax and legs are completely dry, when it is ready to be placed in the collection. Spiders thus prepared are as easily and conveniently examined as insects impaled in the usual manner; but if one has besides a collection in spirits, so much the better. Very few species (e. g. some of the genus *Xysticus*) lose a little of their colour in hardening: nearly all others, if rightly manipulated, remain entirely unchanged.

1) Originally diagnosis was looked upon as synonymous with "differentia specifica", i. e. a *definition* comprising the marks necessary and sufficient to distinguish the species from all other species belonging to the same genus. But such a definition is possible only when *all* the species of the genus are known, which is far from being always the case: and, in the case of genera containing many species, at any rate such definitions would mostly be too long to be of any great practical utility.

WESTRING has throughout consistently endeavoured to apply the law of the priority of names, and it is therefore only in consequence of his not having had access to certain portions of arachnological litterature, that he has, instead of the oldest and therefore right names, occasionally used newer appellations, not only for a number of species, but even for certain genera. But to this we shall have occasion hereafter to return.

The remarks we have to make against WESTRING'S work are not many. It may be mentioned as an imperfection, that the author has paid no attention to those characteristics, the examination of which requires the aid of the microscope, and some of which, e. g. those derived from the structure of the spinners and the claws, are by no means unimportant either in classification or specific description. A somewhat more detailed account of the different species' of spiders occurence, economy, industry, etc., than what the author has furnished, would have been acceptable, and might also certainly by him, who for so long a series of years has devoted his attention to that group of animals, easily have been supplied.

As regards the *families* into which WESTRING has distributed the Swedish spiders, they are, as corresponding with the Latreillean family-groups (by me considered as sub-orders) very natural, but might perhaps at least in part be resolved with advantage into several, as is particularly the case with the *Drassidæ* WESTR., which most modern authors divide into three or more separate families. With regard to the division of the families into *genera*, the author appears in general to have hit upon the right mean course between too strict an adherence to the views of older systematizers and the occasionally over minute subdivision of genera, such as has been introduced into the territory of arachnology by for instance MENGE; nevertheless it appears to us, that some of the older genera preserved unchanged by WESTRING, e. g. *Theridium*, *Philodromus*, *Lycosa*, *Attus*, might well have borne with some division, as well as *Epeira*, *Clubiona*, *Drassus*, etc., which he has divided into several smaller generic groups.

To facilitate comparison between the Spider-fauna of the Scandinavian peninsula and that of Great Britain and Ireland, as they appear in the

---

We accordingly find in descriptive works of moderate bulk the diagnoses generally so expressed, that they serve to distinguish only those species of the genus, that are immediately under treatment, and have therefore no other object than to *facilitate the determination of an unknown species.* But for that purpose — the only one which in a diagnosis needs be considered — it needs not be very verbose, not even in very large genera, if nota bene these genera are duly subdivided into smaller easily distinguishable groups.

works of WESTRING and BLACKWALL here referred to, a tabular view is here given of the number of species belonging in these countries to the different families and genera of the order of spiders, in which I have followed WESTRINGS system, and endeavoured to aggregate to the genera and families adopted by him, such species as by BLACKWALL have been otherwise classified. In the case of certain species among these, to me unknown, I have however been unable with full certainty to determine the corresponding genus in WESTRING'S system. This has been especially the case with several of the species comprehended by BLACKWALL, in the genus *Neriene*. Most of the species in that genus belong indeed to WESTRING'S *Erigone;* nevertheless it is probable, that some more than the few (6) species that I have reckoned to *Linyphia* WESTR., ought to be referred to this genus: perhaps also one or two *Neriene*-species belong to WESTRING'S *Theridium*. Being unable to come to any certain conclusion in this matter, I have aggregated to the genus *Erigone* WESTR. all the species (about 20 in number) of the genus *Neriene*, of whose place in WESTRING'S system I felt uncertain.

	Sweden and Norway.		Gr. Britain and Ireland.	
EPEIRIDÆ WESTR. (= *Epeiridæ* BLACKW. et *Ciniflonidæ* BLACKW. ad part.				
Epeira WESTR. = *Epeira* BLACKW. [1]) ad maximam partem	19	—	19	—
Singa WESTR. = *Epeira* BLACKW. ad partem	5	—	4	—
Zilla WESTR. = *Epeira* BLACKW. ad partem	3	—	3	—
Meta WESTR. = *Epeira* BLACKW. ad partem	4	—	5	—
Tetragnatha WESTR., BLACKW.	2	—	1	—
Veleda BLACKW. (*Uloborus* LATR.)	—	—	1	—
Mithras WESTR.	1	34	—	33
THERIDIDÆ WESTR. (= *Therididæ*, *Linyphiidæ* et *Scytodidæ* BLACKW.)				
Linyphia WESTR. = *Linyphia* BLACKW. [2]) ad max. partem + *Neriene* BLACKW.[2]) ad part. + *Theridion* BLACKW.[2]) ad part.	30	—	38?	—
Tapinopa WESTR. = *Linyphia* BLACKW. ad part.	1	—	1	—
Pachygnatha WESTR., BLACKW.	3	—	3	—
Ero WESTR. = *Theridion* BLACKW. ad part.	2	—	1	—
Theridium WESTR. = *Theridion* BLACKW. ad max. part.	24	—	25?	—
Episinus WESTR. = *Theridion* BLACKW. ad part.	1	—	1	—
Erigone WESTR. = *Walckenaera* BLACKW.[2]) + *Neriene* BLACKW. ad max. part.	53	—	74?	—

1) BLACKWALL mentions 31 species of the genus *Epeira*.

2) The genus *Linyphia* has in BLACKWALL 33, *Theridion* 27, *Neriene* 48 and *Walckenaera* 32 species.

	Sweden and Norway.		Gr. Britain and Ireland.	
Pholcus WESTR., BLACKW.	1	—	1	—
Scytodes BLACKW.	—	115	1	145
DRASSIDÆ WESTR.				
(= *Drassidæ*, *Ciniflonidæ* ad max. part., *Agelenidæ* et *Dysderidæ* BLACKW.				
Segestria WESTR., BLACKW.	2	—	2	—
Dysdera WESTR., BLACKW.	1	—	3	—
Schænobates BLACKW.	—	—	1	—
Oonops BLACKW.	—	—	1	—
Tegenaria WESTR. = *Tegenaria* BLACKW. ad max. part.[1])	3	—	3	—
Agelena WESTR. = *Agelena* BLACKW. ad part.[1])	1	—	3?	—
Textrix WESTR., BLACKW.	1	—	1	—
Agroeca WESTR. = *Agelena* BLACKW. ad part.	1	—	1	—
Hahnia WESTR. = *Agelena* BLACKW. ad part. + *Tegenaria*			—	—
BLACKW. ad part.	3	—	4	—
Apostenus WESTR. = *Agelena* BLACKW. ad part.	1	—	2	—
Cœlotes BLACKW.	—	—	1	—
Zora WESTR. = *Hecaërge* BLACKW.	1	—	1	—
Phrurolithus WESTR. = *Drassus* BLACKW. ad part.[1])	2	—	1	—
Micaria WESTR. = *Drassus* BLACKW. ad part.	4	—	2	—
Drassus WESTR. = *Drassus* BLACKW. ad part.	7	—	3?	—
Pythonissa WESTR. = *Drassus* BLACKW. ad part.	4	—	1	—
Melanophora WESTR. = *Drassus* BLACKW. ad part.	5	—	3	—
Drassodes WESTR. = *Drassus* BLACKW. ad part.	4	—	4	—
Argyroneta WESTR., BLACKW.	1	—	1	—
Anyphæna WESTR. = *Clubiona* BLACKW. ad part.[1])	1	—	1	—
Amaurobius WESTR. = *Ciniflo* BLACKW.	2	—	5	—
Cheiracanthium WESTR. = *Clubiona* BLACKW. ad part.	2	—	2	—
Dictyna WESTR. = *Ergatis* BLACKW.	5	—	3	—
Clubiona WESTR. = *Clubiona* BLACKW. ad max. part.	10	—	9	—
Sparassus WESTR., BLACKW.	2	63	1	59
THOMISIDÆ WESTR., BLACKW.				
Thomisus WESTR., BLACKW.	18	—	19	—
Philodromus WESTR., BLACKW.	12	30	9	28

1) *Tegenaria* in BLACKWALL comprehends 4, *Agelena* 9, *Drassus* 13, and *Clubiona* 12 species.

	Sweden and Norway.		Gr. Britain and Ireland.	
LYCOSIDÆ WESTR., BLACKW.				
Lycosa WESTR., BLACKW.	32	—	16	—
Dolomedes WESTR. = *Dolomedes* BLACKW. ad part.[1]	1	—	2	—
Ocyale WESTR. = *Dolomedes* BLACKW. ad part.	1	—	1	—
Sphasus WESTR., BLACKW.	1	35	1	20
ATTIDÆ WESTR. (= *Salticidæ* BLACKW.)				
Salticus WESTR. = *Salticus* BLACKW. ad part.[2]	1	—	1	—
Attus WESTR. = *Salticus* BLACKW. ad max. part.	30	—	16	—
Eresus BLACKW.	—	31	1	18
MYGALIDÆ BLACKW.				
Atypus BLACKW.	—	—	1	1
	—	308	—	304

A glance at the foregoing table shows, what is remarkable enough, that the number of species of spiders observed in Sweden and Norway on the one side and Great Britain and Ireland on the other is very nearly the same, a little above 300. As however the last mentioned countries, from their more southerly position and warmer climate, ought probably to possess a richer spider-fauna than our peninsula, one may perhaps asssume, that this latter has been more carefully scrutinized with respect to its arachnology than the British Isles, where accordingly a rich after-harvest of new species probably remains to be made.[3]

The number of species that compose the different families, is on the contrary very unlike within the faunistic districts in question. The *Theraphosoidæ* (*Mygalidæ* BLACKW.), which in England are represented by *Atypus piceus*, are altogether absent from Scandinavia. Unless we aggregate *Pholcus* to the *Scytodoidæ*, that family is also unrepresented in Scandinavia. While WESTRING has in his *Theridiidæ* (*Theridioidæ* + *Scytodoidæ* NOB.)

1) *Dolomedes* in BLACKWALL comprehends 3 species.

2) *Salticus* in BLACKWALL comprehends 17 species.

3) In some papers that have since appeared, CAMBRIDGE and BLACKWALL have indeed considerably augmented the number of known spiders in these islands. In Sweden also several new forms have in the meantime been observed, but not yet published; some of these we shall have occasion in the following pages to mention.

but 115 species, BLACKWALL has described 144 species of the same family or rather sub-order — 145, if we consider *Scytodes* as belonging to the *Theridiidæ* WESTR. It is the genera *Linyphia* and *Erigone* alone (comprehending together 112 species in BLACKWALL and but 83 in WESTRING) that determine that family's great preponderance in the British compared with the Scandinavian fauna. On the other hand *Lycosidæ* and *Attidæ* are far more numerous here than within the boundaries of Britain, the first of these families showing 35 to 20, the latter 31 to 18 species; of the genus *Lycosa* WESTRING takes up double as many (32) species as BLACKWALL, of *Attus* nearly double (30 to 16).

The families *Epeiridæ*, *Drassidæ* and *Thomisidæ* of WESTRING are about equally numerously represented in both countries. As regards his *Drassidæ* (*Tubitelariæ* NOB.) it should however be observed that, whereas the *genuine* Drassidæ (*Drassoidæ* NOB.) are far more numerous in our country (41 to 29 species — *Sparassus* being referred to the *Thomisoidæ*, *Agrœca*, *Argyroneta* and the *Ciniflonidæ* BLACKW. to the *Agalenoidæ* —), the number of species of the *Dysderoidæ* and *Agalenoidæ* is not so great in the Scandinavian peninsula as in Great Britain and Ireland (the respective proportions being 3 to 7 and 17 to 22).

Among the forms described by BLACKWALL, we find several belonging also to the southern and middle parts of the European continent, which are entirely absent here: such for ex. are *Uloborus Walckenaerii* LATR. (*Veleda lineata* BLACKW.), *Scytodes thoracica*, *Segestria florentina*, *Dysdera punctoria*, *Cœlotes saxatilis*, *Eresus cinnaberinus* and *Atypus piceus*. Besides *Uloborus*, *Scytodes*, *Cœlotes*, *Eresus* and *Atypus*, the genera *Oonops* and *Schœnobates* of the family *Dysderoidæ*, (each of one species), are entirely unrepresented in the Scandinavian peninsula, whereas again — since *Hyptiotes (Mithras) paradoxus* has been found in England [1]), and if we have been right in referring BLACKWALL's *Agelena gracilipes* and *celans* to *Apostenus* WESTR. — all the genera that occur in our country have their representatives in the fauna of Great Britain and Ireland.

Among the species common throughout a great part of Sweden, but which seem to be absent in Great Britain and Ireland, we mention (by the names given them in WESTRING's work) the following: *Zilla montana*, *Theridium castaneum*, *albo-maculatum*, *Melanophora nocturna*, *Philodromus margaritatus*, *formicinus*, *Lycosa nemoralis*, *monticola*, *tarsalis*, *lignaria*, *paludicola*, *tæniata*, *cuneata*, *Attus hastatus*, *medius*, v-*insignitus* and *arcuatus*.

---

1) BLACKWALL, Notice of the capture of *Mithras paradoxus* in England.

2. "*A History of the Spiders of Great Britain and Ireland, by John Blackwall*", (Part I. 1861, Part II. 1864), is the title of the second of the works with the examination of which we are occupied. The work is published by *The Ray Society*, and is a costly work, a small folio, with 384 pages of text and 29 coloured plates. The author, who has long borne an honoured name among the zoologists of the present age, has not only by numerous essays of a descriptive character on the order of spiders, but also by important discoveries relative to these animals' economy, their industry and their (outward) structure, laid this branch of zoology under great obligations. Since however the greatest part of BLACKWALL'S previous works are scattered over a series extending to many years of English journals and other periodical works, they are not so easily or generally accessible as were to be desired, and accordingly several of the continental arachnologists seem not to be aware of them. We are therefore so much the more thankful for the work before us, which unites to a whole in an independent treatise and worthily completes the author's previous labours in illustrating the spider-fauna of Great Britain and Ireland.

As we have already seen (p. 25), this work contains descriptions of 304 species, distributed into the following 12 families: *Mygalidæ* (1 species) *Lycosidæ* (21), *Salticidæ* (18), *Thomisidæ* (29), *Drassidæ* (27), *Cinifloridæ* (9), *Agelenidæ* (15), *Theridiidæ* (28), *Linyphiidæ* (116), *Epeiridæ* (32), *Dysderidæ* (7) och *Scytodidæ* (1). It is preceeded by an introduction, in which the author gives a short general account of the external and internal structure of the spiders, their economy, the construction of their webs and their manner of living, which is so much the more valuable, as being founded on BLACKWALL'S own observations and discoveries. [1]) This is perhaps the ground on which the respected author labours with most success: as a systematizer he does not appear to us to be always quite so fortunate. We cannot, for example, accept the author's method, proposed by him in 1841 [2]), and since then constantly maintained, of dividing the spiders into 3 tribes *distinguished by the number of the eyes: Octonoculinæ* with 8, *Senoculinæ*

1) BLACKWALL seems not to have witnessed a peculiarly important fact ascertained many years ago by MENGE (Ueb. die Lebensw. d. Arachn., p. 36), viz. that the male spider, before the act of copulation, emits from the sexual aperture situated under the base of the abdomen, a drop of sperma on a kind of small web made for the purpose, which drop he then takes up in the genital bulb of the palpi. This process has newly been observed also by AUSSERER (Beob. über die Lebensw. der Spinnen, pag. 194 et sequ.)

2) BLACKW., The differ. in the numb. of eyes etc., p. 632.

with 6, and *Binoculinæ* with 2 eyes [1]). Now not only is a *fourth* tribe wanting for the spiders, which have *no* eyes, as e. g. *Stalita* SCHIÖDTE and *Hadites* KEYSERL., of which genera the first is nearly connected with *Dysdera* and the other is, so to say, a blind *Agalena*; but this whole system of classification lies open to the objection, that it is entirely *artificial*. By a one-sided adherence to a single feature not correlated with an aggregate of characters or intimately affecting the whole organism of the animal, nearly related forms are, as is well known, almost always widely sundered, and others, which are really far removed from one another, united in the same division, — and this is also the case when spiders are grouped according to the number of their eyes. As proof of this assertion we need no more than to refer to the genera *Pholcus* and *Spermophora (Rachus)*, of which the former has 8, and the latter 6 eyes. One species of the last named genus was first described by DUGÈS [2]) under the name of *Pholcus senoculatus*, and is in fact so like a *Pholcus*, that WALCKENAER doubted the correctness of DUGÈS' statement as to the number of the animal's eyes, nor was it until LUCAS [3]) also had found a six-eyed *Pholcus*, *(Ph. 4-punctatus* LUC., no doubt identical with the *Pholcus senoculatus)*, and thus confirmed DUGÈS' report, that WALCKENAER formed for these animals the genus *Rachus* [4]). The North American *Spermophora (Oophora) meridionalis* described by HENTZ [5]) is said also to differ from *Pholcus* only in having 6 eyes and

1) WALCKENAER also has made use of the number of eyes as a basis of classification: he however first separated "les Théraphoses" (Mygalidæ) from other spiders ("les Araignées"), and then divided (as early as 1833) these others into two divisions, spiders with 6 and spiders with 8 eyes (Mém. sur une nouv. classif. d. Aran., p. 438). When he afterwards became acquainted with the 2-eyed genus *Nops* MAC LEAY, a third division was added for its accommodation, so that in WALCKENAER'S Ins. Apt., II, p. 510, 511 (1841) we find "les araignées" divided into *"les Binoculées, les Sénoculées"* and *"les Octoculées."* — Whether any *4-eyed* spiders exist, is uncertain: the *Tessarops maritima* RAFIN., which is said to be distinguished by that unusual number of eyes, is so ill described and drawn, that one cannot feel certain even that this animal is a spider at all. (Conf. RAFINESQUE, Descript. d'une araignée qui constitue un nouv. genre, p. 88, Pl. 116, fig. 1).

2) Observations sur les Aranéides p. 160; — CUV., Règne Anim., Arachnides, Atlas, Pl. 9, fig. 7.

3) Exploration de l'Algérie, Arachnides, p. 239, Pl. 15, fig. 2.

4) Hist. Nat. d. Ins. Apt., IV, p. 459.

5) Descr. and fig. of the Aran. of the United States, *in* Boston Journ. of Nat., VI, p. 286.

shorter legs [1]). Even in BLACKWALL'S works the mischievous consequences of the artificial, even if in other respects convenient, system by him adopted are clearly visible: *Segestria* and *Dysdera* are widely separated from the *Drassidæ*, their nearest relations, and placed next to *Scytodes*, the natural place of which is surely in the neighbourhood of *Pholcus*, and which is more nearly related to BLACKWALL'S *Theridiidæ*, than to any other of his families. — BLACKWALL appears to us also to lay too much weight upon an organ which he considers as a 4th pair of spinners grown together, and on the rows of curved hairs or bristles on the metatarsi of the posterior legs, which he calls *calamistrum:* on the always contemporaneous presence of these organs be has based his family *Ciniflonidæ*, in which he brings together forms so widely separated as e. g. *Amaurobius* (*Ciniflo* BLACKW.) and *Uloborus* (*Veleda* BLACKW.), the former of which genera is usually aggregated to the *Tubitelæ* of LATREILLE and the latter to his *Orbitelæ*. The genera *Eresus* and *Dinopis*, which also, as L. KOCH has shown [2]), have the "*calamistrum*" and the above mentioned organ situated immediately under or in front of the spinners (and which we on account of its situation call the *infra-mammillary organ*), [3]) must thus also be referred to the family *Ciniflonidæ*, which accordingly is made to contain a heterogeneous mixture of species belonging to the most widely separated families [4]). However important these characters may be — and we believe that we attribute to

1) Also among the *Theraphosoidæ* (*Mygalidæ*) forms occur with only 6 eyes, e. g. the genus *Pelecodon* DOLESCHALL (Tweede Bijdr. t. d. Kenn. d. Arachn. v. d. Ind. Arch., p. 5) and *Mygale* (*Cteniza*) *hexops* WHITE (Descr. of appar. new spec. of Apt. fr. New Zeal., p. 3). Among the *Thomisoidæ* the *Th. yolophus* DOUMERC has six eyes: among the *Retitelariæ* (*Inæquitelæ*) not only *Spermophora*, but also *Sicarius* WALCK. or *Thomisoides* NICOLET (if this genus really belong to that sub-order and not, as is more probable, to the Thomisoidæ) has also only six eyes. KEMPELEN has lately (Verhandl. d. zool.-bot. Vereins in Wien, XVII) described under the name of *Thysa pythonissæformis* a six-eyed spider from Hungary, which seems to be nearly connected with the genus *Gnaphosa* (*Pythonissa* C. KOCH) belonging to the real *Drassoidæ*.

2) Die Arachn.-fam. d. Drassiden, p. 1; Beschr. n. Arachn. u. Myriap., *in* Verhandl. d. zool.-bot. Gesellsch. in Wien, XVII (1867), p. 231.

3) L. KOCH (Die Arachn.-gatt. Amaurobius, Cœlotes u. Cybæus, p. 1) provisionally calls this organ *cribellum*, because he finds that it has some likeness to a sieve ("Sieb"). The name "Sieb", *colatorium*, has been previously employed by MENGE for the surface of the last joint of the spinners where the spinning-tubes are situated: see MENGE, Preuss. Spinnen, p. 27.

4) BLACKWALL now also includes the *Eresus* among the Ciniflonidæ: see BLACKW., A List of Spid. capt. in the south east reg. of equat. Africa, p. 454.

them sufficient weight, when we set up for the forms which exhibit them, within the family *Epeiroidæ* the sub-family *Uloborinæ*, and within the family *Agalenoidæ* the sub-family *Amaurobiinæ*, and moreover among the *Saltigradæ* reckon the *Eresoidæ* and *Dinopoidæ* as separate families, — they cannot be allowed the importance which BLACKWALL ascribes to them. In the first place it is very uncertain, that the organ, which BLACKWALL considers as a pair of spinners grown together, really is so; I for my part do not think so, for it does not project above the surface of the abdomen, but seems only to consist of a peculiarly modified part of the skin, neither have I been able to discover any spinning-tubes on its surface [1]). But even if BLACKWALL'S explanation of that organ is right, still the family *Mygalidæ* BLACKW. proves, that the number of spinners needs not be the same in all the genera belonging to one and the same family; for to that family BLACKWALL himself reckons genera not only with four but also with six spinners. As regards the *calamistrum*, the purpose of that apparatus in the animal's economy is perhaps as yet too little known to justify the laying of any great weight upon it in classification. If BLACKWALL'S statement, that it is a curling-apparatus used in the construction of the spider's web, [2]) is correct as regards the genera *Amaurobius* and *Dictyna*, which I have no reason to doubt, it can hardly have the same functions in, for ex., the species of *Uloborus* and *Hyptiotes (Mithras)*, which weave regular, so-called geo-

1) It is a matter deserving of investigation, whether the infra-mamillary organ be not connected with *tracheæ*, having their stigmata in or close to that organ. That some spiders *(Dysderoidæ, Argyroneta)* have two tracheal trunks opening on the ventral surface of the abdomen, near its *base*, behind the openings of the two so-called pulmonary sacs, is generally known. In some other species MENGE (Ueb. d. Lebensweise d. Arachn., p. 23; Preuss. Spinnen, p. 81, 189 etc.) has discovered a system of tracheæ opening at the *end* of the abdomen, *immediately in front of the spinners*, with either *two* stigmata (certain *Attoidæ* and the *Erigone-* and *Walckenaera-* or *Micryphantes*-species) or only *one (Cercidia* or *Cerceis prominens)*. But according to v. SIEBOLD (Vergl. Anatom., p. 535) there is in most spiders — he reckons up the different genera *Epeira*, *Tetragnatha*, *Theridion*, *Drassus*, *Clubiona*, *Lycosa*, *Dolomedes*, *Thomisus (Xysticus)* — a fissure *before the spinners*, from which proceed four flattened, band-formed, almost always unramified tracheæ. It seems then that a tracheal system is to be found in all spiders provided with only two "pulmonary" sacs, although it may terminate sometimes with one and sometimes with two very variously situated apertures, and it is certainly not wanting in those genera, which have an infra-mamillary organ and calamistrum.

2) As to BLCKWALL'S beautiful and highly interesting researches on this subject, vid. BLACKW., Notice of sev. rec. discov. in the struct. and œconomy of spiders, p. 472 et sequ.

metrical webs; and as a calamistrum is found in spiders, whose industry is so different, it appears to us that it's importance in the construction of the web cannot always be particularly great.

The 12 families adopted by BLACKWALL all constitute natural groups, with the exception of the *Ciniflonidæ*, of which we have already spoken, and in some degree the *Theridiidæ*, which family in BLACKWALL comprises only the Walckenaerian genera *Theridium* and *Pholcus;* for the remaining genera of SUNDEVALL'S *Theridides* he has formed the family *Linyphiidæ.* This division of the old family *Theridides* is certainly unnecessary: BLACKWALL does not mention a single character of the animals themselves, whereby the families may be distinguished. *Pholcus* appears to us rather to belong to the *Scytodoïdæ*, and that family should immediately follow the Theridioïdæ. — The families are very briefly, often insufficiently or not at all, characterized: the genera also very briefly, but in general with sufficient detail for practical behoof in the examination of an unknown form.

BLACKWALL, as regards the number and extent of the genera he adopts, is much more conservative than WESTRING; he acknowledges but a small number of new generic groups over and above those already established by WALCKENAER, like whom, in determining the limits of the genera, he appears to fix his attention almost exclusively on the characteristics of the organs of the mouth, the position of the eyes, and the relative length of the legs. The greatest part of the genera proposed by others, for ex., C. KOCH, and the distinctions of which are founded also on characteristics deduced from other parts of the body, are rejected by BLACKWALL, although they, if often in a more or less modified form, appear to have been pretty generally acknowledged by the arachnologists of the Continent.

The author's remarks on the instincts, haunts and general economy of the species described, their manner of constructing their webs etc., are particularly valuable and interesting. The descriptions of the species are themselves, with few exceptions, very fully detailed, and, in combination with the figures, sufficient for the recognition of the species. Especial attention has been very properly paid to the form of the palpi of the males, to the construction of the spinners and other finer details of structure, except as regards the *spines* wherewith the legs and palpi are armed, which are only superficially touched upon. The descriptions are however often occupied in a great measure by characters, which, as common for the whole genus or most of the species comprised by it, are of little or no use in determining the species. We wish to call attention to this as an impediment in the use of the work, as also, and that especially, to the *absence of diagnoses*, a want which is felt as much the more, as the author seldom separately gives any hints as to the characters by which the species most resembling each other may best be distinguished. The *unit of length* for expressing the dimensions of the animals is also, as it seems to us, not very well chosen. Instead of stating, as most zoologists do, the size of such small animals in millimeters or lines and decimals of one or other of these units, BLACKWALL measures the length and breadth of the spider's cephalothorax, abdomen etc. by fractions of an inch difficult to compare, so that one is sometimes obliged to submit the given measure to an arithmetical reduction, before it is possible to form a clear idea of the relative size of the parts described. — The figures are, with a few exceptions, (as e. g. some in Pl. I and II) good. Not only are coloured entire figures of both the male and female of almost every species given, but also outlines of the male's palpi, frequently also of the position of the eyes, the different organs of the mouth, and so forth. The large and difficult genera *Neriene* and *Walckenaera* (= *Erigone* WESTR.) are treated with especial care, and indeed there scarcely exists in arachnological literature anything surpassing the superb, highly magnified figures, that BLACKWALL has furnished of these remarkable little spiders.

BLACKWALL in his synonyms cites but few arachnological works; excepting his countryman LISTER he mentions none of the older authors, non even CLERCK or FABRICIUS: he generally follows, and only with a too implicit confidence, the determinations and nomenclature of WALCKENAER. Accordingly, as we shall hereafter see, his specific names will be frequently obliged to give place to other, older denominations. In other respects he appears in his nomenclature to have conscientiously observed the law of priority.

3. "*Histoire Naturelle des Araignées (Aranéides) par Eugène Simon*" is a work differing in many respects from the two preceeding, and which we here submit to examination only on account of the "*Catalogue synonymique des espèces européennes*" appended to it. The work contains a description of the internal and external structure of spiders, followed by a detailed account of their division into families, sub-families and genera, a list of the species belonging to each separate genus, as also an account of the principal species' haunts and economy. The plan and object of the work are clearly set forth by the author in the following words: "résumer . . . . dans un cadre restreint tous les travaux anciens et modernes qui ont été publiés sur cette classe d'animaux, en y joignant les observations qui me sont propres, tel est le but que je me suis proposé en publiant ce traité."

We are however afraid that the author has not executed his work with sufficient care and accuracy to be said to have solved so comprehensive a problem in a satisfactory manner: he does not even possess the acquaintance with arachnological literature indispensably necessary for a work of this nature [1]). Accordingly a very large number both of genera and species made known before the publication of SIMON'S work by German, English, American, Swedish and other authors, are in that work wanting.

SIMON divides the spiders into 9 families: *Scytodiformes*, *Mygaliformes*, *Drassiformes*, *Théridiformes*, *Épéiriformes*, *Salticiformes*, *Lycosiformes*, *Thomisiformes* and *Myrméciformes*. The first of these families, *Scytodiformes*, includes the genera *Scytodes*, *Omosites*, *Rachus*, *Pholcus* and *Artema*, which union appears to us fully justified. On the other hand the combination of two so widely different generic forms as *Myrmecium* and *Chersis (Palpimanus)* in the family *Myrméciformes*, does not appear to be a happy step, nor does the union of *Agelenidæ* BLACKW. with *Theridides* SUND. to one family, *Théridiformes*, seem more reasonable. *Agalenoidæ* and *Theridioidæ* are by SIMON not even separated as *sub-families* or tribes: when breaking up his *Théridiformes* into 3 such divisions *(Clothéiens*, *Théridiens*, and *Linyphiens)*, he removes e. g. the genera *Linyphia* and *Micryphantes* from his *Théridiens* (*Theridium*, *Erigone*, etc.) and unites them with the *Agalenoidæ* in the sub-family *Linyphiens*. — Whether SIMON has been right in removing *Eresus* from the *Attides* SUND. and aggregating that genus to his *Épéiriformes*, it must also be permitted to doubt.

Most of the 93 *genera* adopted by SIMON appear to us founded in nature; indeed they correspond, with the exception of *Oxyptila* (for *Thomisus claveatus* WALCK.) and *Phrynoides* (for *Th. rugosus* WALCK. and *Th.*

1) This is evidenced already in the introduction, where the author has attempted to give a brief account of the development of arachnology: thus for ex., having mentioned WALCKENAER'S *Tableau des Aranéides* and *Histoire Naturelle des Insectes Aptères*, the former printed in 1805 and the latter 1837—1847, he immediately continues: "Cependant, *presque en même temps*, LISTER, en Angleterre, donnait son *Histoire des Araignées* trop courte et trop incomplète; CLERCK et DE GEER, en Suède, poursuivaient des études sur les moeurs de quelques espèces", etc. LISTER'S classical work, *"Historiæ Animalium Angliæ tres tractatus. Unus de Araneis"* etc. here referred to, was however printed in 1678, CLERCK'S *"Svenska Spindlar, Aranei Suecici"*, in 1757, and the volume (Tom. VII) of DE GEER'S *"Mémoires pour servir à l'Histoire des Insectes"*, which treats of spiders, in 1778. The works in question were then by no means published, as SIMON states, nearly at the same time, but during the course of three successive centuries.

foka VINS.) to genera already received under either the same or other names. But if at first sight the number of genera adopted by SIMON appears not to be unreasonably great, but rather the reverse, the fact is nevertheless otherwise. Instead of dividing every family or sub-family into a number of groups comparable with each other, viz. the *genera*, and merely applying to these a generic name, he has followed the altogether objectionable example set by some authors, of forming so-called *sub-genera;* and not content with dividing a large number of genera into *"sous-genres"* with separate names, he goes farther and divides these "sous-genres" into *"groupes"*, also loaded with names of a similar kind, whereby the number of generic names used by SIMON becomes very considerable. The "sous-genres" and "groupes" are often very vaguely distinguished, and the characters attributed both to them and to the genera properly so called, nay even to the families themselves, are in general by no means trustworthy and indeed not unfrequently erroneous. [1])

SIMON has endeavoured to give the etymological derivation of every generic name; his services however in this respect are greatly depreciated

---

1) This may be sufficiently shown by a couple of examples. Of the family *Scytodiformes* it is said (p. 43), that their physionomy has "quelque chose de particulier, qui est dû à la *forme globuleuse du corselet, élevé surtout en arrière*" etc. But of the genera belonging to that family, *Scytodes* is the only one to which this description is applicable; for *Omosites* has the cephalothorax "déprimé" and *Rachus* has its "parties latérales et postérieures déprimées", *Pholcus* has it "déprimé", and *Artema* "déprimé en arrière", according to SIMON's own account. — According to SIMON (p. 256) the genus *Singa* differs from *Epeira* "par une forme particulière et charactéristique de l'abdomen; . . . . il s'élève et s'élargit graduellement jusqu'à sa partie postérieure, dont la portion supérieure est un tubercule et dont la portion inférieure est tronquée obliquement"—a description which may very well suit for *S. conica* (which however WESTRING and MENGE, as it seems to us with good reason, do not aggregate to the genus *Singa*), but which is quite inapplicable to e. g. *S. hamata*, which is typical of the genus, as well as to *S. Herii* and others.— *Epeira marmorea* and *pyramidata (scalaris)*, which are perfectly similar both in form and economy, and distinguished only by colour, are referred by SIMON to different "groupes" of the "sub-genus" *Epeira:* the former is a *Neopora* SIM., which group is said (p. 261) to have the abdomen "globuleux faiblement anguleux", and the species of which are "araignées vivant . . . dans les jardins, ne construisant pas de coques", whereas the last belongs to the group *Neoschœna* [*Neoscona*] SIM., the species of which have the abdomen "tout-à-fait globuleux et oviforme", and are "araignées vivant sur les bords des eaux, se renfermant dans des coques" etc. — The sub-genera, into which the genus *Micryphantes* is subdivided, are distinguished by characters belonging only to one sex, and one of them, *Viderius* SIM., is characterized (p. 196) by a peculiarity ("les deux

by the mistakes that he not unfrequently commits [1]). Many names he alters — in consequence probably of their, according to his notions, erroneous etymology — in a manner which it appears to me difficult to justify. Thus for ex. he changes *Theridion* or *Theridium* into *Theridio*, *Erigone* to *Erygona*, *Neriene* to *Nerieneus*, *Textrix* to *Tectrix*, *Hersilia* to *Herselia*, and so forth. A large number of generic names have their terminations arbitrarily altered: many for ex. with the termination *es* or *us* terminate in SIMON'S work in *a*, so that we there read *Scytoda*, *Eresa*, *Atta*, *Thomisa*, *Sparassa*, *Philodroma*, etc. instead of *Scytodes*, *Eresus*, *Attus*, *Thomisus*, etc. Neither does he observe any consistency in this, for he preserves the names *Dolomedes*, *Pholcus*, *Drassus*, *Uloborus* etc. unaltered, nor does he appear to remark, that, by making or adopting such changes, he applies names already appropriated to other genera of animals, as for ex. in the case of the names *Atypa*, *Myrmecia* and *Atta* (instead of *Atypus* WALCK., *Myrmecium* LATR. [2])

---

yeux latéraux de la ligne supérieure sont placés chacun à l'extrémité d'un long pédicule horizontal"), which cannot be recognized in either of the two species *(M. cucullatus* and *M. tibialis)* which SIMON looks on as belonging to the sub-genus *Viderius*, nor in any other European spider, that I know of.

1) *Omosites* (ὠμόσιτος, eating raw food, from ὠμός, raw and σιτέομαι, eat) SIMON derives from "ομος, même; σιτος, nourriture." *Anyphæna* (ἀνυφαίνω, unravel a web) he derives from "ανυω, tuer; φοινος, rouge ou sanglant"; — *Theridium* (θηρίδιον, little animal) from "θηρα, chasse; ειδω, voir"; — *Phrurolithus* (φρουρέω watch, and λίθος, stone) from "φρεω, creuser", and λίθος; — *Erigone* (Ἠριγόνη, mythol. prop. name) from "ἐρύω défendre; γόνος, progéniture"; *Micryphantes* (μικρός, small, ὑφάντης, weaver) from μικρός and "φανθεις, brillant"; *Neriene* (prop. name of the wife of Mars) from "νευρις, corde, fil; ενη, le soir"; — *Meta* (Μῆτα, mythol. proper name) from "μῆτις, sagesse, prévoyance." — *Uloborus* (οὐλοβόρος, deadly biting — of οὖλος fatal, deadly, and βιβρώσκω, eat) is by SIMON derived from "ὕλη, broussailles; βορὸς, qui dévore"; — *Argyrodes* (ἄργυρος, silver, εἶδος, appearance) from ἄργυρος and "οἰδος, gonflé"; — *Singa* (Σίγγα, geogr. prop. name) from "σύν, marque similitude; γᾶ ou γῆ, terre"; *Micrathena* ("μικρα, parva, et Ἀθηνα, nomen Græcum Minervæ": SUNDEV.) from μικρός and "θείνω aiguillonner"; — *Hersilia* (proper name of Romulus' Sabine wife) from "ἔρσις, action d'enlacer", etc.

2) LATREILLE formed the genus *Myrmecium* in 1824 (Notice sur un nouveau genre d'Aranéides, p. 23): afterwards, in 1829, he changed the name to *Myrmecia* (in CUVIER'S Règne Animal, 2ᵉ Édit., p. 261). In consequence of the too great similarity of the name with that of *Myrmecia* already employed by FABRICIUS: DALMAN (Årsberättelse, 1826, p. 59.) proposed to change *Myrmecium* to *Myrmidea*, which however to us appears unnecessary. — *Myrmecium* or *Myrmecion* is the classical name of a species of spider, "formicæ similis capite, alvo nigro, guttis albis distinguentibus": PLIN. Hist. Nat., L. XXIX, c. 27.

and *Attus* WALCK.), of which, as is well known, the first name has long belonged to a Hemipterous, and the last two again to two Hymenopterous genera [1]). The names which through these and similar alterations have come into SIMON'S work, I have not, in the giving of synonyms and determinations of priority, considered myself bound to treat as new names formed by him, but only as varied orthographies of those which he has altered.

Very many of the generic denominations used by SIMON were already appropriated to other animals before they were used as names for spiders [2]). They must accordingly be rejected and, where necessary, be replaced by others, either by already existing synonyms or by new formed

---

1) More reasonable grounds for altering the orthography of certain names in the classification of spiders are not wanting, and SIMON would no doubt have won the approbation of many, had he written *Chiracanthium* instead of *Cheiracanthium*, *Hyptiotes* for *Uptiotes*, *Philœca* for *Philoica*, etc. Several of the names which SIMON himself desires to introduce into the science, and in the formation of which he has used Greek words, which he has specified, stand in sore need of such correction. Thus it seems to us tolerably evident that the name formed by the combination of *μελία* and *κέρας* should not be *Melicertus*, but perhaps *Melicerus*, or rather *Melioceros*; of *ἄῤῥην* and *κέρας* one would form not *Arrecerus*, but e. g. *Arrhenoceros*; of *νέω* and *σχοῖνος* one may form *Neoschœna*, but not *Neoscona*; of *τρι* and *ἄκανθα* *Triacantha*, not *Tricantha*; of *σύν* and *αἷμα* *Synæma*, not *Synema*; of *κιῤῥός* (light yellow) and *φέρω* *Cirrhophora*, not *Cirrofera*, etc. In a couple of the names formed by SIMON, the letters *ζ* and *ξ* have been confounded: he writes *Pesionyx* instead of *Pexionyx* (from *πῆξις* and *ὄνυξ*), *Ozyptila* instead of *Oxyptila* (from *ὀξύς* and *πτίλον*); in two others he has confounded *υ* and *ου*, viz. in *Nuctobia* (*νυκτόβιος*) and *Nuctenea* (*νύξ*, *νέω*), which should be written *Nyctobia* and *Nyctinea*. Fortunately most of the names formed by SIMON will probably be found supererogatory.

2) Such is the case with at least the following names. *Artamus* C. KOCH 1837 (*Artamus* VIEILL. [Aves] 1816); — *Clotho* WALCK. 1809 (*Clotho* FAUJ. [Moll.] 1808); — *Cyrtocephalus* LUCAS 1845 (*Cyrtocephalus* AUD. [Coleopt.] 1834); — *Cyrtonota* SIMON 1864 (*Cyrtonota* CHEVR. [Coleopt.] 1834); — *Dia* C. KOCH. 1850 (*Dia* DEJ. [Coleopt.] 1834); — *Diana* C. KOCH 1850 (*Diana* RISSO [Pisc.] 1826); — *Eucharia* C. KOCH 1836 (*Eucharia* HÜBN. [Lepidopt.] 1816); — *Eurysoma* C. KOCH 1839 (*Eurysoma* GISTL [Coleopt.] 1829); — *Isacantha* SIM. 1864 (*Isacantha* HOPE [Coleopt.] 1833); — *Janus* C. KOCH 1846 (*Janus* STEPH. [Hymenopt.] 1835); — *Ino* C. KOCH 1850 (*Ino* LEACH [Lepidopt.] 1814); — *Lachesis* SAV. et AUD. 1825—27 (*Lachesis* DAUD. [Rept.] 1802); — *Leimonia* C. KOCH 1848 (*Leimonia* HÜBN. [Lepidopt.] 1816); — *Macaria* C. KOCH 1835 (*Macaria* CURT. [Lepidopt.) 1826); — *Melicertus* SIM. 1864 (*Melicertus* RAFIN. [Crust.] 1814); — *Monastes* LUC. 1847? (*Monastes* NITZSCH [Aves] 1840); — *Mygale* WALCK. 1802 (*Mygale* CUV. [Mammal.] 1800); — *Pachyptila* SIM. 1864 (*Pachyptila* ILLIG. [Aves] 1811); — *Pales* C. KOCH 1850 (*Pales* ROB. DESV. [Dipt.] 1830); — *Pandora* C. KOCH 1850 (*Pandora* BRUG. [Moll.] 1791); — *Parthenia* C. KOCH 1850

names [1]). As regards the greater number of SIMON's "coupes génériques", which will thus become nameless, I have not been able to persuade myself that they are sufficiently founded in nature to deserve, at least for the present, to be preserved: the case seems however to be otherwise with the following genera, for which I accordingly consider that new names ought to be formed:

Instead of	*Artamus* C. KOCH	we propose	*Artanes* ('Αρτάνης, proper name);
„	*Cyrtocephalus* LUC.	„	*Cyrtauchenius* (κυρτός, arched, bent; αὐχήν, neck);
„	*Dia* C. KOCH	„	*Ælurops* (αἴλουρος, cat; ὤψ, face);
„	*Diana* C. KOCH	„	*Diæa* (Διαῖος, prop. name);
„	*Lachesis* SAV. et AUD.	„	*Laches* (Λάχης, prop. name, masc.);
„	*Monastes* LUC.	„	*Monæses* (Μοναίσης, prop. name);
„	*Philia* C. KOCH	„	*Philæus* (Φιλαῖος, prop. name);
„	*Phrynoides* SIM.	„	*Phrynarachne* (φρύνη, toad; ἀράχνη, spider);
„	*Rhanis* C. KOCH	„	*Rhene* ('Ρήνη, prop. name, fem.) [2]).

(*Parthenia* ROB.-DESV. [Dipt.] 1830; — *Philia* C. KOCH 1846 (*Philia* SCHIÖDTE [Hemipt.] 1842); — *Phœbe* C. KOCH 1850 (*Phœbe* SERV. [Coleopt.] 1835); — *Phrynoides* SIM. 1864 (*Phrynoidis* FITZ. [Rept.] 1843); — *Potamia* C. KOCH 1848 (*Potamia* ROB.-DESV. [Dipt.] 1830); — *Pyrophorus* C. KOCH 1837 (*Pyrophorus* ILLIG. [Coleopt.] 1809); — *Rhanis* C. KOCH 1848 (*Rhanis* DEJ. [Coleopt.] 1834); — *Sphodros* WALCK. 1837 (*Sphodrus* CLAIRV. [Coleopt.] 1806); — *Trivia* C. KOCH 1850 (*Trivia* GRAY [Moll.] 1832). — Such of the above notices as do not refer to spiders, are for the most part taken from AGASSIZ' Nomenclator Zoologicus.

1) *Clotho* WALCK. ought to be replaced with *Uroctea* DUF.; *Eurysoma* C. KOCH with *Eurycorma* THOR. and *Cærostris* THOR. (Eugenie's Resa, Arachn. 1, p. 3, 4); *Janus* C. KOCH with *Synemosyna* (HENTZ); *Macaria* C. KOCH with *Micaria* WESTR.; *Mygale* WALCK. with e. g. *Theraphosa* WALCK.; *Pyrophorus* C. KOCH with *Salticus* (LATR.) SUND.

2) Instead of some generic names already previously appropriated, not, it is true, adopted by SIMON, but applied by other arachnologists, and which appear to me to indicate good and well characterized genera, I avail myself of this opportunity to propose the following appellations:

For	*Ariadne* DOLESCH. (1857)	I propose	*Ariamnes* ('Αριάμνης, prop. name, masc.);
„	*Cerceis* MENGE (1866)	„	*Cercidia* (κερκίς, shuttle);
„	*Cyphagogus* GÜNTH. (1862)	„	*Cyphonethis* (κυφός, curved, νηθίς, female spinner);
„	*Galena* C. KOCH (1845)	„	*Gelanor* (Γελάνωρ, prop. name myth.);
„	*Latona* L. KOCH (1866)	„	*Lampona* (Λάμπων, prop. name);
„	*Pasithea* BLACKW. (1858)	„	*Peucetia* (Πευκέτιος, prop. name myth.);
„	*Triclaria* C. KOCH (1848)	„	*Trechalea* (τρηχαλέος, rough, savage).

To point out and correct the numerous erroneous or contradictory statements we have met with in SIMON's work, excepting where they concern the synonymies of the European spider-fauna, would be foreign to the problem we are endeavouring to solve. It would moreover be an unnecessary and thankless task, for these errors are generally of such a character that they are readily seen by any one tolerably versed in arachnology. — SIMON appears to have no idea, that a name, once imposed, ought to be respected, and not arbitrarily changed for another. He rechristens LATREILLE's *Cteniza* (= *Nemesia* SAV. et AUD.) *Mygalodonta*, merely because he imagines himself to have discovered, that the former apellation "est restée inconnue." The name *Aulonia* C. KOCH he rejects without assigning any reason at all, and gives to that genus the new name *Lycosina*. It is therefore evident that many changes must be made in his nomenclature: the name *Actinopus* PERTY has right of priority before *Sphodros* WALCK., [1]) which is adopted by SIMON, *Palpimanus* DUF. before (*Platyscelum* SAV. et AUD., and) *Chersis* WALCK., SIM., etc. Several corrections connected with this we shall have occasion to make in the following pages.

II.

VIEW OF THE GENERA OF EUROPEAN SPIDERS.

It is customary to begin the series of spiders with the *Epeiroidæ* or the *Orbitelariæ*. and in the following pages I have conformed to this custom, because it appears to me that the properties that distinguish the whole Order of Spiders, are most strikingly seen in that family, and the Epeiroidæ may therefore be considered as including the very type itself of the Order.

(*Ariadna* SAV. et AUD. [Aran.] 1825—7; — *Cerceis* MILNE-EDW. [Crust.] 1840; — *Galene* DE HAAN [Crust.] 1835; — *Latona* SCHUM. [Moll.] 1817; — *Pasithea* LAMOUROUX [Polypi] 1812; — *Triclaria* WAGN. [Aves] 1838]. — The name *Cyphagogus* has been given to a genus among the *Curculiones* [Coleopt.] long before 1862, according to GERSTÄCKER (Bericht üb. die wissensch. Leistungen im Gebiete d. Entomologie währ. d. Jahres 1862, p. 560)).

1) *Sphodros* was, it is true, already in 1833 mentioned by WALCKENAER in his Mém. s. une nouv. classif. d. Aranéides, but all that is there communicated about that genus, is, that it has "les yeux écartés" like *Missulena (Eriodon)*, and that it is "intermédiare entre les Missulènes et les Mygales", which assuredly cannot be considered as a characterization of the genus. — (*Sphodrus* CLAIRV. [Coleopt.] 1806).

I begin then with that family, not because I consider it to stand higher than others, but for the same reason, for which, when systematically treating, for instance, the Class of Fishes, one usually goes out from the *Teleostei* or Bony fishes, and not from the undoubtedly far higher organized *Selachiæ* [and *Dipnoi*]. I am far from persuaded that the family Epeiroidæ really includes the highest forms within the Order Araneæ. If we had only to take account of the development of the *instincts*, we might, in consideration of the more artistic construction of the webs of the Epeiroidæ, place that group above the other families of the Order; but then again, if we consider, as we reasonably ought to do, more the harmonious development of the body's various parts, the superior development of the organs of sense, and suchlike, we soon see that the Epeiroidæ, with their weak cephalothorax and heavy abdomen, their slow and clumsy motions, their comparatively small eyes, etc., are surpassed by more than one of the other families, usually looked upon as lower. Generally speaking, the opinion that spiders which build a web, are higher animals than those which hunt their prey, seems to be unfounded. Those which are *most perfectly organized* ought to have the higher rank assigned them, and it appears to me difficult to show, that in that respect the *weavers* in any way take precedence of the *hunters*. The family *Theraphosoidæ* or *Mygalidæ*, which surpass all other spider-families in *magnitude*, form through *Liphistius desultor* SCHIÖDTE, which is destitute of spinners and has the back of the abdomen covered with jointed horn-shields, a connexion with the *Phrynoidæ* and *Scorpions*, which I believe must be considered as more highly organized animals than spiders [1]). The *Lycosoidæ*, and in a still

1) I do not however consider the remarkable agreement between *Liphistius* and *Phrynus* as proving that that genus has any nearer *affinity* (depending on a closer propinquity of descent) to *Phrynus* than other spiders have, but I only consider it as an example of the *analogy* that can exist between groups of animals not intimately related. In the case of spiders this is not a solitary example. *Anetes cœletrum* MENGE (Verz. d. Danziger Spinnen, p. 71), which is said to stand in near relation to *Arcys* among the *Thomisoidæ*, is without spinners, like *Liphistius*. The relationship which VINSON believes to exist between the Epeiroid genus *Arachnura* VINS. (Aran. d. Iles de la Réunion etc., p. 289) — in which genus the abdomen is drawn out into a kind of tail — and the Scorpions, depends upon a similar, only still weaker analogy. MAC LEAY mentions (On some new forms of Arachn., p. 5) "a singularly flat and minute, hard-shelled, six-eyed spider with a *sessile* abdomen", which is met with in Cuba, and which he considers as forming the connecting link between spiders and *Acari*. Here perhaps we have an instance of true *affinity* between spiders and a lower group: indeed the Order of spiders is undoubtedly no

higher degree the *Attoidæ*, distinguish themselves by their well proportioned forms, their powerfully developed cephalothorax, by the quickness and force of their movements, highly developed organs of sight, and the *Attoidæ* also by an expression of intelligence, which cannot escape even the most casual observer, and which, among other lower invertebrate animals, is only to be found in that Order of Insects which comprises undeniably the most highly developed animals of this Class, the Hymenoptera. As regards the other reasons that have been adduced in support of the assumption of the preeminence of the Epeiroidæ before all other spiders, such as the numerousness and beauty of the species, the small number of transition-forms, etc., they hold equally true of the *Attoidæ*, which form a unit quite as close, compact and rich in species as the Epeiroidæ; in the brilliancy and variety of their colours they surpass both these and the other families of spiders, and may even be compared with the most showy families of Coleoptera, so distinguished for beauty and brilliancy of colour.

If it is difficult to agree on, which group of spiders is to be considered as the *highest*, it is on the contrary easy enough to determine which of the sub-orders received by us occupies the *lowest* rank. We without hesitation assign that place to the *Tubitelariæ*, among which, it is true, clear and defined transition-forms to lower groups of animals are as little to be met with as in any of the other sub-orders, but which nevertheless show themselves in many respects to stand in a lower stage than the other great subdivisions of the Order. The gradual reduction of the organs of vision is already one evidence of this: most of the spiders, that have only six eyes belong to this sub-order, and it is only within its compass that species have been found having only two eyes *(Nops)*, or even totally

---

mere connecting link between two other orders of Arachnoidea, but appears to have been developed side by side with the so-called *Arthrogastra* (*Solifugæ* SUND.) from an inferior group, probably the *Opiliones*. In the families of the *Scytodoidæ* and *Filistatoidæ* are several features that may be considered to indicate relationship with the last-named order, as for ex. the process at the extremity of the mandible, which in conjunction with its claw almost forms a two-fingered forceps; the extraordinarily long legs of *Polcus*, the tarsus divided into three parts, etc. — HÆCKEL (Generelle Morphologie d. Organismen, II, p. XCVII) believes that spiders were developed from the *Galeodoidæ*, independently of the other divergent branches which, according to his view, go out from the *Galeodoidæ*, viz. *Scorpiones* (including *Phrynus*) and *Opiliones;* he considers the *Saltigradæ* as the spiders, which still stand in the nearest relation to the *Galeodoidæ*, probably on account of the apparent segmentation of the cephalothorax in the genus *Myrmecium*. I can however in no wise accede to this opinion.

blind forms *(Stalita, Hadites)* [1]). In contradistinction to the *Saltigradæ*, *Citigradæ*, *Orbitelariæ* etc., the sub-order *Tubitelariæ* is extremely *polymorphous*, and forms only a loosely connected combination of very heterogeneous elements: it must be divided into many families and a great number of genera, and but few of these last seem to contain more than a very limited number of species. Transition-forms to almost all the other sub-orders are also to be found among the *Tubitelariæ*, which form as it were the chaos, from which the other more sharply defined and clearer types have been gradually developed. The forms are frequently coarse, ugly and clumsy, the colour dark and dusky; even their generally concealed and nocturnal habits indicate the lower rank of these animals. Among the different families, into which this sub-order is divided, the first place must certainly be assigned to the *Agalenoidæ;* the remaining families would appear to be in about the same stage of development, though probably the *Filistatoidæ* are the lowest. With them may be joined, as occupying an equally low position, the family *Scytodoidæ* in the sub-order *Retitelariæ*.

Whether we endeavour to arrange the families and genera of spiders in a continuous series, from that group which is looked upon as the most perfect, down to the lowest, or vice versa, or whether we arrange them after any other principle, we are soon met by the same difficulties which present themselves, whenever we endeavour to arrange in such a manner any class or order whatever of the productions of nature. We are soon obliged to abandon the hope of making the arrangement fully *natural*, i. e. such as to give a clear view of the more near or distant relationships of the various groups, and their thence following mutual similarities and dissimilarities, and in the choice of the various combinations that offer themselves, we have, as Walckenaer (Tabl. d. Aran., p. xii) happily expressed himself, often enough only "le choix des inconvénients". The arrangement of the series itself is accordingly often enough tolerably unimportant, if one only take care in some other way to account for the natural relations which the various groups have to each other. As regards the larger groups of spiders, the sub-orders and the families, the reasons for the order of arrangement we have chosen will, we hope, easily be seen if one casts one's eye on

---

1) Even the so imperfectly described blind *Antrobia* [*Anthrobia*] *monmouthia* Tellkamp (Beschr. einig. neuen in d. Mammuth-Höhle aufgef. Gliederth., p. 318, Taf. VIII, fig. 13—17) probably belongs to this family, and not to the *Territelariæ* as Tellkamp supposes: Compare his description and figure of the animal's mandibles ("Kieferklauen *eingeschlagen*") and maxillæ.

the accompanying diagram, which gives a view of the connexion founded on real *affinity* which the families of the spiders adopted by us, according to our opinion, have with each other [1]). That connexion is more easily shown in a plane than in a series, but of course even that method of representation still leaves much to be desired. Thus e. g. the line (13) that marks the *Thomisoidæ* ought to be supposed drawn in another plane, so that the distance between the *Theraphosoidæ* (10) and the *Lycosoidæ* (14) may not be greater than between the first of these and the *Thomisoidæ* (13).

We first divide the order of spiders (Araneæ) into 7 *sub-orders:* I. *Orbitelariæ*, II. *Retitelariæ*, III. *Tubitelariæ*, IV. *Territelariæ*, V. *Citigradæ*, VI. *Laterigradæ*, and VII. *Saltigradæ*, corresponding to the old, almost similarly named Latreillian families (*Retitelariæ* nob. = *Inæquitelæ* Latr.) [2]). These subdivisions moreover fall asunder into the 22 families inserted in the figure. If we begin with the *Tubitelariæ* as the lowest sub-order, we might also consider the others as radiating from it in *three* principal branches, whereof one is composed of the *Retitelariæ* and *Orbitelariæ*, another of the *Territelariæ*, the third of the *Laterigradæ*, *Citigradæ* and *Saltigradæ:* the 16 families represented in the European fauna would perhaps then be most accurately conjoined in the following manner:

1. *Epeiroidæ.*
2. *Theridioidæ.*
3. *Scytodoidæ.*
4. *Enyoidæ.*

16. *Attoidæ.*
15. *Eresoidæ.*
14. *Oxyopoidæ.*
13. *Lycosoidæ.*
12. *Thomisoidæ.*

5. *Urocteoidæ.*
6. *Hersilioidæ.*
7. *Agalenoidæ.*
8. *Drassoidæ.*
9. *Dysderoidæ.*
10. *Filistatoidæ.*

11. *Theraphosoidæ.*

---

1) I believe with Darwin, "that *propinquity of descent* — the only known cause of the similarity of organic beings — is the bond, hidden as it is by various degrees of modification, which is partially revealed to us by our classifications" (On the Origin of Species, 4th Ed., p. 489).

2) The old division of the spiders into two great coordinate groups, *Théraphoses* and *Araignées:* Walck. (*Mygalées* and *Aranées:* Dugès), or *Quadripulmonaires* and

ARANEÆ.

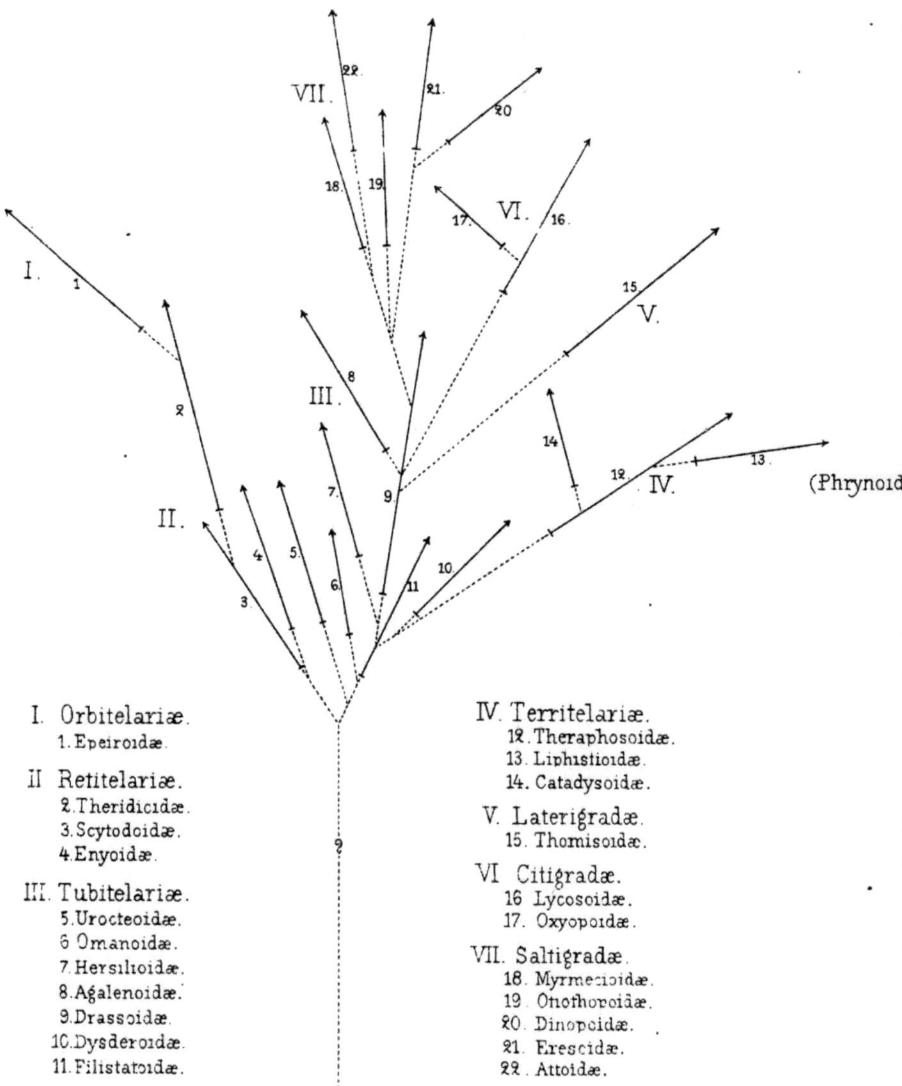

Lith. o tr. h. Schlachter & Seedorff, St.

Concerning the *exotic* families accepted by us a few words appear necessary. The family *Myrmecioidæ* stands almost half-way between the *Attoidæ* and the *Drassoidæ*, and, though forming a continous series with them, cannot well be aggregated to either. — The wonderful genus *Otiothops* MAC LEAY [1]), which differs from all other known spiders in the form of the first pair of legs, which are palpiform, short and thick, without claws, and composed of only 6 joints, certainly appears to stand nearest to *Palpimanus* among the *Eresoidæ*, but cannot easily be united either with that or any other as yet formed family, on which account we have been obliged to set up a new family especially for it, the *Otiothopoidæ*. — Whether the *Dinopoidæ* really ought to occupy the place I have allotted to them, is uncertain: L. KOCH [2]) unites them with the *Eresoidæ*, to which they appear to me to approach nearest: DOLESCHALL [3]) however says of a species described by him, *Dinopis Kollari* from Amboina, that "it builds a very long irregular web between trees, in the middle of which web it sits, with its frontmost pair of legs stretched out." DOLESCHALL places *Dinopis* between *Dolomedes* and *Oxyopes;* even MAC LEAY, who was the first to make known this genus, reckons it to the *Citigradæ*. — As regards the *Catadysoidæ*, I have thought it necessary to form that new family for the North American *Catadysas* [*Katadysas*] *pumilus* HENTZ [4]), which, by having the palpi inserted near the extremity of the maxillæ, and by the longitudinal direction of the mandibular claw, is related to the typical *Theraphosoidæ*, but in other respects seems to approach very near to the *Lycosoidæ:* like them it is said to have only two tracheal ("pulmonary") sacs. — The family *Liphistioidæ* has been

---

Bipulmonaires: DUFOUR (*Tetrapneumones* and *Dipneumones:* LATR.), is now, I think, pretty generally abandoned.

1) On some new forms of Arachn., p. 13. — The name is no doubt formed of *ὠϑέω* or *ὠϑίζω*, run against, push, and *ὤψ*, eye, with reference to the circumstance, that the two posterior intermediate eyes, which are of different size, stand so close together, that the larger seems to push the smaller out of the way. The name in this case ought to be quite otherwise written — one might at least alter it to *Othiotops*. But, as some may perhaps prefer to derive the word from *ὠτίον*, a little ear, and *ϑώψ*, flatterer, however impossible that etymology may appear to me, I have not considered myself at liberty to alter the usually received way of writing the word. In AGASSIZ' Nomencl. Zool. it is derived from *ὠτίον*, auris and *ὤψ* facies!

2) Beschr. neuer Arachn. u. Myriap. (1867), p. 59 (231).

3) Tweede Bijdr. t. de Kennis d. Arachn. v. d. Ind. Arch., p. 11.

4) Araneides of the United States, *in* Boston Journ. of Nat. Hist., Vol. VI, p. 287, Pl. X, fig. 16.

formed for the remarkable genus *Liphistius* SCHIÖDTE [1]) — from the Island of Pinang — which, as we have already observed, differs from all other known spiders in having the back of the abdomen covered with a row of horny plates jointed into one another, and which is destitute of spinners, but in other respects is similar to the *Theraphosoidæ*. — Lastly, the family *Omanoidæ* is identical with the *Œcobiidiæ* BLACKW. [2]), a family, which BLACKWALL has formed for a six-eyed spider with calamistrum and infra-mammillary organ and two claws on the tarsi, and which he has described under the name of *Œcobius navus;* this spider does not however belong to the genus *Œcobius* LUCAS (as will be shown under that genus), and we have therefore changed its name into *Omanus*, and that of the family into *Omanoidæ*. — For the new families formed by us, that have representatives in the fauna of Europe, we shall have occasion fully to account in the following pages.

GENERA ARANEARUM EUROPÆARUM.

Sub-ordo I. ORBITELARIÆ.

Fam. I. EPEIROIDÆ.

Sub-fam. I. EPEIRINÆ.

Gen. 1. *Argiope* SAV. et AUD.
2. *Epeira* (WALCK.).
3. *Cyrtophora* (SIM.).
4. *Singa* (C. KOCH).
5. *Cercidia* THOR.
6. *Zilla* (C. KOCH).
7. *Meta* (C. KOCH).
8. *Tetragnatha* (WALCK.).

Sub-fam. II. ULOBORINÆ.

9. *Uloborus* LATR.
10. *Hyptiotes* WALCK.

Sub-ordo II. RETITELARIÆ.

Fam. I. THERIDIOIDÆ.

Gen. 1. *Pachygnatha* SUND.
2. *Formicina* CANESTR.
3. *Episinus* WALCK.
4. *Argyrodes* SIM.
5. *Tapinopa* WESTR.
6. *Linyphia* (LATR.).
7. *Erigone* SAV. et AUD.
8. *Walckenaera* (BLACKW.).
9. *Nesticus* THOR.
10. *Ero* (C. KOCH).
11. *Phyllonethis* THOR.
12. *Dipœna* THOR.
13. *Theridium* (WALCK.).
14. *Steatoda* (SUND.).

---

1) Om en afvigende Slægt af Spindlernes Orden, p. 5 (621).

2) Descr. of newly disc. spid. fr. the isl. of Madeira, p. 382.

15. *Lithyphantes* THOR.
16. *Lathrodectus* WALCK.
17. *Euryopis* (MENGE).
18. *Asagena* SUND.
19. *Pholcomma* THOR.

Fam. II. SCYTODOIDÆ.

Sub-fam. I. PHOLCINÆ.

Gen. 1. *Pholcus* WALCK.
2. *Spermophora* HENTZ.

Sub-fam. II. SCYTODINÆ.

3. *Scytodes* LATR.
4. *Loxosceles* HEIN. et LOWE.

Fam. III. ENYOIDÆ.

Gen. 1. *Zodarium* WALCK.
2. *Enyo* SAV. et AUD.

Sub-ordo III. TUBITELARIÆ.

Fam. I. UROCTEOIDÆ.

Gen. 1. *Uroctea* DUF.
2. *Œcobius* LUC.

Fam. II. HERSILIOIDÆ.

Gen. 1. *Hersiliola* THOR.

Fam. III. AGALENOIDÆ.

Sub-fam. I. AMAUROBIINÆ.

Gen. 1. *Dictyna* SUND.
2. *Argenna* THOR.
3. *Titanœca* THOR.
4. *Amaurobius* (C. KOCH).
5. *Lethia* (MENGE).

Sub-fam. II. AGALENINÆ.

6. *Cybæus* L. KOCH.
7. *Cœlotes* BLACKW.
8. *Tegenaria* (LATR.).
9. *Cryphœca* THOR.
10. *Hahnia* (C. KOCH).
11. *Agalena* (WALCK.).
12. *Histopona* THOR.
13. *Textrix* SUND.
14. *Hadites* KEYSERL.
15. *Agrœca* WESTR.

Sub-fam. III. ARGYRONETINÆ.

15. *Argyroneta* LATR.

Fam. IV. DRASSOIDÆ.

Gen. 1. *Zora* (C. KOCH).
2. *Apostenus* WESTR.
3. *Trachelas* L. KOCH.
4. *Liocranum* L. KOCH.
5. *Anyphæna* SUND.
6. *Clubiona* (LATR.)
7 *Chiracanthium* C. KOCH.
8. *Phrurolithus* (C. KOCH).
9. *Micaria* WESTR.
10. *Drassus* (WALCK.).
11. *Melanophora* C. KOCH.
12. *Gnaphosa* (LATR.).
13. *Thysa* KEMP.

Fam. V. DYSDEROIDÆ.

Gen. 1. *Segestria* LATR.
2. *Schœnobates* BLACKW.
3. *Ariadne* SAV. et AUD.
4. *Dysdera* (LATR.).
5. *Harpactes* TEMPL.
6. *Oonops* TEMPL.
7. *Stalita* SCHIÖDTE.

Fam. VI. FILISTATOIDÆ.

Gen. 1. *Filistata* LATR.

Sub-ordo IV. TERRITELARIÆ.

Fam. I. THERAPHOSOIDÆ.

Gen. 1. *Atypus* LATR.
2. *Cyrtauchenius* THOR.
3. *Nemesia* SAV. et AUD.

4. *Diplura* (C. KOCH).
5. *Trechona* (C. KOCH).
[6. *Avicularia* (LAM.).]

Sub-ordo V. LATERIGRADÆ.

Fam. I. THOMISOIDÆ.

Sub-fam. I. PHILODROMINÆ.

Gen. 1. *Micrommata* (LATR.).
2. *Sparassus* (WALCK.).
[3. *Heteropoda* (LATR.)].
4. *Selenops* DUF.
5. *Artanes* THOR.
6. *Philodromus* (WALCK.).
7. *Thanatus* C. KOCH.

Sub-fam. II. THOMISINÆ.

8. *Monœses* THOR.
9. *Thomisus* (WALCK.).
10. *Misumena* (LATR.).
11. *Diœa* THOR.
12. *Xysticus* (C. KOCH).
13. *Coriarachne* THOR.

Sub-fam. III. ANETINÆ.
14. *Anetes* MENGE.

Sub-ordo VI. CITIGRADÆ.

Fam. I. LYCOSOIDÆ.

Gen. 1. *Aulonia* C. KOCH.
2. *Lycosa* (LATR.).
3. *Tarentula* (SUND.).
4. *Trochosa* (C. KOCH).
5. *Pirata* SUND.
6. *Dolomedes* (LATR.).
7. *Ocyale* SAV. et AUD.
[8. *Ctenus* (WALCK.).]

Fam. II. OXYOPOIDÆ.

Gen. 1. *Peucetia* THOR.
2. *Oxyopes* LATR.

Sub-ordo VII. SALTIGRADÆ.

Fam. I. ERESOIDÆ.

Sub-fam. I. ERESINÆ.

Gen. 1. *Eresus* WALCK.

Sub-fam. II. PALPIMANINÆ.

2. *Palpimanus* DUF.

Fam. II. ATTOIDÆ.

Gen. 1. *Salticus* (LATR.).
2. *Leptorchestes* THOR.
3. *Epiblemum* (HENTZ.).
4. *Heliophanus* C. KOCH.
5. *Ballus* (C. KOCH).
6. *Marpissa* (C. KOCH).
7. *Menemerus* (SIM.).
8. *Dendryphantes* (C. KOCH).
9. *Euophrys* (C. KOCH).
10. *Philœus* THOR.
11. *Attus* (WALCK.).
12. *Ælurops* THOR.
13. *Yllenus* (SIM.).

Ordo ARANEÆ.

Sub-ordo I. ORBITELARIÆ.

Syn.: Vide infra sub Fam. *Epeiroidæ.*

The European spiders belonging to this sub-order, which comprises only one family, the *Epeiroidæ*, are, as is known, most easily distinguished from their nearest relations, the *Retitelariæ*, by a very low, not transversally impressed forehead, where the distance between the margin of the clypeus and the intermediate of the anterior eyes is *less*, or at any rate not greater, than that between the anterior and posterior intermediate eyes (except in the case of some males with a strongly projecting forehead); in the *Retitelariæ* on the contrary (with the exception of *Tapinopa*) the former distance is *greater* than the latter. They all have *three* strong, genuine claws on the tarsi, of which the two superior are pectinated, and the inferior armed with *two* close and (except in the case of *Hyptiotes*) blunt comb-teeth; in *Uloborus* and *Cyrtophora conica* alone I have observed on that claw only *one* well developed tooth. Moreover the extremity of the tarsus is always previded with two or more *accessory* (or auxiliary) claws. The palpus-claw in the female is always well developed, pectinated or serrated. The first pair of legs is longer than the others, except in *Cercidia* (*Cerceis* MENGE). The spinners are short, the superior and inferior about equal in length.

Fam. I. EPEIROIDÆ.

Syn.: 1817. **"Orbitèles"** LATR. *in* CUVIER, Règne Anim., R. III, p. 86.
1823. **Retiariæ** SUND., Gen. Aran. Suec., p. 15.
1825. **Orbitelæ** LATR., Fam. Nat. du Règne Anim., p. 315.
1833. **Epeirides** SUND., Consp. Arachn., p. 13.
1833. **Araneæ Orbitelariæ** PERTY, Delect. Anim. Art. Bras., p. 193.
1850. **Epeirides** C. KOCK + **Mithraides**, Uebers. d. Arachn.-Syst., 5, p. 8 et 15. ID.

In this family we include all the spiders already reckoned as belonging to it by LATREILLE, who called the family *Orbitelæ*, and by SUNDEVALL, who introduced the appellation *Epeirides*, accordingly all those that spin regular, so-called "*geometrical*" webs. Amongst C. KOCH'S spider-families, besides his *Epeirides*, the *Mithraides*, including the genera *Poltys* and *Mithras* (*Hyptiotes* WALCK.), belong to this family. That *Poltys* C. KOCH is an Epeiroid spider, has been shown by KEYSERLING: [1]) as to *Mithras* C. KOCH or *Hyptiotes*, I have in a previous paper [2]) endeavoured to show its near relationship to *Uloborus* LATR. and consequently its connexion with the Epeiroidæ, whereof more farther on under the Gen. *Hyptiotes*. — WESTRING gives to this family the same limits that we have assigned to it; BLACKWALL on the contrary separates *Uloborus* (*Veleda* BLACKW.) and *Hyptiotes* from the Epeiroidæ and includes them among his *Ciniflonidæ*, on which step we shall presently have occasion to express our opinion (when treating the sub-family *Uloborinæ*) — SIMON'S arrangement of this family, which he calls "*Épéiriformes*" [3]), appears to me not particularly happy. He divides it into four "tribus": 1:o "*Nuctobiens*" or "*Théridio-Épéires*" (for the genus *Nyctobia* SIM. = *Meta* C. KOCH + *Zilla* ID.); 2:o "*Tétragnathiens*" (with the genera *Uloborus*, *Zosis*, *Tetragnatha* and *Argyrodes* SIM. = WALCKENAER'S "Linyphies épéirides"); 3:o "*Epéiriens*" (including *Singa*, *Epeira*, *Nephila*, *Gasteracantha*, *Acrosoma*, *Arachnura* and *Dolophones)*, and 4:o "*Érésiens*" (the genus *Eresus*). — The species of *Argyrodes* SIM. live, according to VINSON, [4]) who assigns them a place among the *Linyphiæ*, parasitically in the "toiles" of certain Epeiroidæ, in which they weave "leurs petits réseaux." VINSON does not say that these "réseaux" have a form different from those of other Linyphiæ, which I think he would not have omitted to do if such had been the case, and it is therefore not probable that the webs of these spiders are like those of the Epeiroidæ, or that *Argyrodes* belongs to that family. — That there is no near connexion between *Eresus* and the Epeiroidæ, it is probably unnecessary to prove. — *Uloborus* and *Zosis* (= *Orithyia* BLACKW.) undoubtedly deserve in combination with *Hyptiotes* to be classed as a separate "tribus" or sub-family on account of the presence of an infra-mammillary organ [5]) and calamistrum: *Tetragnatha* on the contrary is closely related to *Meta*, and that genus, as also *Zilla*, is so intimately connected with *Epeira* and *Singa*, that these 5 genera can hardly be distributed among different sub-families. — MENGE, whose "tribus" *Orbitelæ* corresponds to our Epeiroidæ, divides [6]) the spiders of this tribus described by him, (all of which belong to the sub-family

1) Beschr. neuer etc. Orbitelæ, p. 1 and 22.

2) Till känned. om Mithras och Uloborus, p. 202 et seq.

3) Hist. Nat. des Araignées, p. 233.

4) Aran. de la Réun., Maur. et Madag., p. 259.

5) See above, p. 29.

6) Preuss. Spinn., I, p. 40 and 90.

Epeirinæ NOB.) into two "families", *Epeiridæ* and *Tetragnathidæ*, which however differ only in the position in which the animals place themselves in their webs, and in their methods of capturing their prey and of copulating, but not in any point of bodily formation. — We can admit of only *two* European sub-families, *Epeirinæ* and *Uloborinæ*.

We accordingly arrange the European Epeiroidæ in the following manner:

§ Organum infra-mamillare nullum; metatarsi postici calamistro carent. . . . . I. EPEIRINÆ.

† Maxillæ breves, latitudine non vel parum longiores.

* Series oculorum postica, desuperne visa, fortiter procurva [1]). Cephalothorax sub-planus, parte cephalica parva, humili. . . . 1. *Argiope*.

** Series oculorum postica, desuperne visa, sub-recta vel recurva. Cephalothorax modice convexus, parte cephalica (in feminis saltem) sat magna.

A. Oculi laterales postici a mediis posticis multo longius distantes quam hi inter se.

I. Pedes 1^mi^ paris reliquis longiores.

a. Oculi laterales antici ab anticis mediis sæpissime evidenter, plerumque dimidio — duplo longius distantes quam hi inter se. Abdomen plerumque ovatum, interdum subglobosum, sæpe antice tuberculatum. . . . . . . 2. *Epeira*.

b. Oculi laterales antici ab anticis mediis non vel paullo tantum longius distantes quam hi inter se. Series oculorum posticorum, desuperne visa, evidenter recurva; frons ultra basin mandibularum plus minus prominens.

1. Oculi laterales sæpissime sat late disjuncti. Abdomen postice in formam coni productum vel ibi tuberculatum. . . . . 3. *Cyrtophora*.

2. Oculi laterales sub-contingentes, ab anticis mediis vix vel non longius distantes quam hi inter se. Abdomen cute molli tectum, cylindrato-ovale, cum cephalothorace parce pilosum et sub-nitidum. . . . . . 4. *Singa*.

II. Pedes 4^ti^ paris reliquis longiores. Abdomen cute duriuscula tectum. . . . . 5. *Cercidia*.

B. Oculi laterales postici non vel (in ♂) parum longius a mediis posticis distantes quam hi inter se. Series oculorum posticorum,

---

1) The series is said to be *procurva*, when its *convexity* is directed *backwards*, *recurva*, when the convexity is directed *forwards* (towards the mouth). — The curvature of the anterior series is determined as seen from before, and that of the posterior as seen from above.

desuperne visa, sub-recta; frons non ultra basin mandibularum prominens. Cephalothorax parvus, debilis. Abdomen breviter ovale, sub-depressum . . . . 6. *Zilla.*

†† Maxillæ dimidio — duplo longiores quam latiores. Series oculorum anticorum fortiter recurva. Oculi laterales sub-contingentes, a mediis non multo longius distantes quam hi inter se. . . . . 7. *Meta.*

††† Maxillæ latitudine duplo vel ultra longiores. Series oculorum anticorum sub-recta. Oculi laterales disjuncti. . . . . 8. *Tetragnatha.*

§§ Organum infra-mamillare adest; metatarsi postici calamistro instructi. . . . . II. ULOBORINÆ.

1. Series oculorum antica margini frontis proxima, procurva, postica recurva. . . . . 9. *Uloborus.*
2. Oculi a margine frontis longe remoti, spatium magnum occupantes; series antica procurva, postica recurva, longa. . . . . . . 10. *Hyptiotes.*

Whether *Dolophones notacanthas* (QUOY et GAIM.) [1]) really belong to the *Orbitelariæ* and *Epeiroidæ*, as SIMON thinks, [2]) appears to me tolerably doubtful.

The New-Holland genus *Celænia* THOR., [3]) distinguished by its high clypeus and its long slender anterior pairs of legs developed into a sort of *pedes raptorii*, the two upper tarsal claws of which are of very different size, ought certainly to form a separate sub-family (perhaps a particular family). This genus has 3 claws upon the tarsi, which is said not to be the case with the, as it seems, else so nearly related genus *Thlaosoma* CAMBR., the species of which, according to CAMBRIDGE, [4]) have only *two* tarsal claws, and make no web, but live "beneath folded leaves." CAMBRIDGE assigns *Thlaosoma* to the *Thomisoidæ*, to which family *Celænia* at any rate cannot be aggregated.

WHITE has [5]) under the name of *Linyphia* (*Leucauge*) *argyrobapta* described a spider taken by DARWIN in Brazil, which probably belongs to the *Epeiroidæ;* at least there is reason to suppose so, in consequence of the following notice of the remarkable method pursued by this species in the construction of its web, which WHITE communicates from DARWIN'S MSS. "Web very regular, nearly horizontal, with concentric circles: beneath, but sometimes above, the concentric web, there is an irregular or thin tissue of network; the animal rests in the centre, on the inferior surface."

---

1) Conf. WALCK., Hist. Nat. d. Ins. Apt., I, p. 382.
2) Hist. Nat. d. Araignées, p. 297.
3) Eugenies Resa omkr. Jorden, Zool., Arachn. 1, p. 1.
4) Descr. of a new gen. and six new spec. of Spid., p. 273, 274.
5) Descr. of new or little known Arachn., p. 473.

Sub-fam. I. EPEIRINÆ.

Syn.: 1841. **Epeiridæ** Blackw., The differ. in the numb. of eyes etc., p 668.
1866. **Epeiridæ** Menge + **Tetragnathidæ** id., Preuss. Spinn., I, p. 40, 90.

This sub-family includes all the *genuine* European Epeiroids, i. e. those that are destitute of infra-mamillary organ and calamistrum. The Epeirinæ belonging to the fauna of Europe may, we think, be united under the genera *Argiope*, *Epeira*, *Cyrtophora*, *Singa*, *Cercidia*, *Zilla*, *Meta* and *Tetragnatha*, all which, with the exception of the last, constitute portions of the old Walckenaerian genus *Epeira*.

Gen. 1. ARGIOPE Sav. et Aud. 1825—7.

Deriv.: Ἀργιόπη, mythol. proper name.

Syn.: 1825—7. **Argiope** Sav. et Aud., *in* Descript. de l'Égypte, (2:e Ed.:) XXII, p. 328.
1829. **Argyopes** Latr., *in* Cuv., Règne Anim., Nouv. Éd., V, p. 528.
1831. **Argyope** id., Cours d'Entomol., p. 529.
1835. **Miranda** C. Koch, *in* Herr.-Schæff., Deutschl. Ins., 128, 14.
1839. **Nephila** id., Die Arachn., V, *(ad partem:)* p. 33.
1864. **Argyopes** Sim., Hist. Nat. d. Araignées, p. 281 *(ad max. partem)*.
1864. **Nephila** id., ibid., p. 275 *(ad partem)*.

Type: *Argiope lobata* (Pallas).

We take this genus, which has been separated from Walckenaer's *Epeira* by Savigny and Audouin, in its original compass, which indeed seems to be the same as that assigned to it by Sundevall [1]) and Keyserling [2]), whereas some of the species attributed by C. Koch and Simon to the genus *Argiope*, e. g. *A. tridentatus* and *gonygaster*, hardly seem rightly to belong to it. Sav. and Aud. include (loc. cit. p. 329) *Aranea Brünnichii* Scop. (*Ar. fasciata* Oliv.) in *Argiope*, which is perfectly right. By C. Koch it was first erroneously assigned to *Miranda*, and afterwards to *Nephila* Leach, which is not a more fortunate disposition, as the characteristics that mark this last genus (the anterior part of the cephalothorax elevated, broad; maxillæ considerably longer than they are broad, etc.) by no means apply to *A. Brünnichii* [3]). The genus *Nephila* must, for the present at least, be removed from the list of the European genera of spiders. We suspect in fact that *Epeira ambagiosa* Walck. also is an *Argiope*. Simon indeed classes it with the genus *Nephila* (p. 276), and, following Walckenaer, gives

1) Consp. Arachn., p. 15.
2) Beschr. neuer etc. Orbitelæ, p. 2 (64).
3) Conf. Leach, Zool. Misc., II, p. 133; — Keyserling, loc. cit.

"Espagne" for its country, although on the page immediately following he says of *Arg. Brünnichii* (*Nephila fasciata* SIM.) that it is "la seule espèce Européenne" of the genus *Nephila*, and also omits to include *E. ambagiosa* in his "Catalogue Synonymique"; but it is referred by WALCKENAER himself [1]) to the same (1e) Race of the genus *Epeira's* 4th family ("les Décorées"), to which *E. fasciata*, *aurelia*, *ætherea* and the other species of *Argiope* with unlobated or unnotched abdomen belong.

With reference to the orthography of the name *Argiope*, the following remarks may be reasonably made [2]). Both in the passage of the *Descr. de l'Égypte*, where that genus is described by AUDOUIN (T. XXII, p. 328 of the 2nd Edit.) and in the index to that volume (p. 466), its *Latin* name is *Argiope*, but in *French* he calls it *Argyope* ("Genre *Argyope*, *Argiope*" just as he writes "Genre *Tégénaire*, *Tegenaria*", "Genre *Pholque*, *Pholcus*" etc.). This latter orthography has moreover since been used not only in the French but also in the Latin names [3]) of the species, and this has caused several subsequent authors to write *Argyope* instead of *Argiope*. As however AUDOUIN *first*, and in *characterizing the genus*, wrote *Argiope*, that orthography must be preserved, especially as it is that which is etymologically right (see the derivation of the word above), and the unreasonable spelling *Argyope*, which has been received by LUCAS, WALCKENAER and others — including myself, [4]) before I had the opportunity of consulting the *Descr. de l'Egypte* — must be abandoned. LATREILLE first (loc. cit. in Syn.) changed *Argiope* into *Argyopes*, in which he has been followed by SUNDEVALL, C. KOCH, KEYSERLING and others, but shortly after wrote *Argyope* (loc. cit.). — The genus will, it is to be hoped, hereafter retain its original and proper name: *Argiope* SAV. et AUD.

---

1) Hist. Nat. d. Ins. Apt., II, p. 113.

2) Conf. THORELL, Om *Aranea lobata* PALL., p. 596.

3) The confounding of *i* and *y* in names borrowed from the Greek is not very rare among French writers. SIMON for ex. has in his Hist. Nat. des Araignées, p. 433 formed a genus that he calls *Pachyptila*, but p. 526, where he reckons up the European species of that genus, he calls it *Pachyptyla*.

4) Nya exot. Epeirider, p. 299.

Genus 2. EPEIRA (Walck.). 1805.

Deriv. unknown [1]).

Syn.: † 1804. **Aranea** Latr., *in* Nouv. Dict. d'Hist. Nat., XXIV, p. 135 *(ad partem).*
1805. **Epeira** Walck., Tabl. d. Aran., p. 53 *(ad partem).*
1837. **Miranda** C. Koch, Uebers. d. Arachn.-Syst., 1, p. 4 *(ad max. part.).*
1837. **Epeira** id., ibid. p. 1.
1837. **Atea** id., ibid., p. 3.
1861. **Epeira** Wesrtr., Aran. Suec., p. 20 *(ad max. part.).*
1864. „ Blackw., Spid. of Gr. Brit., II, p. 323 *(ad max. part.).*
1864. „ Sim., Hist. Nat. d. Araignées, p. 259 *(ad max. part.).*

Type: *Epeira diademata* (Clerck).

Among modern authors only Lucas, Blackwall, Vinson and a few others have preserved Walckenaer's genus *Epeira* (= *Aranea* Latr. 1804) undivided. Separate groups of species have from time to time been taken from it, and formed into particular genera, and although several of these (especially those formed by C. Koch) are very imperfectly characterized, they have yet, though often with modified limits, been more or less generally accepted. The greater part of the species included by Koch in his genera *Miranda* and *Atea* we refer to *Epeira* strictly so called: his *Meta*, *Zilla* and *Singa* (with the limits assigned to the two first by Westring and to the last by Menge and ourselves) may on the other hand be suffered to retain their place as independent genera. The determination of *Epeira* sensu strictiori given by us in Rec. crit. (p. 106) has been adopted by Westring (loc. cit.); we now however think that *E. conica* were better separated from *Epeira* and referred to a separate genus, *Cyrtophora* (Sim.). Simon's view of the genus *Epeira* agrees with that of Westring, except that, in conformity with Koch, he reckons *E. conica* to *Singa.* He moreover divides *Epeira* into three sub-genera, *Miranda*, *Atea* and *Epeira*, and this last sub-genus he farther divides into six "groups", *Nyctinea [Nuctenea]*, *Eriophora*, *Neoschœna [Neoscona]*, *Neopora*, *Epeira* and *Cyrtophora*, most of which appear to be even more unnecessary and still worse defined than

---

1) In Agassiz' *Nomencl. Zool.* it is derived from "ἐπείρομαι, examinor", a derivation which appears to me destitute of all grounds. According to Simon, *Epeira* comes from "ἐπείρω, faire un tissu", which verb I have not been able to find in any Greek Lexicon to which I have access. — May not the name perhaps be formed of ἐπί, *on*, and εἶρος, *wool* (with reference to the circumstance of the female's being usually found, after laying her eggs, sitting beside or upon the wool-like cocoon)? — As however all this is but conjecture, I have not ventured, like Erichson and Schiödte, to change the received orthography *Epeira* into *Epira.*

the sub-genera proposed by KOCH. — Of KOCH'S above named new genera KEYSERLING adopts only *Meta* [1]).

MENGE [2]), in dividing WALCKENAER'S genus *Epeira*, has adopted KOCH'S *Singa*, *Zilla*, *Zygia*, *Miranda* and *Meta*, as also *Epeira*, but not *Atea*, and has moreover formed two new genera, *Cyclosa* (for *E. conica*) and *Cerceis* (for *Singa prominens* WESTR.) But as he, in marking the characteristics of these genera, fixes his attention almost exclusively on differences in the organs of generation, differences which do not seem to me sufficient to distinguish genera, I have not for the present considered it advisable to adopt his division entirely unaltered, more especially as, in many of the species, only one sex is known, and MENGE has therefore in some cases been uncertain whether the species really belong to the genus to which he has aggregated them or not. To *Miranda* he assigns, besides certain species referred to it by C. KOCH, *Ep. acalypha* WALCK.; whereas, according to the characteristics given by MENGE, the following for example do not belong to his *Miranda: Epeira armida* SAV. and AUD. and *M. ceropegia* C. KOCH (which is not identical with *E. ceropegia* WALCK., as we shall hereafter, when we come to examine WESTRING'S *E. ceropegia*, show), although they are very nearly related to *Epeira ceropegia* WALCK. and *E. adianta* ID., which MENGE assigns to *Miranda*. — The greatest part of the species, that compose KOCH'S *Atea*, are referred by MENGE, as also by us, to *Epeira*. — *Zygia (Z. atrica)* differs only by modifications of the organs of copulation from *Zilla*, which genus in MENGE'S work embraces only *Zilla montana* WESTR. (non KOCH), described by MENGE under the name of *Z. calophylla*. The true *Z. calophylla* (WALCK.) C. KOCH does not, according to the characteristics given by MENGE, belong either to *Zygia* or *Zilla*. — *Singa prominens*, which has the 4th pair of legs longer than the others, well deserves to be received as the type of a new genus, *Cercidia* NOB. (= *Cerceis* MENGE). — *Ep. conica* (and probably also *Ar. citricola* FORSK., *Ep. oculata* WALCK. and some others) ought to be united with *Ep. opuntiæ* DUF., which by the greater distance between the lateral eyes, and by the peculiar form of its abdomen, seems to form the type of a separate genus, *Cyrtophora* SIM. ad part.

OHLERT [3]) has latterly endeavoured more accurately to determine and fix the genera reckoned by C. KOCH to the Epeiroidæ, which are represented in the Prussian fauna. He first divides them into three groups. In the first group (including the genera *Epeira*, *Singa*, *Miranda* and *Atea*)

1) Beitr. z. Kenntn. d. Orbitelæ, p. 2 (800).
2) Preuss. Spinn., p. 41.
3) Die Aran. d. Prov. Preussen, p. 20—21.

the 3rd pair of legs is *more* than half as long as the first pair; in the second group (*Zygia* and *Zilla*) the length of the 3rd pair is *less* than half that of the 1st; and lastly in the third group (*Meta*) the length of the 3rd pair is *equal* to half that of the first. The first group is also distinguished from the second and third by the form of the 4th joint of the male's palpi, which is short, broad and bowl-formed, not, as in the case of the others, cylindrical and of greater length than breadth — a characteristic accordingly, that holds good only for one sex. The distinctions derived from the relative length of the 1st and 3rd pairs of legs do not however hold even for the few species that OHLERT has treated: in his *Zilla acalypha* (at least ♀) for ex. the 3rd pair of legs is not at all shorter but on the contrary considerably longer than half the 1st, and that species ought therefore not to have been referred to *Zilla* but to *Epeira* (to which genus I have aggregated it) or *Singa*. Moreover this relation is sometimes different in the different sexes of the same species.[1]) — Within the first group, *Epeira* and *Singa* on the one hand are distinguished from *Miranda* and *Atea* on the other by the 4 intermediary eyes of the first named genera forming a *trapezoid*, of which the posterior side is shorter than the anterior, and being all of about the same size, whereas in the latter they form a *rectangle*, and the hindermost of them are sensibly larger than the anterior ones. Of how little consequence however these differences are, will doubtless be immediately seen by anybody who has examined a larger number of not only European Epeiroidæ. If suchlike trifling differences in the position and size of the eyes are to be considered as decisive in the formation, within this family,

1) In a large full-grown female of *Meta Menardi* (LATR.), I find the length of the 1st and 3rd pairs of legs, reckoned from the edge of the cephalothorax, respectively 32 ½ and 19 ½, or, if the coxæ be taken into account, 34 and 21 millimeters; accordingly the 3rd pair of legs *more* than half as long as the 1st. This is also certainly the case in the male *M. Menardi:* in the only full-grown specimen I have of this spider, the tarsi of the 1st pair are wanting, but if these be considered as only half the length of the metatarsi, the 3rd pair in this specimen will still be longer than half the first. — In a moderate-sized ♂ of *Meta Merianæ* (SCOP.) I have indeed found the 1st pair of legs just double the length of the third, when the coxæ are included, but in ♀ the case is otherwise: in a small, but fullgrown female specimen I found the first pair of legs 14mm. and the 3rd pair 8 ½mm., reckoned from the edge of the cephalothorax; reckoned from the bases of the coxæ these pairs were respectively 15mm. and 9 ½mm. long. — In ♂ of *Zilla reticulata* C. KOCH or *Meta segmentata* (CLERCK) the first pair is more than double, nearly 3 times, as long as the 3rd, but in the *female* scarcely double: if the coxæ are included, the 3rd pair is at least *sometimes* longer than half the first.

of generic groups, the genera will be in the highest degree artificial, and species in all other respects very nearly related will be torn from each other and referred to different genera: e. g. *E. angulata* will be separated from *E. bicornis*, which must be an *Atea* Ohl., *E. sclopetaria* from *E. sollers*, which would also, according to this system, belong to the genus *Atea; Singa pygmæa* (*S. Herii* Westr.) and *S. albo-vittata* from *S. hamata*, and so forth. — *Atea* [1]) according to Ohlert is distinguished from *Miranda*, by the lateral eyes being in the last named genus *more* than double, whereas in *Atea* they are *less* than double so far removed from the intermediary eyes, as these are from each other, and the anterior series being in *Atea* straight and in *Miranda* slightly curved backwards. (According to C. Koch the anterior intermediary eyes in *Atea* are farther apart than the posterior, whereas according to Ohlert these four eyes are placed rectangularly, — and yet both these authors refer the same species, *Ep. agalena* Walck., to the genus *Atea.*) — Ohlert distinguishes *Singa* from *Epeira* by the side-eyes being in *Singa* at the *same* distance, and in *Epeira* at *double* the distance from the intermediary eyes, which the anterior intermediary eyes are from each other. If there were no transitions (as however there are both to *Atea* and *Miranda*), this would be a very good characteristic; and the genus *Singa* is perhaps the only one of Koch's new genera here mentioned, which can in the present state of the science be retained, not so much however on account of anything characteristic in the position of the eyes, as for the animal's general form and appearance, somewhat resembling that of certain Theridioidæ. (Conf. Westring, Aran. Suec., p. 56). — In addition to the above-mentioned really trifling and moreover not even constant difference in the proportion of the length of the 1st and 3rd pairs of legs, which, according to Ohlert, exists between *Zygia* and *Zilla* on the one side and *Meta* on the other, the two first of these genera are stated to have their eyes *rectangularly* placed, and the lateral *little more* distant from the intermediary eyes than these latter from each other, whereas in the case of *Meta* the anterior intermediary eyes are *somewhat nearer to each other* than the posterior, and the side eyes at the *same* distance from the intermediate as these latter from each other. All these are, as we have already said, extremely weak and insignificant distinctions, which have not even the merit of being reliable, for e. g. Koch's and Ohlert's *Zilla reticulata*

---

1) *Atea melanogaster* C. Koch (*Dipæna melanogaster* nob.) is not an *Epeiroid* at all, but belongs to the family *Theridioidæ*. Vid. infra: Gen. *Dipæna* of that family.

has the intermediary eyes placed precisely in the same manner as those of their *Meta fusca*. — Concerning these three genera, vid. infra: Gen. *Zilla* and *Meta* (p. 59, 61).

Genus 3. CYRTOPHORA (SIM.). 1864.

Deriv.: κυρτός, crooked, bowed; φέρω, bear.

Syn.: 1837. **Singa** C. KOCH, Uebers. d. Arachn.-Syst., 1, p. 6 *(ad partem)*.
1861. **Epeira** WESTR., Aran. Suec., p. 20 *(ad partem)*.
1864. „ BLACKW., Spid. of Gr. Brit., II, p. 323 *(ad partem)*.
1864. **Singa** Hist. Nat. d. Araignées, p. 255 *(ad partem)*.
1864. **Epeira**: «groupe» **Cyrtophora** ID., ibid., p. 262 *(ad partem)*.
1866. **Cyclosa** MENGE, Preuss. Spinn., I, p. 73.

Type: *Cyrtophora opuntiæ* (DUFOUR).

The spiders which, in consequence of peculiarities in the form of the abdomen, and usually also in the position of the eyes (conf. p. 49), we assemble under this generic name, are chiefly exotic: the only species found in northern Europe is *Ar. conica* PALLAS, which, sometimes referred to *Epeira*, sometimes to *Singa*, sometimes elevated to the rank of a separate genus, appears to us to stand in very close relation to *Epeira opuntiæ* DUF. [1]), a spider which we take as typical of the genus *Cyrtophora*, and which seems to be so considered by SIMON (loc. cit.). For *C. conica*, MENGE has formed the genus *Cyclosa*, which name, being younger, must be rejected in favour of that given by SIMON. SIMON however aggregates to *Cyrtophora* several species, which surely have little or no relationship with *C. opuntiæ:* such are *Epeira mexicana* LUCAS, *E. paradoxa* ID. — which would seem to be a *Cyrtarachne* THOR. (*Cyrtogaster* KEYSERL.) — and *E. mitralis* VINSON, belonging to the genus *Cærostris* THOR., of the other species of which genus one, *C. tuberculata* (VINS.), is included by SIMON in the "groupe" *Eriophora* of the genus *Epeira*, and another, *C. imperialis* (WALCK.), in the sub-genus *Eurysoma* of the genus *Gasteracantha*. [2]) Further on in his work (p. 494, 495) he assigns even the species of the "groupe" of the genus (and sub-genus) *Epeira*, for which he had first specially reserved that name, (*E. angulata*, *E. bicornis*, etc.), to *Cyrtophora*, whereby this latter group is certainly not made more natural.

---

1) DUFOUR, Descr. de six Arachn. nouv., p. 359, Tab. LXIX, fig. 3.

2) SIMON, loc. cit., p. 261 and 284.

In *C. opuntiæ*, as in most Epeiroidæ, the inferior tarsal claw has two comb-teeth in close juxtaposition, but in *C. conica* I have, as aforesaid, not been able to discover more than one distinctly developed tooth (and sometimes a very small point behind it) on that claw.

Genus 4. SINGA (C. KOCH). 1836.

Deriv.: *Σίγγα*, geogr. prop. name.

Syn.: 1836. **Singa** C. KOCH, Die Arachn., III, p. 42 } (*ad max. part.*).
1837. „ ID., Uebers. d. Arachn.-Syst., p. 6 } (*ad max. part.*).
1861. „ WESTR. Aran. Suec., p. 56 } (*ad max. part.*).
1864. **Epeira** BLACKW., Spid. of Gr. Brit., II, p. 323 (*ad partem*).
1864. **Singa** SIM., Hist. Nat. d. Araignées, p. 255 (*ad max. part.*).
1866. „ MENGE, Preuss. Spinn., I, p. 81.

Type: *Singa hamata* (CLERCK).

This genus, formed by C. KOCH at the expense of WALCKENAER'S *Epeira*, is not ackowledged by BLACKWALL, but received by WESTRING, who however assigns it somewhat different limits from those given it by KOCH (and SIMON), paying more attention to the form of the cephalothorax and abdomen and the consequent peculiarities in the animal's general appearance, than to the position of the eyes, which but slightly deviates from that of certain *Epeiræ*, viz. those in which the lateral eyes, like those of *Singa*, are little, if at all, more distant from the intermediary eyes, than these latter from each other. With MENGE, we assign to *Singa* the same species that are referred to it by WESTRING, excepting *S. prominens* (and *S. scutifera* WESTR.), for which MENGE has formed the genus *Cerceis* (*Cercidia* NOB.). The limits between *Epeira* and *Singa* are however by no means sharply defined, and it is not without some doubts that I have here taken up *Singa* as an independent genus separate from *Epeira*. As its type KOCH himself gives *S. hamata* (CLERCK). Concerning this genus see also under Gen. *Epeira* (p. 53).

Genus 5. CERCIDIA N.

Deriv.: *κερκίς*, shuttle.

Syn.: 1851. **Epeira** WESTR., Förteckn., p. 33 } (*ad partem*).
1861. **Singa** ID., Aran. Suec., p. 56 } (*ad partem*).
1864. **Epeira** BLACKW., Spid. of Gr. Brit., II, p. 323 } (*ad partem*).

1) Uebers. d. Arachn.-Syst., loc. cit.

1865. **Atea** OHLERT, Arachnol. Studien, p. 9.
† 1866. **Cercels** MENGE, Preuss. Spinn., p. 80.

Type: *Cercidia prominens* (WESTR.)

This genus, set up by MENGE under the already engaged name of *Cerceis* [1]) for WESTRING'S *Singa prominens*, differs in sundry important particulars from the other European Epeiroidæ. The hard, almost leather-like skin of the abdomen, and the unusual relative lengths of the legs (the 4th pair being longer than the first) are sufficient to justify the formation of this new genus; but in addition to this there are, according to MENGE, peculiarities in the structure of the organs of generation, and a small transversal opening in front of the spinners, in which 4 fine tubes (supposed by MENGE be air-tubes) terminate [2]). The cocoon is like that of *Ero* but without shaft. Conf. MENGE, loc. cit., p. 80, 81.

Genus 6. ZILLA (C. KOCH). 1834.

Deriv.: *Zilla*, prop. name.

Syn.: † 1834. **Zygia** C. KOCH, *in* HERR.-SCHÆFF., Deutschl. Ins., 123, 17—19.
1834. **Zilla** ID., ibid. (*ad part.:*) 125, 19.
1837. „ ID., Uebers. d. Arachn.-Syst., 1, p. 5 (*ad partem*).
† 1845. **Eucharia** ID., Die Arachn., XII, (*ad part.:*) p. 103.
1861. **Zilla** WESTR., Aran. Suec., p. 68.
1364. **Epeira** BLACKW., Spid. of Gr. Brit., II, p. 323 (*ad partem*).
1864. **Nyctobia** [**Nuctobia**]: sub-gen. **Zilla** SIM., Hist. Nat. d. Araignées, p. 236, 237 (*ad partem*).
1866. **Zilla** MENGE, Preuss. Spinn., I, p. 76.
1866. **Zygia** ID., ibid., p. 77.
1867. „ OHLERT, Aran. d. Prov. Preuss., p. 21.

Type: *Zilla x-notata* (CLERCK.)

The genera *Zilla* and *Meta* appear to us to have with good reason been separated by C. KOCH from *Epeira* WALCK.: they form in many respects a transition to the family *Theridioidæ*, and are distinguished from *Epeira* not only by SIMON, MENGE and OHLERT, but also by WESTRING; BLACKWALL however still includes the species of this genus under *Epeira*. SIMON has even formed of them a separate tribe of Epeiroidæ, which, as we have seen above, he calls "*Nuctobiens*" or "*Théridio-Épéires*"; they form however

1) Vid. sup. p. 37.
2) As to these air-tubes see however p. 30, note 1).

in his work but one genus, *Nyctobia* [*Nuctobia*] SIM., with *Meta* and *Zilla* as sub-genera. (On this subject see further under the article *Meta*). As type for the genus *Zilla*, KOCH [1]), it is true, gives *Z. albimacula* (*Ep. diodia* WALCK.), but as we feel ourselves obliged to refer that species to the genus *Epeira*, we have assumed as typical species *Zygia calophylla* (WALCK.) C. KOCH = *Z. x-notata* (CLERCK), which appears to us best to express the peculiarities of the genus, is the first species of the genus described, and has since been by KOCH himself assigned to *Zilla*, together with the very closely related *Zilla montana* C. KOCH. He accordingly himself combined in one the genera *Zygia* and *Zilla*, and entirely suppressed the former name. In *Zilla* we also include a part of the unnatural genus *Eucharia* C. KOCH — which is formed chiefly of *Theridioidæ* and is synonymous with *Steatoda* (SUND.) NOB. — viz. *E. atrica* C. KOCH. KOCH'S *Z. reticulata* (*Ar. segmentatus* CLERCK.) we refer to *Meta*, his *Z. albimacula* and *Z. acalypha* to *Epeira*. The limits assigned by us to the genus are accordingly the same as those proposed by WESTRING (loc. cit.), which limitation has the right of priority before that adopted by SIMON and OHLERT. In SIMON, *Zilla* is a sub-genus under his *Nyctobia*, with about the same limits as in KOCH, but includes also species [2]), which according to the characteristics given by SIMON himself and derived from the position of the intermediary eyes, ought to be referred to his sub-genus *Meta*. — We have already above in treating of *Epeira* (p. 55, 56) expressed our opinion that the characters whereby OHLERT distinguishes *Zilla* and *Zygia* from *Meta* are utterly insufficient. MENGE and OHLERT readopt the genus *Zygia* abandoned by KOCH: MENGE separates it from *Zilla* in consequence of some difference in the organs of generation in both sexes, OHLERT only in consequence of the greater length, *in the males*, of the 4th joint of the palpi. How nearly related to each other the species are, that compose the genera *Zygia* and *Zilla* of these authors, is evidenced by the fact, that *Zilla atrica* and *Z. x-notata* (*calophylla*), which are so like one another as to be considered both by WALCKENAER and SUNDEVALL as one and the same species, and to have been confounded by OHLERT himself in his synonyms, are the former a *Zygia* OHL. and the latter a *Zilla* OHL. The name *Zygia* was disposed of long before it was applied by KOCH to denote a genus of spiders, and accordingly cannot on any terms be retained [3]). Also OHLERT assigns KOCH'S

---

1) Uebers. d. Arach.-Syst., loc. cit.

2) *Zilla inclinata* and *Z. antriada* (WALCK.) SIM., which latter is synonymous with *Meta Merianæ* C. KOCH, but by no means with *Zilla montana* ID.

3) *Zygia* FABR. [Coleopt.] 1775.

Zilla reticulata (*segmentata*) and *Zilla acalypha* (*Miranda acalypha* MENGE) to the genus *Zilla;* we, in unison with WESTRING and MENGE, refer the first mentioned (as has been already observed) to *Meta*, whereas the other appears to us to be an *Epeira*. *Z. reticulata* in fact agrees in the form of the maxillæ, the position of the eyes, and in its general appearance almost exactly with the species that typify *Meta* (*M. Menardi* (LATR.), *M. Merianæ* (SCOP.) C. KOCH); *Z. acalypha* again in the relative lengths of the legs, and, as far as we can judge from BLACKWALL'S and MENGE'S figures of the male, also in the structure of the palpi of that sex, closely resembles the species of *Epeira* WESTR.: both differ considerably from *Zilla x-notata*, *atrica* and *montana*, which, while by their short maxillæ they approach *Epeira* and *Singa*, by their weak cephalothorax and depressed, short and rounded abdomen exhibit a close analogy with the genus *Steatoda* (SUND.), an analogy, which led KOCH into the error of uniting within his genus *Eucharia* species of *Steatoda* and *Zilla*, which belong to quite different families.

Genus 7. META (C. KOCH). 1836.

Deriv.: *Μήτα*, mythol. prop. name.

Syn.: 1836. **Meta** C. KOCH, *in* HERR.-SCHÆFF., Deutschl. Ins., 134, 12, 13; 135, 14—16.
1837. „ ID., Uebers. d. Arachn.-Syst., 1, p. 6.
1856. „ THOR., Rec. crit. aran., p. 106.
1861. „ WESTR., Aran. Suec., p. 75.
1864. **Epeira** BLACKW., Spid. of Gr. Brit., II, p. 323 (*ad partem*).
1864. **Nyctobia [Nuctobia]** sub-gen. **Meta** SIM., H. N. d. Araignées p. 236, 237.
1864. „ „ : sub-gen. **Zilla** ID., ibid. (*ad partem*).
1864. **Tetragnatha** KEYSERL., Beschr. neuer etc. Orbitelæ, p. 21 (64) (*ad partem*).
1866. **Meta** ID., Beitr. z. Kenntn. d. Orbitelæ, p. 2 (800) (*ad partem*).
1866. „ MENGE, Preuss. Spinn., I, p. 86.

Type: *Meta Menardi* (LATR.).

As *Zilla* (C. KOCH) shows analogy with *Steatoda* (SUND.), so does *Meta* form a transition to *Linyphia* (LATR.).—This genus, which was formed by C. KOCH, and by him referred to the family *Theridioidæ*, notwithstanding that the species cited as its type, *M. fusca* C. KOCH = *M. Menardi* (LATR.), as also *M. Merianæ* C. KOCH and *M. muraria* ID., are, as is probably generally known, true *Epeiroidæ*, is by KOCH so ill defined, that also two real Theridioidæ, *Meta cellulana* C. KOCH and *M. tigrina* ID. = *Linyphia socialis* SUND., are besides by him included in it. It has accordingly been since restricted by me (loc. cit.), and subsequently by WESTRING and MENGE, to those of KOCH's *Meta*-species, which are really *Epeiroidæ*, and some very closely related species, e. g. *Ar. segmentatus* CLERCK (*Ar. reticulata* LINN.), which by KOCH and some others is erroneously referred to *Zilla.* All these species have in their *elongated maxillæ* a common characteristic, which distinguishes them from the other European generic groups separated from WALCKENAER'S *Epeira* (vid. sup.). BLACKWALL refers the species of this genus to *Epeira.* SIMON has, as we have already seen, combined *Meta* and *Zilla* to one genus, which he calls *Nyctobia* [1]), but which however did not require a new name, as there were two older names to choose between. He divides it into two sub-genera, which preserve their Kochian names, but of the species that compose KOCH'S *Meta*, he refers only *M. fusca* KOCH *(M. Menardi)* to *Meta* (see more above under *Zilla*). OHLERT aggregates also *M. Merianæ* C. KOCH to *Meta.* The difference in the position of the eyes assumed by these writers as ground of distinction between the genera *Meta* and *Zilla* is too trifling to be acknowledged as of any decisive weight in the characterization of genera; we have accordingly, as above mentioned, been obliged to transfer one of their *Zilla*-species *(Z. reticulata)* to *Meta.* Also KEYSERLING, who at first united the genera *Meta* and *Tetragnatha*, though he afterwards admitted the independence of the former genus, finds its principal characteristic in the elongated maxillæ, but he assigns to it not only those Epeiroidæ which C. KOCH referred to *Meta*, but also the "1re famille, *Coadunatæ*", of WALCKENAER'S *Tetragnatha* [2]).

Genus 8. TETRAGNATHA LATR. 1804.

Deriv.: *τετρα-*, four, *γνάθος*, jaw [3]).

Syn.: 1804. **Tetragnatha** LATR., *in* Nouv. Dict d'Hist. Nat., XXIV, p. 135.
1805. „ WALCK., Tabl. d. Aran., p. 68.

---

1) SIMON'S reason for this name (from *νύξ*, night, and *βιόω*, live) probably was that he considered VINSON'S "Épeires nocturnes" as corresponding to KOCH'S *Meta* and *Zilla.* At least he says (p. 238) of the species of these genera, or *Nyctobia* SIM.: "M. VINSON les a appelées "*Épeires nocturnes.*" Nevertheless, on the preceding page, where he reckons up the species of the genus *Nyctobia*, SIMON curiously enough has not taken up *a single one* of VINSON'S "Épeires nocturnes." He classes them all under his *Epeira* and *Nephila*, which indeed appears to us right, but is difficult to reconcile with his above quoted words. (Conf. VINSON, Aran. de la Réun., Maur. et Madag., p. 153 et seq.)

2) WALCK., Hist. Nat. d. Ins. Apt., II, p. 219.

3) Among the ancients *τετράγναθον* was the name of a venomous kind of arachnoid animal, probably a *Galeodes.*

1825—7. **Eugnatha** SAV. et AUD., *in* Descr. de l'Égypte (2:e Ed.:) XXII, p. 323 (*ad partem*).
1843. **Dinognatha** [**Deinagnatha**] WHITE, *in* DIEFFENBACH, Trav. in New Zeal., II, p. 271 (*sec.* WHITE [1])).
1861. **Tetragnatha** WESTR., Aran. Suec., p. 83.
1864. **Tetragnatha** BLACKW., Spid. of Gr. Brit., II, p. 367.
1864. „ SIM., Hist. Nat. d. Araignées, p. 248 (*saltem ad part.*).

Type: *Tetragnatha extensa* (LINN.)

To this genus, generally understood as comprising the compass assigned to it by LATREILLE and WALCKENAER, KEYSERLING [2]) also reckons a spider, *Eugnatha filiformis* SAV. and AUD., which by WALCKENAER and SIMON is referred to *Uloborus*, and in WALCKENAER forms the "3me Race" of that genus' "1re famille" [3]). It certainly belongs to the sub-family *Epeirinæ*, and approaches much nearer to *Tetragnatha* than to *Uloborus*, but appears to us, on account of its agreement with *Uloborus* in the relative length of the legs and the position of the eyes, by its shorter maxillæ, its abdomen of almost thread-like dimensions and drawn out in a point, etc., to deserve to be considered as the type of a separate genus. We have elsewhere [4]) proposed to reserve to that genus the name of *Eugnatha*, under which SAVIGNY and AUDOUIN united WALCKENAER'S *Tetragnatha* with that writer's *Uloborus filiformis* (loc. cit.) [5]). — The spiders, which WALCKENAER (loc. cit., p. 219) classes under the genus *Tetragnatha's* 2nd family, with the name "*Coadunatæ*", KEYSERLING, as we have just seen, refers to *Meta* KOCH. It seems to us that they ought rather to form a separate genus betwen *Meta* and *Tetragnatha*. The "3me Famille" of *Tetragnatha* WALCK., "*Lézardiformes*", (loc. cit. p. 224) might also well be separated from *Tetragnatha* LATR., and that genus be thus restricted to the compass, which it originally had in LATREILLE and in WALCKENAER'S *Tableau des Aranéides*. — The di-

---

1) WHITE, Descr. of a new gen. of Arachn., w. notes on two other spec. of spiders, p. 13.

2) Beiträge zur Kenntniss der Orbitelæ, p. 38 (836).

3) Hist. Nat. d. Ins. Apt., II, p. 230.

4) Till kännedomen om slägtena Mithras och Uloborus, p. 194.

5) Near this genus *Eugnatha* (SAV. and AUD.) stands DOLESCHALL'S *Ariadne* (Bijdr. tot de Kenn. d. Arachn. v. d. Ind. Arch., p. 410) by the relative lengths of the legs and the form of the abdomen (which is still more drawn out behind the spinners and thread-like), but the position of the eyes is quite different. As the name *Ariadne* had been already in 1825—7 applied by SAV. and AUD. for another genus of spiders, we have (p. 37) proposed to exchange the name *Ariadne* DOLESCH. for *Ariamnes*.

stinctive marks, on which WHITE formed the genus *Dinognatha*, are too trifling to warrant the separation of the spiders thus characterized from *Tetragnatha*.

Sub-family II. ULOBORINÆ.

Under this denomination I include those Epeiroidæ, which are provided with *infra-mammillary organ and calamistrum.* The spiders of this sub-family known to me all agree in at least one more respect, viz. that their *legs are not armed with spines.* Of the two European genera that belong to this sub-family, *Uloborus* and *Hyptiotes*, the first named had, ever since it was first so classed by LATREILLE, been considered as an *Epeiroid*, until BLACKWALL discovered in *U. Walckenarii (Veleda lineata* BLACKW.) the agreement, which *Uloborus*, by the presence of an infra-mammillary organ and calamistrum, exhibits with *Amaurobius* (C. KOCH) = *Ciniflo* BLACKW. and *Dictyna* SUND. = *Ergatis* BLACKW. That agreement induced BLACKWALL to refer *Uloborus* to the family *Ciniflonidæ*, formed by him in 1841 for the two above named genera[1]), and placed between his *Drassidæ* and *Agelenidæ*, so that by BLACKWALL *Uloborus* is widely separated from the Epeiroidæ. KEYSERLING[2]) assents to the opinion of BLACKWALL. — We have already in the preceding pages (p. 29) drawn attention to the unnatural character of the family *Ciniflonidæ*, in that it not only brings together forms so widely separate as e. g. *Uloborus* and *Amaurobius*, but even includes *Eresus* and *Dinopis*, for also these genera have an infra-mammillary organ and calamistrum. As regards especially *Uloborus*, it appears to me that its agreement with *Epeira* and *Tetragnatha* as well in the form of the cephalothorax and abdomen as in the structure of the parts of the mouth and the extremities etc. must more than compensate the differences, which are found, and which we have indicated above; that agreement is so complete as scarcely to require the additional evidence of this genus' belonging to the Orbitelariæ or Epeiroidæ, which is furnished by the circumstance, that its species all spin regular, *circular* nets. We may also allow ourselves to call attention to a commonly overlooked characteristic, which is found in *Uloborus*, as also in *all* other *Epeiroidæ* that I know of, and in a part only of the *Theridioidæ* and *Scytodoidæ*, but which is absent in the *Tubitelariæ* (even *Amaurobi-*

— Another nearly related genus is *Oxysoma* NICOLET (GRAY, Hist. fis. y pol. de Chile, Zool., III, p. 511).

1) BLACKWALL, The differ. in the number of eyes, etc., p. 606.

2) Beschr. neuer etc. Orbitelæ, p. 2 (64).

us and *Dictyna)* in the *Laterigradæ*, *Citigradæ* and *Saltigradæ*, as well as in the *Territelariæ* that I have had the opportunity of examining, namely, the presence of *accessory claws*, together with the (three) ordinary genuine claws, at the end of the tarsi [1]). These accessory claws are perhaps of as much importance for the animal's industry and for the determination of its systematic position, as the infra-mammillary organ and calamistrum; the presence of these claws in *Uloborus* may surely therefore be considered as an additional reason for referring that genus to a family where they always occur, and separating it from forms, in which I have never observed them. By considering *Uloborus* as the type of a separate sub-family of the Epeiroidæ, sufficient notice is certainly taken of the deviations of that genus from the *typical* Epeiroidæ. — Besides the genera *Uloborus*, *Hyptiotes* and *Zosis*, it is probable that also *Cyllopodia* Hentz (Aran. of the United States, *in* Bost. Journ. of Nat. Hist., V, p. 466), which is reported to have only six eyes, belongs to the sub-family *Uloborinæ*.

Genus 9. ULOBORUS Latr. 1806.

Deriv.: οὐλοβόρος, with deadly bite (οὖλος, deadly, βιβρώσκω, eat).

Syn.: 1806. **Uloborus** Latr., Gen. Crust. et Ins., I, p. 109.
1841. „ Walck., H. N. d. Ins. Apt. II, p. 227 (*ad partem*).
† 1855. **Phillyra** Hentz, Aran. of the United States, *in* Bost. Journ. of Nat. Hist., VI, p. 25.
1859. **Uloborus** Thor., Till känned. om Mithras och Uloborus, p. 194.
1859. **Veleda** Blackw., Descr. of six recently disc. spec. etc., p. 95.
1864. „ id., Spid. of Gr. Brit., I, p. 150.
1864. **Uloborus** Sim., H. N. d. Araignées, p. 244 (*ad max. part.*).

Type: *Uloborus Walckenaerii* Latr.

---

1) By accessory claws (*ungues spurii*, *secundarii*), I mean those unguiform or pectiniform appendages, which in the spiders here spoken of are to be found at the end of the tarsus, and occasionally also near the ordinary claw at the extremity of the palpus of the ♀. They are posited generally under or immediately beside, though occasionally even above, the genuine claws (*ungues veri*), from which they are easily distinguished by not being curved downwards, but directed straight forward (outwards), sometimes slightly upward. Generally they are slightly curved in the manner of an ∽; often however almost straight. They are in general smaller, especially slenderer, than the genuine claws, and, like them, are on the under side (though finer) dentated or serrulated, the serrulation being sometimes of extreme fineness. They are not always equally developed on the tarsi of the different pairs of legs. Their number varies greatly: generally there are 2 or 4, sometimes 6 or even more (as in the case of *Pholcus)* on each tarsus, arranged symmetrically near the

In the above-cited passage I have more accurately determined the genus *Uloborus* so as to include only the species, which can be referred to the 1st and 2rd Races of the 1st Family ("Les Divergentes, *Deflectentes*") of that genus in WALCKENAER (loc. cit.). The 1st Family's 3rd Race is the genus *Eugnatha* (SAV. et AUD.), of which more has been said above under the article *Tetragnatha.* The 2nd Family ("les Ecartées, *Divaricatæ*") of WALCKENAER'S *Uloborus* forms the genus *Zosis* WALCK.

DUFOUR'S statement, cited by LATREILLE [1]) and LUCAS [2]), that *U. Walckenaerii* has but *one* claw upon the three hinder pairs of legs, is entirely unfounded. The claws in that, as well as in the other species of this genus, are three in number on *each* tarsus. In *U. Walckenaerii* the tarsal claws are very small, but rather powerful; of the two superior claws the inner is considerably thicker at the base than the outer; on the 1st pair of legs the former has at least 5 somewhat curved comb-teeth, gradually increasing in length, the points of which, together with that of the claw, form an almost straight line: the outer has about 4 teeth, the innermost shortest, the others gradually increasing, and the last considerably longer than the rest, issuing from about the middle of the claw. The inferior claw is somewhat shorter than the superior, short and thick, strongly curved downwards. In *U. Latreillii* THOR. there is on this claw *one* long sharp tooth; in *U. Walckenaerii* I have with certainty observed such a tooth on the 3rd pair of legs, and I think I have seen one also on the 1st pair. On the 3rd pair the claws are shorter, curved more abruptly downwards, and provided with fewer teeth than on the 1st pair.

The female's palpus-claw is weak and of uniform thickness, slightly curved, with some few (in *U. Latreillii* about 5, in *U. Costæ* THOR. 2 or 3) weak teeth pointing forwards, nearer the tip. The claws of the palpi thus

---

end of the joint. They cannot be put in motion by muscles of their own, as is the case with the genuine claws, and are in fact to be considered merely as transformed bristles or spines. Beside in the *Epeiroidæ*, I have met with these accessory claws in *Pachygnatha*, in many, especially the larger, species of *Linyphia*, *Theridium*, *Steatoda* etc., as also in *Pholcus*. (Conf. THORELL, Till kännedomen om slägtena *Mithras* och *Uloborus*, p. 200).—The accessory claws of *Epeira diademata* have been described and figured already in RÖSEL'S Ins. Belust., IV, p. 252, Tab. XXXIX, fig. 8; BLACKWALL has also described and figured these claws in the *Epeiroidæ* (Notice of sev. rec. disc. in the struct. and œc. of spid., p. 476, Tab. XIII, fig. 4).

1) For inst. in Cours d'Entomol., p. 527.

2) Hist. Nat. d. Crust., d. Arachn. et d. Myriap., p. 443.

exhibit a considerable similitude of appearance to those of the genus *Linyphia* in the following family.

Genus 10. HYPTIOTES. (WALCK.). [1833] 1837.

Deriv.: *ὑπτιάω*, *ὑπτιάζω*, to be *ὕπτιος*, leaning back, easy, careless.

Syn.: 1833. **Hyptiotes [Uptiotes]** WALCK., Mém. s. une nouv. classif. d. Aran., p. 438.
† 1834. **Mithras** C. KOCH, *in* HERR.-SCHÆFF., Deutschl. Ins., 123, 9.
1837. „ ID., Uebers. d. Arachn.-Syst., 1, p. 6.
1837. **Scytodes** WALCK., H. N. d. Ins. Apt., I, p. 275 (*ad partem*).
1837. **Hyptiotes [Uptiotes]** ID., ibid., p. 277.
1860. **Mithras** THOR., Till känned. om Mithras och Uloborus, p. 198.
1861. „ WESTR., Aran. Suec., p. 87.
1864. **Hyptiotes [Uptiota]** SIM., H. N. d. Araignées, p. 184.

Type: *Hyptiotes paradoxus* (C. KOCH).

This genus is first mentioned by WALCKENAER 1833 in his above cited *Mémoire sur une nouvelle classification des Aranéides* under the name of *Uptiotes*, but not otherwise characterized, than that it is erroneously stated to have six eyes, like *Scytodes* and *Omosites*, together with which genera it is placed in the family "*Cellulicoles*". It was not till in 1837 [1]) that WALCKENAER gave a recognizable, though inaccurate, diagnosis of the genus "*Uptiotes*". But during this interval, 1835, the animal, which was the type of WALCKENAER'S *Uptiotes*, had been described by C. KOCH under the name of *Mithras paradoxus* (see Synonym.) Under such circumstances it may seem dubious, which generic name ought to be preserved; I for my part should have unreservedly given the preference to the name *Mithras*, if it had not been previously engaged for another genus of animals; but that being the case [2]), the Walckenaerian name must be adopted.

As *Uptiotes* is unquestionably formed from *ὕπτιος*, *resupinus*, the name, as has been already observed in AGASSIZ' *Nomenclator Zoologicus*, must be written *Hyptiotes*.

The systematic position of this genus has been, and still is, a subject of much dispute. It is remarkable that as long as WALCKENAER and KOCH believed it to have only 6 eyes, they acknowledged its intimate connexion with the *Epeiroidæ*. KOCH even introduced it into that family [3]).

---

1) Hist. Nat. d. Ins. Apt., loc. cit.
2) *Mithras* HÜBN. [Lepidopt.] 1816. — *Mythras* HALID. [Hymenopt.] 1829.
3) Uebers. d. Arachn. Syst., 1, p. 6.

WALCKENAER, it is true, says (probably on the strength of the imaginary agreement in the number of the eyes) that his *Uptiotes* is most nearly related to *Scytodes*, and he takes up KOCH'S *Mithras paradoxus* — which he considers different from *Uptiotes anceps* — under the name of *Scytodes mithras;* but he nevertheless remarks, that *U. anceps* approaches very near to the genus *Uloborus* "in the position of the eyes and the form of the cephalothorax" [1]). Afterwards, in 1847 [2]), he however maintains, that "the genus *Hyptiotes* in its cephalothorax, maxillæ and abdomen (?) is intermediate between the genera *Theridium* and *Argus*", and says not a word of its relationship to *Uloborus.* He accordingly classes it with the *Theridioidæ* ("les *Rétitèles*": loc. cit. p. 527), and is here followed by SIMON, who has given the genus a place between *Ero* and *Dictyna.* Excepting the abnormally great distance between the anterior row of eyes and the base of the mandibles, and the more sharp-pointed teeth on the inferior tarsal claw, I find nothing in *Hyptiotes* that approaches more to the *Theridioidæ* than to the *Epeiroidæ.* — KOCH maintains [3]), that "this genus, by the position of the eyes, the structure of the body, and by its habits in general (?) belongs to a family of spiders, of which as yet no other genus is known to exist". Of the *habits* of this genus KOCH seems nevertheless to have known nothing. In his Uebers. d. Arachn.-Syst., 5 (1850), he calls this new family *Mithraides*, and refers to it, together with *Mithras*, the genus *Poltys* (C. KOCH). The family *Mithraides* takes its place between *Epeirides* and *Theridides.* (Conf. THORELL loc. cit. p. 192).

BLACKWALL (loc. cit.) and KEYSERLING [4]) refer *Hyptiotes*, together with *Uloborus*, to the *Ciniflonidæ* BLACKW. on account of the infra-mammillary organ and calamistrum: even AUSSERER [5]) places that genus next to *Dictyna* and *Amaurobius* (reckoned by him to the family *Agalenoidæ*), which is so much the more remarkable, as he is acquainted with the form, in which *Hyptiotes paradoxus* makes its web. What we have above (see p. 64) said on the matter with respect to *Uloborus*, holds good also of *Hyptiotes.* By OHLERT this genus was first [6]) and rightly assigned to the *Epeiroidæ;* afterwards [7]) he included it in the family *Thomisoidæ*, with which

1) Hist. Nat. d. Ins. Apt., I, p. 279.
2) Ibid., IV, p. 388.
3) Die Arachn., XII, p. 98.
4) Beschr. neuer etc. Orbitelæ, p. 3 (65).
5) Die Arachniden Tirols, I, p. 150.
6) Beitr. z. Diag. u. Rev. d. Preuss. Spinnengattungen, p. 2; — Beitr. z. einer auf d. Klauenbildung gegr. Diagn. u. Anordn. d. Preuss. Spinnen, p. 238.
7) Die Aran. d. Prov. Preuss., p. 110 and 125.

Hyptiotes, as far as I am aware, does not possess a single characteristic in common.

As early as 1856 [1]) I classed *Hyptiotes* or *Mithras* among the *Epeiroidæ*, and have in a later paper [2]) developed and expounded the grounds for that view. As I there endeavoured to show, the genus *Hyptiotes* approaches more nearly to *Uloborus* than to any other known genus of spiders, while at the same time by the shorter and robuster form of its body, its short and broad maxillæ, its only slightly tapering extremities, its stouter tarsal and palpal claws, its two teeth on the inferior tarsal claw, it stands in nearer relation to the *typical* Epeiroidæ than do the species of *Uloborus*. The deviations from them, which *Hyptiotes* exhibits, such as the presence of the infra-mammillary organ and calamistrum, the distribution of the eyes into two rows *diverging* at the ends, it has also almost all in common with *Uloborus*. A remarkable analogy between *Hyptiotes* and the species of *Uloborus*, with which I am acquainted, is displayed in the fact that the hairy covering on the sides of the back of the abdomen are conglomerated into *fascicles*, *arranged in two rows along the back*. In *Hyptiotes*, as in *Uloborus*, the 4th pair of legs is longer than the second, and the legs are destitute of spines. A pair of accessory claws appear at the extremity of the tarsus in *Hyptiotes*, as well as in *Uloborus* and other *Epeiroidæ*. The only character of any consequence, in which *Hyptiotes* deviates at once from *Uloborus* and the *Epeirinæ*, appears to me to lie in the great extent of the eye-area, and its considerable distance from the fore-edge of the cephalothorax. But a similar relation is also observed in *Poltys* C. KOCH (*Pleuromma* DOLESCH.), especially as regards the unusually far back located position of the posterior side-eyes [3]), and that genus seems in this respect to occupy the same relation to *Epeira*, as *Hyptiotes* to *Uloborus*. C. KOCH united, as has before been said, the genera *Poltys* and *Hyptiotes* in the same family: the former belongs indisputably to the *Epeirinæ*, whither KEYSERLING subsequently referred it, and the latter must with equal certainty be placed in the most intimate relation to *Uloborus*.

That even its *habits* and *industry* claim for *Hyptiotes* a place among the Orbitelariæ, will be evidenced by the following lines which we cite from our above-mentioned paper:

---

1) Rec. crit. Aran. Suec., p. 107.

2) Till kännedomen om slägtena *Mithras* och *Uloborus*, p. 202 et seq.

3) Conf. KOCH, Die Arachn., X, p. 97, fig. 821.—KEYSERL., Beschr. neuer etc. Orbitelæ, p. 23, Tab. III, fig. 1—3.

"In the summer of 1855 I first met with *Mithras paradoxus*, in the neighbourhood of Stockholm, the only part of this country, in which it has been observed. July, August and September are the months in which it is met with full-grown. The males are extremely rare, and I have as yet not found more than one fully developed. It was taken Aug. 5. The female on the contrary is pretty common, and is met with principally in woods of trees of the fir kind, especially in pine woods. Between the dry bare branches of two neighbouring trees, she spins a strong thread in a horizontal direction, from a point of which she afterwards draws obliquely downwards three other threads, which form equal angles with the original thread and each other and lie in the same vertical plane. These four threads form the radii of the web; over them are laid concentric cross-threads, 16—22 in number, and tolerably wide apart. The loose net thus constructed forms a circular sector of about 45 degrees with a radius of a foot or more. It is therefore very large in proportion to the spider itself. The animal does not build itself any shelter or nest near the web, but hangs on the first-named horizontal thread that bears the web, near one of the twigs to which it is fastened, and at a considerable distance from the common point of intersection of the radii. The identity of colour between the animal and the dry branches causes it not to be so easily perceived: if disturbed, it draws in its legs and lets itself down to the ground. Its movements are slow and sluggish: the prey, which has fastened in the web, is spun into an envelope of silk, before it is devoured — a process employed, as far as I am aware, only by the *Epeiroidæ* (according to LUCAS also by *Uloborus*)."

"Although the web made by *Mithras paradoxus* is so peculiar and so unlike that of every other known species of spider, it is easily seen from the description, that it cannot be looked upon as any separate and independent form of web, but must be classed under the head of the known so-called geometrical nets of the Epeiroidæ. Here, as with them, it consists of radii diverging from a point, united by threads running concentrically; the difference is simply that, whereas with the other species belonging to the family it forms a closed circle, with *Mithras* it is but a circular *sector*. A transition to this latter form may in a certain sense be looked for in the case, of which one sometimes meets with examples, where, in the common circular net, the interval between two radii is left open, by the circular threads being terminated at these radii [1]). Not

1) Another more evident transition is described by DARWIN (Journal of Researches etc. during the voyage of the Beagle, p. 42) in the following words: "In a

only then in external character, but also in habits and form of web is the greatest likeness visible between *Mithras* and the Epeiroidæ. Hitherto, and with good reason, the habits of spiders, and especially the form they give their webs, have been considered as affording the surest basis for a natural grouping and classification of these animals; and as, in all probability, all the species that belong to the family Epeiroidæ distinguish themselves by their power of spinning regular geometrical webs [1]) — on which account that family received from LATREILLE the name of Orbitelæ — and in short one never assigns to any other family a species, which is known to spin such a net [2]), it seems evident that the genus *Mithras* ought to be included in the Epeiroidæ, although it must be placed last among them, nearest to the genus *Uloborus*, with which it also best agrees in the looseness of its web." (Loc. cit. p. 203—204).

Sub-ordo II. RETITELARIÆ.

Syn.: 1817. **"Inéquitèles"** LATR., *in* CUV., Régne Anim., T. III, p. 84.
1823. **Laquearîæ** SUND., Gen. Aran. Suec., p. 13.
1825. **Inæquitelæ** LATR., Fam. Nat. du Règne Anim., p. 314.
1833. **Theridides** SUND., Consp. Arachn., p. 15.

The limit between this and the next following sub-order, *Tubitelariæ*, is difficult to determine with sharpness. The genera *Dictyna*, *Titanœca*,

---

lofty valley of the Cordillera, near Mendoza, I found another spider with a singularly formed web. Strong lines radiated in a vertical plane from a common centre, where the insect had its station; but only two of the rays were connected by a symmetrical meshwork, so that the net, instead of being, as is generally the case, circular, *consisted of a wedge-shaped segment.* All the webs were similarly constructed."

1) We should perhaps except the genus *Dolophones*, if that genus really belong to the Epeiroidæ (Conf. WALCK. H. N. d. Ins. Apt., I, p. 383), and, according to SUNDEVALL (Consp. Arachn., p. 13), an East Indian species of *Epeira*, which he calls *E. abnormis*, but does not describe: it is said to spin an irregular net. Of *Argyrodes* SIM. vid. p. 48.

2) "I take no notice of the curious classification of the family *Therididæ*, in N:o 5 of KOCH's Uebers. d. Arachn.-Systems, where such genera appear as for example *Meta*, of the five cited species of which three are *Epeiroidæ (M. fusca = M. Menardi* (LATR.), *Merianæ = M. fusca* (DE GEER), and *muraria)*, one belongs to the genus *Linyphia (M. tigrina = Lin. socialis* SUND.) and the fifth *(M. cellulana)* appears to be a *Theridium;* or *Eucharia*, of the three species of which two, *E. bi-*

Uroctea for ex. might with almost as good reason be referred to the one as to the other. — Of the characteristic features of this sub-order the following ought to be observed: the extremities are, with few exceptions, fine and slender, and provided with in general weak, serrated or pectinated tarsal claws: there is almost always an inferior tarsal claw, sometimes toothless, sometimes armed with one or two, *never more*, teeth. The palpal claw in the females is generally weak, serrated or pectinated, but often without teeth: sometimes it is rudimentary or even entirely absent. The hairy covering is thin; the markings of the abdomen depend upon the colour of the skin, not upon that of the hairy covering. The eyes (in all European species) form typically two transversal rows, but sometimes the intermediary eyes, especially of the males in certain genera *(Argyrodes, Linyphia, Walckenaera)* of the family *Theridioidæ*, are drawn, on account of the unusual development of the pars cephalica of the cephalothorax, out of their ordinary regular position. There is no infra-mammillary organ or calamistrum. The spinners are — excepting in the *Enyoidæ*, in which the *inferior* spinners are considerably longer than the others — short and of almost equal length, and consist of *only two* joints. — See also above p. 47.

We divide the spiders that compose this sub-order, which very nearly corresponds to LATREILLE'S *Inæquitelæ* or SUNDEVALL'S *Theridides*, into three families, *Theridioidæ*, *Scytodoidæ* and *Enyoidæ*, which may be thus distinguished:

I. Mamillæ inferiores (anteriores) reliquis non vel parum longiores.

1. Tarsi articulo unguifero libero carentes. Labium liberum. Mandibulæ non ad basin coalitæ. . . . . I. *Theridioidæ*.
2. Tarsi articulo libero unguifero aucti. Labium cum sterno plerumque sine sutura coalitum. Mandibulæ versus basin plerumque inter se unitæ. . . . . II. *Scytodoidæ*.

II. Mamillæ inferiores reliquis multo longiores. . . . . . . III. *Enyoidæ*.

---

punctata and *castanea* are typical of the genus *Steatoda* SUND., the third, *E. atrica*, is an Epeiroid spider, belonging to KOCH's genus *Zilla*, and standing so near his *Z. montana* and *Z. calophylla* = *Z. x-notata* (CLERCK) both in form, colour and habits, that all three are considered by WALCK. and SUNDEV. as belonging to the same species." — C. KOCH's mistake with respect to the systematic position of the Epeiroidæ here mentioned has in fact been observed and corrected by almost all subsequent writers.

Fam. I. THERIDIOIDÆ.

Syn.: 1837. **Therididea** C. KOCH, Uebers. d. Arachn.-Syst., 5, p. 6 (*ad max. part.*).
1861. **Theridiidæ** WESTR. (*excl.* Pholco), Aran. Suec., p. 90.
1864. „ BLACKW. (*excl.* Pholco) + **Linyphiidæ** BLACKW., Spid. of Gr. Brit., II, p. 175, 210.
1866. **Retiariæ (Pachygnathidæ + Linyphidæ + Therididæ)** MENGE, Preuss. Spinn., I, p. 94, 100, 146 (*ad max part.*).

The European spiders belonging to this family, as well in their habits, the construction of their webs and the form they give their cocoons, as in their general appearance, exhibit very considerable differences, but yet form a perfectly compact and natural group, which it is not easy to resolve even into tolerably well-defined sub-families. Nevertheless BLACKWALL has divided them into two separate *families*, *Linyphiidæ* and *Theridiidæ*, but without indicating any difference of *bodily form* between these two families. They are said to be distinguishable however by the different manner in which the species composing them construct their webs: the *Linyphiidæ* fabricate a fine sheet of web, the snares of the *Theridiidæ* on the contrary are said to consist of lines intersecting one another in different planes and at various angles, and to present the appearance of being constructed without any regular plan [1]). Even if this basis for their classification could be admitted, which seems to me dubious, it not being taken from the animals themselves, I still think it does not, if strictly applied, lead to a natural grouping. Its application would probably in many cases be impossible, for the webs of many species belonging to BLACKWALL'S *Walckenaera* and *Neriene* are unknown, and it seems to me probable, that these do not all construct webs resembling those of *Linyphia*, nay that some of them fabricate none at all. *Steatoda bipunctata* and *castanea*, *Lithyphantes corollatus* and others, belonging to BLACKWALL'S *Theridion*, weave nets not indeed so close and fine as those of the Linyphiæ, but which yet have the form of a sheet, and are quite as regular as theirs, and these species surely no one would think of separating from the Theridiidæ. Other authors also have endeavoured to separate between Theridiidæ and Linyphiidæ, but it is vain to look for any agreement as to which genera are to be assigned to the one or other of these groups. SIMON refers *Micryphantes* to the *Linyphiidæ*, but *Erigone* to the *Theridiidæ:* by BLACKWALL on the other hand the species composing both these genera are referred to the *Linyphiidæ*. MENGE, who

1) Conf. BLACKW., Spid. of Gr. Britain., II, p. 175 and 210.

divides his "Netzspinnen, *Retiariæ*" into three families, *Pachygnathidæ*, *Linyphidæ* and *Therididæ*, of which the two latter are distinguished by the form of their webs, just as in Blackwall, refers to the *Linyphidæ* those spiders only, which belong to Westring's *Linyphia* and *Tapinopa*, thus excluding both the *Erigone*- and *Micryphantes*-species, which he makes *Therididæ*. According to Menge, the *Pachygnathidæ* do not make any web, but both Westring [1]) and Blackwall [2]) speak of their webs as being *irregular*, and the latter says of *Pach. Clerckii*, that it is related with *Theridium* "by the irregularity of the scanty web which it spins": nevertheless Blackwall aggregates this genus to the *Linyphiidæ*, not to the *Theridiidæ*. It having thus been found impracticable to assign any certain line of separation between Theridiidæ and Linyphiidæ either in the form of the body or the appearance of the web, I have (in company with Westring and Ohlert) not considered the latter as either a family or sub-family separate from the former. Also Walckenaer, who had at first distinguished them under the names of "*Rétitèles*" (Theridiidæ) and "*Tapitèles*" (Linyphiidæ), subsequently united them under the denomination *Rétitèles* [3]). Neither needs *Pachygnatha* be separated from them, although that genus certainly deviates a little from the typical Theridioidæ, as is the case also with *Episinus*, which genus is by some (e. g. Simon and Ohlert) included in the family *Thomisoidæ*.

C. Koch's division of his Theridides [4]) into 5 sub-families ("*Beutelspinnen*", *Wandspinnen*", "*Eigentliche Webspinnen*", "*Strickerspinnen*" and "*Bodenspinnen*") is altogether impracticable and full of gross errors. Epeiroidæ and Drassoidæ occur there mixed up with real Theridioidæ in a manner, which is utterly unaccountable. — Simon divides his "*Théridiformes*" into three "tribus": "*Clothéiens*", "*Théridiens*", and "*Linyphiens*"; the first of these appears to us to form two separate families, which we call *Enyoidæ* and *Urocteoidæ*, and of which we only refer the former to the *Retitelariæ*, whereas the *Urocteoidæ* may perhaps better be united with the next suborder, the *Tubitelariæ;* the *Théridiens* and *Linyphiens* together correspond very nearly with our *Theridioidæ* and *Agalenoidæ*, which last Simon has united with his *Linyphiens*. Of *Hyptiotes* and *Dictyna*, which he aggregates to his *Théridiens*, we include the former genus in the *Uloborinæ* of the family *Epeiroidæ*, and the latter in the *Amaurobiinæ* of the family *Agalenoidæ*. — Menge [5]) refers not only *Dictyna* (and *Lethia*) but also *Hahnia* to his Thericidæ; we unite this latter genus with the genuine *Agaleninæ*.

---

1) Aran. Suec., p. 144.
2) Spid. of Gr. Brit., II, p. 320.
3) Hist. Nat. d. Ins. Apt., IV, p. 527.
4) Uebers. d. Arachn.-Syst., 5, p. 15–24.
5) Preuss. Spinn., III, p. 244, 249, 251.

Among the exotic genera related to our Theridioidæ, we may mention the genus *Mimetus* HENTZ [1]), remarkable for its long mandibles and its roving habits, and which seems to form a link between the Epeiroidæ and Theridioidæ; as also *Thalamia* HENTZ [2]), which has its eyes posited in four transversal rows, the 2nd pair of legs longest, the 1st pair shortest, and which forms a tubular dwelling of silk in the crevices of walls. These genera cannot easily be inserted among our Theridioidæ (sub.-fam. *Theridiinæ*) otherwise than as types of separate sub-families. The exotic genus *Phoroncidia* WESTW., which by its spiny abdomen so closely resembles *Acrosoma* (PERTY) and *Pycnacantha* BLACKW., and which by WESTWOOD [3]) is referred to the *Epeiroidæ*, is by CAMBRIDGE [4]), in all probability rightly, assigned to the *Theridioidæ*. Another (Brazilian) genus, with a three-spined abdomen and belonging to the Theridioidæ, has lately been described by SIMON under the name of *Trithena* [5]). A somewhat similar spider appears to belong to the European fauna: for O. G. COSTA has, in Fauna del Regno di Napoli, Aracn., Tav. II, fig. 8, figured a *Theridium*-like species, the abdomen of which shows three small pointed processes, one on each side and one behind. That spider has however never been described nor named.

The European genera adopted by us may be thus distinguished:

§. Abdomen petiolo longo, nodoso cum cephalothorace unitum. 2. *Formicina.*

§§. Petiolum brevissimum, angustum (ut in araneis plerisque).

† Oculi non in tres turmas dispositi, neque inter se valde inæquales.

A. Oculi laterales inter se spatio minore disjuncti, quam quo distant medii antici a mediis posticis.

* Pedes aculeis sparsis armati. Oculi laterales sæpissime contingentes. (Palpus feminæ ungue sæpissime instructus).

A. Oculi medii in trapezium antice angustius dispositi: maxillæ paralleliter porrectæ vel in labium paullo tantum inclinatæ.

I. Oculi medii antici a margine clypei spatio non breviore distantes, quam quo a mediis posticis distant. Cephalothorax modice convexus. . . . . 6. *Linyphia.*

II. Oculi medii antici a margine clypei multo minus distantes, quam quo a mediis posticis distant. Cephalothorax robustus, convexior. . . . . 5. *Tapinopa.*

---

1) On North Amer. spiders, p. 104; — Aran. of the United States, *in* Bost. Journ. of Nat. Hist., VI, p. 31.

2) Aran. of the United States, loc. cit. p. 34.

3) Insect. Arachnoidumque nov. dec. duo, p. 452.

4) Descr. of a new gen. and six new species of spid., p. 270.

5) Sur trois Araignées nouv., p. 9.

B. Oculi medii in quadratum dispositi. Maxillæ angustæ, in labium fortiter inclinatæ. Cephalothorax alte convexus, sub-hemisphæricus. 10. *Ero.*

** Pedes aculeis carentes (rarissime serie aculeorum subtus instructi).

A. Mandibulæ femore plus duplo crassiores, usque a basi divergentes. . . . . 1. *Pachygnatha.*

B. Mandibulæ non vel apice tantum divergentes.

A. Mandibulæ non vel parum crassiores quam femora anteriora. Maxillæ oblongæ, lateribus sub-rectis, paralleliter porrectæ. Pedes tenues valde, prop. 1, 2, 4, 3. Oculi medii fere in quadratum dispositi, anteriores in tuberculo forti, prominenti positi. (Caput maris valde prominens. Palpus feminæ ungue instructus). . . . . 4. *Argyrodes.*

B. Mandibulæ plerumque femore crassiores et apice divergentes: pedum prop. plerumque 4, 1, 2, 3. (Species minutæ, nigræ, fuscæ vel rufescentes, abdomine sæpissime unicolore. Palpus feminæ ungue sæpissime carens).

a. Maxillæ sub-parallelæ vel in labium inclinatæ, ad basin non vel parum latiores. . . . . . . 8. *Walckenaera.*

b. Maxillæ ad basin valde dilatatæ. . . . . 7. *Erigone.*

C. Mandibulæ sæpissime femore angustiores et sub-cylindratæ. Maxillæ plerumque in labium fortiter inclinatæ. (Palpus feminæ ungue pectinato instructus).

a. Oculi laterales contingentes.

a. Oculi medii trapezium antice duplo angustius formantes. Maxillæ latæ, ovato-sub-quadratæ. Pedes prop. 1, 4, 2, 3, longi et graciles valde. . 9. *Nesticus.*

b. Oculi medii aream antice non vel parum angustiorem quam postice occupantes. Maxillæ in labium fortiter inclinatæ vel circa labium curvatæ, plerumque angustæ et sub-lineares.

α. Spatium inter oculos posticos medios et laterales duplo circiter majus, quam spatium, quo distant oculi medii inter se. Maxillæ extus sub-dilatatæ. Abdomen longius ovatum, modice convexum. Pedes valde longi et graciles. . . . 11. *Phyllonethis.*

β. Spatium inter oculos posticos medios et laterales non vel paullo tantum majus, quam quo distant illi inter se.

I. Series oculorum postica, desuperne visa, procurva vel sub-recta.

1. Oculi minores: postici medii a lateralibus posticis spatio distantes, quod oculi maximi diametrum superat vel saltem æquat. Abdomen (♀) plerumque altum, valde convexum, globosum vel formâ fere pyri. 13. *Theridium.*

2. Oculi majores: postici medii a lateralibus posticis spatio distantes, quod oculi maximi diametro brevius est. Abdomen (♀) sæpissime breviter ovatum vel ovale, supra sub-depressum. . . . . 14. *Steatoda.*

II. Series oculorum posticorum, desuperne visa, evidenter recurva.

1. Pedes 1mi paris reliquis longiores. Abdomen sub-globosum. . . . . . 12. *Dipæna.*

2. Series oculorum posticorum, desuperne visa, fortiter recurva. Pedes 4ti paris reliquis longiores. Abdomen ovatum (postice interdum acuminatum, mamillis superioribus reliquis plerumque multo robustioribus). 17. *Euryopis.*

b. Oculi laterales disjuncti.

a. Spatium inter oculos anticos medios et laterales vix majus, quam spatium, quo distant laterales inter se. Pedes tenues 1mi et 4ti paris longi. Abdomen postice latius. . . . . 3. *Episinus.*

b. Spatium inter oculos anticos medios et laterales multo majus, quam quo distant hi inter se.

1. Clypeus humilior, altitudine circiter dimidiæ mandibulæ. Oculi medii in rectangulum dispositi. Pedes 1mi paris sæpissime (num semper?) reliquis longiores. . . . . 15. *Lithyphantes.*

2. Clypeus altus, altitudine fere mandibulæ longitudinem æquanti. Oculi medii aream antice paullo angustiorem occupantes. Pedes 1mi paris reliquis longiores. . . . . 18. *Asagena.*

B. Oculi in duas series sub-parallelas dispositi. . . 16. *Lathrodectus.*

†† Oculi in tres turmas dispositi, duas laterales ex oculis trinis magnis constantes, tertiam ex oculis duobus minutissimis inter illas sitis. 19. *Pholcomma.*

Gen. 1. PACHYGNATHA SUND. 1823.

Deriv.: παχύς, thick; γνάθος, jaw.

Syn.: 1823. **Pachygnatha** SUND., Gen. Aran. Suec., p. 16.
182.. **Theridium** HAHN, Monogr. Aran. (*ad part.:*) 4, Tab. 4, fig. *B.*

1833. **Manducnlus** BLACKW., Charact. of some undescr. gen. and spec. of Aran., p. 110.
1841. **Linyphia** WALCK., H. N. d. Ins. Apt., II, p. 233 (*ad partem*).
1861. **Pachygnatha** WESTR., Aran. Suec., p. 144.
1864. „ BLACKW., Spid. of Gr. Brit., II, p. 318.
1864. „ SIM., H. N. d. Araignées, p. 229.
1866. „ MENGE, Preuss. Spinn., I, p. 94.

Type: *Pachygnatha Clerckii* SUND.

WESTRING, BLACKWALL, MENGE and others have already remarked the close analogy between this genus and *Tetragnatha* among the Epeiroidæ in the structure of the male's palpi, the form of the mandibles, etc.: it was even referred by SUNDEVALL, when he first described it, to his "Retiariæ" i. e. the *Epeiroidæ*. — The superior tarsal claws are in this genus rather large and powerful, yet slender, pretty uniformly curved, strongly pectinated, with several (12 or less) long, straight teeth; the inferior claw is small, with one very small blunt tooth. The palpal claw in ♀ is under the middle provided with a few close-set teeth gradually increasing in length.

A remarkable genus, which by the form of its large diverging mandibles seems to be related to *Pachygnatha*, is *Prodidomus* HENTZ (Aran. of the United States, *in* Bost. Journ. of Nat. Hist., V, p. 466): it is said to show "some of the characters of *Clubiona* and *Theridium*".

* Gen. 2. FORMICINA CANESTR. 1868.

Deriv.: *Formica*, ant.

Syn.: 1868. **Formicina** CANESTR., Nuovi Aracn. Ital., p. 197.

Type: *Formicina Mutinensis* CANESTR.

This to me unknown genus is said by CANESTRINI to be related to *Pachygnatha:* it has however not the mandibles diverging almost at a right angle, but only slightly diverging at the extremities. The most distinguishing feature of the genus is, that the petiolum, which unites the cephalothorax and abdomen, is long and nodose, thus giving these spiders a certain resemblance with ants. The 4 intermediary eyes form a trapezium: the posterior pair are farther distant from the side-eyes than from each other; the anterior are so close as almost to touch each other. The side-eyes are contiguous, the cephalothorax tolerably long and narrow; the relative length of the legs: 1, 2, 3, 4. See CANESTR., loc. cit.; Aran. Ital., p. 118.

Gen. 3. EPISINUS Walck. 1809.

Deriv.: Probably ἐπισινής, hurtful (σίνομαι, plunder, injure).

Syn.: 1809. **Episinus** Walck., *in* Latr., Gen. Crust. et Ins., IV, p. 371.
1861. „ Westr., Aran. Suec., p. 193.
1864. **Theridium** Blackw., Spid. of Gr. Brit., II, p. 175 (*ad partem*).
1864. **Episinus [Episina]** Sim., H. N. d. Araignées, p. 404.

Type: *Episinus truncatus* Walck.

The natural place of this genus is still a matter of controversy. Latreille included it among his *Inæquitelæ:* also Walckenaer[1]) and Lucas[2]) place it near the genus *Theridium*, and they are here followed by Westring and Blackwall; the last-named author does not even look upon it as generically differing from *Theridium* Walck. C. Koch also at first [3]) referred it to his "Theridides", but subsequently [4]) gave it a place among the *Epeiroidæ*, probably on account of a certain similitude of appearance with *Tetragnatha.* Both its industry and the form of its extremities however remove *Episinus* both from the *Epeiroidæ* and the *Thomisoidæ*, to which latter this genus is referred not only by Simon (loc. cit.) but also by Ohlert [5]), who had nevertheless previously [6]) declared, on the ground of the number and structure of the tarsal claws, his conviction, that it ought to be classed among the *Theridioidæ*, which *all*, like the Epeiroidæ, have *three* claws at the extremity of the tarsus, whereas the Thomisoidæ have only *two.* From this last family *Episinus* differs also, and that essentially, in not being in the least *laterigrade.* A certain *analogy* with the *Philodrominæ* of the family Thomisoidæ we will not deny that it exhibits, especially as regards the form of the abdomen; but the extremely fine and weak, tapering extremities clearly show that *Episinus* is a genuine sedentary [7]), not, like the Thomi-

1) H. N. d. Ins. Apt. II, 375.
2) Explor. de l'Algér., Arachn., p. 269.
3) Uebers. d. Arachn.-Syst., 1, p. 10.
4) Ibid., 5, p. 14.
5) Die Aran. d. Prov. Preuss., p. 110.
6) Ohlert, Klauenbildung d. Preuss. Spinnen, p. 10.
7) Ohlert indeed (with Latreille) includes also the *Thomisoidæ* in the list of *Araneæ sedentes*, but he does not give to that expression the sense we think it ought to bear. Only such spiders ought to be called "sedentes" or "sedentary", as await their prey in a web or nest, in contradistinction to those wich wander about in search of it. In this sense "Sedentes" are perhaps only the *Orbitelariæ, most Retitelariæ* and *some Tubitelariæ (Filistatoidæ*, certain *Dysderoidæ*, and most, if not all, *Agalenoidæ)*: the others, and especially *Laterigradæ, Citigradæ* and *Saltigradæ*, with the exception *perhaps* of the *Eresoidæ* and *Dinopoidæ*, are "*Araneæ vagantes*". Conf. Walckenaer, Ins. Apt., I, p. 187 et seqq.

soidæ, a wandering spider, whose web however appears to consist merely of a few irregular threads, on which it sits with its legs stretched straight forwards and backwards, in much the same position as *Tetragnatha.* — The male by his large palpal clava resembles the *Hyptiotes paradoxus* ♂.

The superior tarsal claws are slender, slightly curved, with about 5 curved saw-teeth between their basis and middle, gradually increasing in length (on the 1st pair of legs); the inferior claw is very small, with the extremity curved a little outwards, and has two blunt teeth; the female's palpus-claw is small, more strongly curved, and has about 6 close-set saw-teeth of about the same form as on the superior claws of the tarsi.

Gen. 4. ARGYRODES SIMON. 1864.

Deriv.: *ἄργυρος*, silver; *εἶδος*, form, appearance.

Syn.: 1841. **Linyphia** WALCK., H. N. d. Ins. Apt., II, p. 233 (*ad part.:* »3e Fam. Les Épéirides, *Epeirides*»).

1864. **Argyrodes** SIM., H. N. d. Araignées, p. 253 (*saltem ad partem*).

Type: *Argyrodes Epeiræ* SIMON.

This genus, formed with good reason by SIMON at the expense of *Linyphia* WALCK., is especially remarkable for its living parasitically in the webs of larger *Epeiroidæ*. SIMON has lately [1]) under the name of *Argyrodes Epeiræ* described a species from Spain, which he considers as new, but which certainly is identical with *Linyphia argyrodes* WALCK., of which I have seen specimens from Brazil, and which is also met with in the Ile de la Réunion (Bourbon) [2]) and probably also in Java [3]) and in Georgia in North America[4]). It appears to be as widely geographically distributed as *Cyrtophora opuntiæ* (DUF.), in the web of which, according to SIMON, it in Spain resides. — On this genus see also above p. 48.

The tarsal claws of *A. Epeiræ* are somewhat weak (as in *Linyphia*); the two superior differ considerably from each other in size, and are provided with 2 or 3 teeth of unequal size, pointing obliquely forward (on the 1st pair of legs). The inferior claw is equally large with the greater of the superior claws, more powerful, with a long, pointed tooth. The female's tarsal claw is fine and slender, slightly curved, with two teeth of different size pointing obliquely forward.

---

1) Sur quelques Araignées d'Espagne, p. 281.

2) VINSON, Aran. d. Iles de la Réun., Maur. et Madag., p. 318.

3) VAN HASSELT, Notice of VINSON's Aran. de la Réun. etc., p. 17.

4) WALCKENAER, Hist. Nat. d. Ins. Apt., II, p. 283.

Gen. 5. TAPINOPA WESTR. 1851.

Deriv.: *ταπεινός*, low; *ὤψ*, face.

Syn.: 1834. **Linyphia** REUSS., Zool. Misc., Arachn., (*ad part.:*) p. 264 (270)).
1851. **Tapinopa** WESTR., Förteckn., p. 38.
1864. **Linyphia** BLACKW., Spid. of Gr. Brit., II, p. 210 (*ad partem*).
1864. „ SIM., H. N. d. Araignées, p. 222 (*ad partem*).
1866. **Tapinopa** MENGE, Preuss. Spinn., I, p. 143.

Type: *Tapinopa longidens* (REUSS).

This remarkable genus, which was formed by WESTRING, approximates, in the very small distance between the anterior central eyes and the margin of the clypeus, to the preceding family, the *Epeiroidæ.* The lateral eyes are however far distant from that margin, as in other *Theridioidæ*, to which family the whole appearance of the only yet known species clearly indicates it as belonging. The form of the web of *T. longidens* is also identical with that of *Linyphia.*

The claws are of the same form as in *Linyphia:* the superior claws of the tarsi are slender, strongly bent, with about 3—5 small saw-teeth under their first half; the inferior claw is comparatively large, with two close-set teeth of unequal size. The female's palpal claw is still more slender, less curved, with a couple of small teeth nearer its base.

Gen. 6. LINYPHIA (LATR.). 1804.

Deriv.: *λίνον*, flax; *ὑφαίνω*, weave.

Syn.: 1804. **Linyphia** LATR., *in* Nouv. Dict. d'Hist. Nat., XXIV, p. 134 (*ad partem*).
1805. „ WALCK., Tabl. d. Aran., p. 70 (*ad max. part.*).
1833. **Erigone** SUND., Sv. Spindl. Beskr., *in* Vet.-Akad. Handl. f. 1832, p. 259 (*ad partem*).
1833. **Neriene** BLACKW., Charact. of some undescr. gen. and spec. of Aran., p. 187 (*ad partem*).
1837. **Bolyphantes** C. KOCH, *in* Uebers. d. Arachn.-Syst., 1, p. 9 (*ad partem*).
1861. **Linyphia** WESTR., Aran. Suec., p. 90.
1864. „ BLACKW., Spid. of Gr. Brit., II, p. 210 (*ad max. part.*).
1864. **Neriene** ID., ibid., p. 248 (*ad partem*).
1864. **Linyphia** SIM., H. N. d. Araignées, p. 222 (*ad max. part.*).
1864. **Bolyphantes** ID., ibid., p. 231.
1866. **Linyphia** MENGE, Preuss. Spinn., I, p. 101.
1866. **Bathyphantes** ID., ibid., p. 111.
†1866. **Pedina** ID., ibid., p. 125.

· 1866. **Helophora** MENGE, Preuss. Spinn., I, p. 126.
†1866. **Stylophora** ID., ibid., p. 128 [1]).
1866. **Lephthyphantes** [**Leptyphantes**] ID., ibid., p. 131.
1866. **Bolyphantes** ID., ibid., p. 134.
1866. **Stemonyphantes** ID., ibid., p. 138.
1866. **Drapetisca** ID., ibid., p. 140.

Type: *Linyphia triangularis* (CLERCK).

Now that SUNDEVALL has separated *Pachygnatha*, WESTRING *Tapinopa*, and SIMON *Argyrodes* from LATREILLE'S *Linyphia*, it appears to me that all the remaining European species referred by WALCKENAER to this genus, excepting only *L. crypticolens* WALCK. (*Nesticus cellulanus* (CLERCK) NOB.), may be retained under that generic name. C. KOCH has removed from it and made a separate genus of *Bolyphantes*, in which he at first also included a couple of species of the genus *Chiracanthium*, which he however since removed from that place. I cannot persuade myself that the genus *Bolyphantes* is sufficiently distinctly characterised to deserve to be retained, although it has been adopted by OHLERT and some others. The number of teeth on the palpal claw, on which OHLERT appears to lay the principal stress in determining the limits of *Linyphia* and *Bolyphantes*, affords a characteristic peculiar only to the female, and moreover that number is too different in the most nearly related species (8 in e. g. *Bol.* or *Lin. alticeps*, 3 or 4 in *Bol. alpestris* or *Lin. luteola* BLACKW.) to be allowed any great importance; neither can I find in the position of the eyes any constant difference between these two genera.

BLACKWALL has combined certain species of WALCKENAER'S *Linyphia* with some forms of that writer's *Argus* or WESTRING'S *Erigone* into a separate genus *Neriene*, which appears to me to include elements to different to be natural. Even had it been natural, there was no occasion for a new name, for it contains species, which had already received from SAVIGNY and AUDOUIN the generic name of *Erigone*. Judging from the characters assigned by BLACKWALL to *Linyphia* and *Neriene*, it would seem that the latter genus differs from the former principally in having the 1st and 4th pairs of legs *equally long or nearly so*, whereas in *Linyphia* the 1st pair is *longer* than the rest. The maxillæ are *dilated at the extremity* and inclined or curved towards the lip in *Neriene*, straight or slightly inclined towards the lip and *somewhat quadrate* in *Linyphia*. The genus *Walckenaera*, according to BLACKWALL, is distinguished from *Neriene* by having the maxillæ

1) *Pedina* AGASS. [Echin.] 1840. — *Stylophora* ROB.-DESV. [Dipt.] 1830.

dilated at the base. On the difference in the form of the maxillæ BLACKWALL appears however to have laid no especial weight, for his *Neriene longipalpis* for ex. has unquestionably the maxillæ strongly dilated at the base, and the maxillæ of *Neriene trilineata* do not, as far as I can see, differ in form from for ex. those of *Linyphia montana* (CLERCK). The distinction then must be founded on the difference in the relative length of the legs; and the consequence is, that species so heterogeneous as for ex. *Neriene (Erigone) longipalpis* and *N. trilineata (Lin. bucculenta)* have been united in one genus with *N. marginata (Lin. clathrata* SUND.), which accordingly is made not to belong to the same genus as for ex. *Linyphia pratensis* BLACKW.! That too much weight ought not in these genera to be laid on the characteristics derived from the relative length of the legs and a somewhat different form of the maxillæ, has nevertheless not escaped the observation of BLACKWALL, for he himself says in his account of his *Neriene dubia* (p. 289) that even "the sexes of the same spider frequently differ in the relative length of the legs and in the form of the maxillæ". — The genus *Neriene* ought, in my opinion, to be wholly suppressed and its species distributed among *Linyphia*, *Walckenaera* and *Erigone*, in such wise that the species that have scattered *spines* on their legs be assigned to *Linyphia*, and of the others those, of which the maxillæ are greatly dilated at the base, to *Erigone*, and the rest to *Walckenaera*.

The numerous genera which MENGE has separated from *Linyphia*, I cannot adopt, as they appear to me almost exclusively based upon minute modifications of the organs of copulation [1]), frequently only discoverable by means of the microscope. Compare what we have said on this subject in our remarks on the genus *Epeira*, p. 54.

---

1) This of course does not prevent our acknowledgement of the great importance of MENGE's researches on the subject of the more detailed structure of the sexual organs, which previous to him had been almost entirely unknown, and whereby a new and highly interesting field for investigation has been opened. It is probable that the results obtained by this species of research will have a certain influence on the classification of spiders, as soon as they have obtained a wider compass, and a richer store of materials of observation has been accumulated. But as yet it is perhaps too early to attempt to deduce from the modifications of these organs the decisive characteristics of *genera*, for we have as yet too little knowledge of what connexion these modifications have with the different form or arrangement of other organs of acknowledged systematic weight, e. g. the parts of the mouth, the extremities and the eyes. It appears to us, that similarities or dissimilarities in the structure of the organs of copulation is far from always indicating a corresponding similitude or dissimilitude in the organisms generally; at least species which are very

Linyphia passes gradually into *Walckenaera (Micryphantes)*, and only a tolerably artificial limit can be established between these two genera, however unlike each other they on the whole may be. We admit, with Westring, that the *presence of scattered spines* on the legs is the mark which distinguishes *Linyphia* from *Walckenaera* (and *Erigone*), in preference to the distinction proposed by Ohlert, the *presence of palpal claws* in *Linyphia*, and the absence of them in *Walckenaera* (and *Erigone*). In fact, the character deduced from the spines on the legs is easily verified, whereas the palpal claw in some of the smaller Linyphiæ is so fine and slender as to be very difficult to distinguish from an ordinary bristle, and furthermore the presence or absence of a claw on the palpus is a characteristic applicable only to *one sex*, the female. Besides, *Erigone vagans* Sav. et Aud. is said to be, unlike other species of that genus, provided with a palpal claw [1]), and of the genus *Ceratina* Menge, which we unite with *Walckenaera* or *Micryphantes*, *C. brevis*, according to Menge (as well as Westring's *Eri-*

closely related in every thing else (e. g. the species of the genus *Zilla* (Koch) Westr.) show very considerable differences with respect to the organs in question, and this has caused newer arachnologists with predilection to deduce the distinctions of *species* from the almost endlessly varying forms of the palpi of the males and the "epigyne" or vulva of the females. — Science however must necessarily gain by every endeavour after an improved classification and a sharper distinction of the genera of spiders. It is indisputable, that Arachnology stands in a much lower position than most other branches of Entomology, and that especially the scientific diagnosis of families and genera is as yet very defective. More than one arachnologist of the present time content themselves with Walckenaer's system, and follow in their descriptions the method of that *princeps arachnologorum*, whose honour one by no means depreciates by not believing, that the science ought always to remain stationary at the point, to which he carried it and where he left it. Only a few of the few, who devote themselves to this branch of zoology, labour to promote it by other means than by increasing the number of better or worse descriptions of species. Of late years however signs of a better spirit have shown themselves, and among the works which rise above the ordinary level, those of Menge undoubtedly occupy a particularly high place. But many workmen are still wanted upon that so slightly cultivated field, and in order to obtain these, it is necessary in the first place to make faunistic and descriptive works in general as easy to use and as practical as possible. As long as good and sufficient marks of distinction can be found by the aid of the simple magnifying lens, one must not make the *compound microscope* an indispensable instrument for any one, who may wish to determine the name and systematic position of an unknown spider. That the microscope is not necessary in order with certainty to distinguish even the smallest species of spiders, is fully evidenced by the works of Westring and Ohlert.

1) Descript. de l'Égypte, Arachn., (Éd. 2:) XXII, p. 320.

gone phæopus, which belongs to *Ceratina*) is destitute of palpal claws [1]), whereas other species of *Ceratina* described by MENGE are provided with them.

The superior tarsal claws in the *Linyphiæ* are slender, slightly curved, and have generally 6—12, sometimes even up to nearly 20, pointed teeth, gradually increasing in length; the inferior claw on the contrary has only *one* or *two pointed* teeth. The female's palpal claw has usually one, two or no teeth; sometimes, but rarely, 3—8 short saw-teeth. In many of the smaller species it is very slender, and sometimes entirely absent.

Gen. 7. ERIGONE SAV. et AUD. 1825—27.

Deriv.: *Ἠριγόνη*, *Erigone*, mythol. proper name.

Syn.: 1825—27. **Erigone** SAV. et AUD., Descr. de l'Égypte, (Éd. 2:) XXII, p. 319.
1833. „ SUND., Sv. Spindl. Beskr., *in* Vet.-Akad. Handl. f. 1832, p. 259 (*ad partem*).
1833. **Neriene** BLACKW., Charact. of some undescr. gen. and spec. of Aran., p. 187 (*ad partem*).
1837. **Micryphantes** C. KOCH, Uebers. d. Arachn.-Syst., 1, p. 11 (*ad partem*).
† 1841. **Argus** WALCK., H. N. d. Ins. Apt., II, p. 344 (*ad part.:* "1e Fam., 1e Race, Les Érigonides, *Erigonides*") [2]).
1861. **Erigone** WESTR., Aran. Suec., p. 195 (*ad partem*).
1864. **Neriene** BLACKW., Spid. of Gr. Brit., II, p. 248 (*ad partem*).
1864. **Erigone** [**Erygona**] SIM., H. N. d. Araignées, p. 191.
1867. „ OHLERT, Aran. d. Prov. Preuss., p. 34.
1868. „ MENGE, Preuss. Spinn., II, p. 195.
1868. **Tmeticus** ID., ibid., p. 184 (*ad partem*).

Type: *Erigone vagans* SAV. et AUD.

Of the species included by WESTRING in the genus *Erigone*, we preserve under that generic name only those, which are distinguished by having the *maxillæ considerably dilated at the base*, short, and inclined towards the lip. *Erigone* NOB. accordingly corresponds to the 1st race ("les *Erigonides*") of the 1st family of *Argus* WALCK. In that compass, which appears to be what by SAVIGNY and AUDOUIN was originally intended, it corresponds very nearly with C. KOCH'S, SIMON'S and OHLERT'S *Erigone*, although the last mentioned author gives as the special characteristic of the genus *the long palpi of the male*, a characteristic then, that applies only to one sex. In WESTRING this genus corresponds to *Erigone* and *Micryphantes* together

---

1) Preuss. Spinn., II, p. 171 et seq.
2) *Argus* TEMM. [Aves] 1815.

of the three last-mentioned writers; in Sundevall it has a still larger compass, in as much as that he has included in it sundry species of the genus *Linyphia.* In Menge again it has been somewhat more confined than in C. Koch and Ohlert. To the form of the maxillæ Menge, in opposition to the majority of writers, seems to give no weight: a couple of species, which we, in consequence of the form of these organs, consider as belonging to *Erigone*, are found in his work attributed to *Tmeticus*, of which again other species belong to *Walckenaera* (Blackw.) nob. or *Micryphantes*. Blackwall reckons the species belonging to this genus to his *Neriene.* — Dugès [1]) places the genus *Erigone* in the family he calls "*Aranées*", which pretty nearly answers to our *Tubitelariæ*.

As to the claws, the species of this genus resemble those of the following: the palpal claw is almost always wanting; the superior tarsal claw is slender, with about 6 teeth, the inferior is armed with *one* long pointed tooth.

Gen. 8. WALCKENAERA (Blackw.). 1833.

Deriv.: Walckenaer, proper name.

Syn.: 1833. **Erigone** Sund., Sv. Spindl. Beskr., *in* Vet.-Akad. Handl. f. 1832, p. 259 (*ad partem*).
1833. **Walckenaera** Blackw., Charact. of some undescr. gen. and spec. of Aran., p. 105.
1833. **Savignia** id., ibid., p. 104.
1833. **Neriene** id., ibid., p. 187 (*ad partem*).
1833. **Micryphantes** C. Koch, *in* Herr.-Schæff., Deutschl. Ins., 121, 19 et seq.
1837. „ id., Uebers. d. Arachn.-Syst., 1, p. 11 (*ad max. part.*).
† 1841. **Argus** Walck., H. N. d. Ins. Apt., II, p. 344 (*ad max. part.*).
1861. **Erigone** Westr., Aran. Suec., p. 195 (*ad max. part.*).
1864. **Micryphantes** [**Micryphantus**] Sim., H. N. d. Araignées, p. 193.
1864. **Walckenaera** Blackw., Spid. of Gr. Brit., II, p. 289.
1864. **Neriene** id., ibid., p. 248 (*ad max. part.*).
1867. **Micryphantes** Ohl., Aran. d. Prov. Preuss., p. 34 (*ad max. part.*).
† 1868. **Ceratina** Menge, Preuss. Spinn , II, p. 170 [2]).
† 1868. **Pachydactylus** id., ibid., p. 176 [3]).
1868. **Platyopis** id., ibid., p. 178.
1868. **Gonatium** id., ibid., p. 180.

---

1) Observ. sur les Aranéides, p. 161.
2) *Ceratina* Latr. [Hymenopt.] 1804.
3) *Pachydactylus* Wiegm. [Rept.] 1834.

1868. **Gongylidium** MENGE, Preuss. Spinn., II, p. 183.
1868. **Tmeticus** ID., ibid., p. 184 (*ad partem*).
1868. **Dicymbium** ID., ibid., p. 193.
1868. **Lophocarenum** ID., ibid., p. 198.
1868. **Lophomma** ID., ibid., p. 209.
1868. **Phalops** ID., ibid., p. 218.
1868. **Dicyphus** ID., ibid., Pl. 43, tab. 121; III, p. 221.
† 1869. **Elaphidium** ID., ibid., III, p. 224[1]).
1869. **Cornicularia** ID., ibid., p. 226.
1869. **Microneta** ID., ibid., p. 227.
1889. **Micryphantes** ID., ibid., p. 236.
1869. **Leptothrix** ID., ibid., p. 240.
? 1869. **Drepanodus** ID., ibid., p. 241.
? 1869. **Pronopius** ID., ibid., p. 243.

Type: *Walckenaera acuminata* BLACKW. (= *Erigone cornuta* (REUSS) WESTR.).

The name *Micryphantes* we find first, and *without characterisation*, applied by C. KOCH in the 121st Number of HERRICH-SCHÆFFER'S Deutschlands Insecten, which appeared in 1833 (or perhaps not before 1834). In 1833 BLACKWALL (loc. cit.) published his genus *Walckenaera* accompanied by a *complete diagnosis:* under this name he united a number of species, which, as it appears from the characterization of *Micryphantes*, *subsequently* given by KOCH (in Uebers. d. Arachn. Syst. 1, p. 11 (1837), and Die Arachn., IV, p. 124—127), also belong to *that* genus. *Walckenaera* and *Micryphantes* are accordingly to be looked upon as synonyms; and from what now has been adduced it is easily perceived, that the first name is to be preferred to the latter, even if not *older than*, but only *contemporaneous with* it.

The genus *Savignia* was formed by BLACKWALL for a spider (*S. frontata* BLACKW. = *Erig. conica* WESTR.) which he erroneously supposed to have but six eyes, but which he has since rightly aggregated to *Walckenaera*. — To *Neriene*, besides many other species, he has referred several, that stand in so close connexion with the spiders that he attributes to *Walckenaera*, that it seems to us, that they ought to be assigned to that genus. C. KOCH referred those of them, with which he was acquainted, to his *Micryphantes*. *Walckenaera* (BLACKW.) NOB. is therefore very nearly identical with *Micryphantes* C. KOCH, which genus has been adopted by

1) *Elaphidion* SERV. [Coleopt.] 1834.

many arachnologists, SIMON and OHLERT among the rest. — Of the value of the distinctions that separate *Neriene* from *Walckenaera* and *Linyphia* we have already delivered an opinion (p. 82—83).

While WESTRING unites into one genus, *Erigone*, the species that compose KOCH'S *Erigone* and *Micryphantes*, the large number of species, of which the latter genus consists, and which makes a division of it desirable, has given occasion to two attempts of the kind, which however do not seem to us to have been attended by fully satisfactory results. SIMON divides *Micryphantes* (loc. cit.) into the sub-genera *Micryphantes*, *Melicertus*, *Pelecopsis*, *Nerieneus*, *Widerius* and *Arrecerus*, in consequence of more or less accurately observed differences in the form of the head and the position of the eyes in the *males*. He however already at the end of the same work discards (*Walckenaera* and) *Nerieneus*. MENGE, in his "Preussische Spinnen", resolves the genus *Micryphantes* or *Walckenaera* into a very large number of new genera, but does not adopt any of SIMON'S or BLACKWALL'S here cited names. Several of these genera are founded on characteristics belonging only to the males, others on modifications, which appear to me of too subordinate importance to be used as the characteristics of genera. Some of them, however, may perhaps deserve to be acknowledged. But as the 3th Part of MENGE'S work, in which many of his new genera are proposed, did not come out till shortly before the present sheet was sent to the press, I cannot now enter into any detailed examination of his classification of the spiders in question, but preserve for the present the genus *Walckenaera* undivided, and of the extent above named.

The weak and slender superior tarsal claws of the species composing the genus *Walckenaera* have usually 6—8 teeth, greatly varying in length; the inferior claw has *one*, rarely two pointed teeth. The palpal claw is *absent*, except in the case of a few species (belonging to the genus *Ceratina* MENGE), in which it has 1—3 teeth. Conf. MENGE, Preuss. Spinnen, II, p. 171 et seq.

Gen. 9. NESTICUS N.

Deriv.: νηστικός, skilful in spinning (νέω, νήθω, spin).

Syn.: 1805. Theridium WALCK., Tabl. d. Aran., p. 72 (*ad part.:* "4e Fam. Les Crypticoles, *Crypticolæ*").

1841. Linyphia ID., H. N. d. Ins. Apt., II, p. 233 (*ad partem*).

1841. Meta C. KOCH, Die Arachn., VIII, (*ad part.:*) p. 123.

1859. Theridium THOR., Om Clercks Origin.-spindelsaml., (*ad part.:*) p. 150.

1861. Theridium WESTR., Aran. Suec., p. 151 (*ad partem*).
1864. Linyphia BLACKW., Spid. of Gr. Brit., II, p. 210 (*ad partem*).
1864. „ SIM., H. N. d. Araignées, p. 222 (*ad partem*).

Type: *Nesticus cellulanus* (CLERCK).

The different genera, to which CLERCK'S *Aran. cellulanus* has been referred by different arachnologists, sufficiently indicate that it does not fully agree with any one of them. By WALCKENAER it was first classed among the *Theridia*, afterwards among the *Linyphiæ*. To the *Epeiroid* genus *Meta*, to which it is assigned by C. KOCH, it of course does not belong. The position of the eyes is the same as in *Linyphia*, but their form, the absence of spines on the legs, and the form of the cephalothorax, seem to vindicate for this spider a place nearer the genus *Theridium* WALCK., and especially near *Steatoda* (SUND.) NOB. In the form of the maxillæ it appears to me to stand midway between *Linyphia* and *Steatoda*. Accordingly, as it cannot properly be united to either of these genera, I have formed a new genus for its reception.

The superior tarsal claws, which are long, slender, and slightly curved, have in *Nesticus cellulanus* about 11—13 closely set, rather short comb-teeth, gradually increasing in length and pointing somewhat forwards; the inferior has two close-set blunt teeth, of which the exterior is considerably thicker than the interior. The palpal claw has about 13 teeth, similar to those of the superior tarsal claws. The armature of the claws is thus very peculiar and different from what we meet with either in *Linyphia*, *Theridium* or *Steatoda*.

Gen. 10. ERO (C. KOCH). 1837.

Deriv. uncertain. *Ero* is probably a proper name [1]).

Syn.: 1805. Theridium WALCK., Tabl. d. Aran., p. 72 (*ad part.:* "8e Fam. Les Tuberculées, *Tuberculatæ*").
1837. Ero C. KOCH, Uebers. d. Arachn.-Syst., 1, p. 8 (*saltem ad part.*).
1861. „ WESTR., Aran. Suec., p. 148.
1864. „ SIM., H. N. d. Araignées, p. 182.
1864. Theridium BLACKW., Spid. of Gr. Brit., II, p. 175 (*ad partem*).
1866. Ero MENGE, Preuss. Spinn., I, p. 146.

Type: *Ero tuberculata* (DE GEER).

1) Had this name been formed from Ἡρώ, C. KOCH would unquestionably have written it with H.

We take this genus, formed by C. KOCH, in the extent which WESTRING has assigned it, and accordingly cannot include in it *Ero saxatilis* C. KOCH, which is a *Theridium* (= *Th. riparium* BLACKW.). — The tarsus in this genus shows at least some appearance of a little separate joint bearing the claws, and by this, as well as by its highly convex cephalothorax, *Ero* shows an evident analogy with *Scytodes*. — The superior tarsal claws are weak, sharply bent downwards, with 3 or 4 small teeth towards the base; the inferior claw appears to me to have only *one* little tooth. The palpal claw has the same form as the superior claws of the tarsi, and 3 or 4 teeth rapidly increasing in length and directed somewhat forward, situated nearly under the middle of the claw (in *E. variegata*).

Gen. 11. PHYLLONETHIS N.

Deriv.: φύλλον, leaf; νηθίς, spinner.

Syn.: 1805. **Theridium** WALCK., Tabl. d. Aran., p. 72 (*ad part.:* "1° Fam. Les Ovales, *Ovatæ*").
1837. **Steatoda** C. KOCH, Uebers. d. Arachn.-Syst., 1, p. 16 (*ad partem*).
1861. **Theridium** WESTR., Aran. Suec., p. 151 (*ad partem*).
1864. „ BLACKW., Spid. of Gr. Brit., II, p. 175 (*ad partem*).
1864. „ sub-gen. **Steatoda** [**Steatodum**] SIM., H. N. d. Araignées, p. 167.
1866. „ MENGE, Preuss. Spinn., II, p. 164.

Type: *Phyllonethis lineata* (CLERCK).

CLERCK'S *Araneus lineatus* differs so largely from the typical *Theridia* (if we consider as such *Th. sisyphium*, *varians*, and their nearest relations) that, unless the great genus *Theridium* WALCK. be preserved undivided, it can hardly be classed among them. WALCKENAER himself formed for this species a special "family" of his *Theridium:* C. KOCH formed for it a separate genus, to which he erroneously affiliated *Th. pictum* WALCK., a spider which accurately agrees with the above-mentioned typical species of *Theridium*. MENGE, who adopts the new genus, restricts it to the species in question, *Ar. lineatus* CLERCK. By C. KOCH it has been denominated *Steatoda*, which name however belongs to quite another group of Theridioidæ (see Gen. 14. *Steatoda*, p. 93). MENGE calls it *Theridium*, but that name had *previously* been by SIMON reserved for *Th. sisyphium* and its nearest allied species (which again are by MENGE taken up under the name of *Steatoda*), and ought, according to the law of priority, to be preserved to them. I have therefore been obliged to give the genus a new name.

The superior tarsal claws are tolerably large, armed with about 5 teeth, of which the two exterior ones are long and divergent; the palpal claw has 5 teeth increasing in length, and of which the 3 exterior are crooked and somewhat diverging. The inferior tarsal claw has a thick, blunt, somewhat crooked tooth and an inconsiderable point behind it.

Gen. 12. DIPŒNA n.

Deriv.: *Δίποινος*, proper name.

Syn.: 1845. Atea C. KOCH, Die Arachn., XI, (*ad part.:*) p. 143.
1863. Theridium CAMBR., Descr. of 24 new spec. of spid., (*ad part.:*) p. 16 (8576).
1864. Epeira: sub-gen. Atea SIM., H. N. d. Araignées, p. 260 (*ad partem*).

Type: *Dipœna melanogaster* (C. KOCH).

We have formed this genus for *Atea melanogaster* C. KOCH (*Ther. congener* CAMBR.), which is not an *Atea* at all, not even an Epeiroid, as C. KOCH maintains (loc. cit.). From *Theridium*, to which this genus approaches more nearly than to any other, and to which the only yet known species is referred by CAMBRIDGE (loc. cit.)[1]), it differs principally by the posterior row of eyes being curved backwards, and by the coarse bristles with which the legs are armed.

In the only known species the height of the clypeus is almost greater than the length of the mandibles, more than $1\frac{1}{2}$ times, nearly double as great as the length of the area occupied by the central eyes. The almost spherical abdomen is slightly emarginate or hollowed out at the base. The tarsal claws are rather strong, but small, especially the inferior, which has *one* tolerably long tooth; the superior are armed with a row of short saw-teeth proceding from the side of the claw and pointing obliquely forwards, which row reaches nearly to the tip of the claw. The female's palpal claw is bent almost exactly to a half-circle, small, and tolerably powerful; I have not been able to see any teeth on it. The construction of the claws is then, in this genus, very unlike that in the genus *Theridium*.

---

1) Even AUSSERER, though he includes it in his genus *Atea*, else consisting only of Epeiroidæ, says, that, on account of its form, it must necessarily be considered a *Theridium*. (Die Arachn. Tirols, I, p. 150). The web of this spider seems to be as yet unknown.

Gen. 13. THERIDIUM (WALCK.). 1805.

Deriv.: *θηρίδιον*, little animal.

Syn.: 1805. **Theridium [Theridion]** WALCK., Tabl. d. Aran., p. 72 (*ad part.*: "3ᵉ Fam. Les Renflées, *Turgidæ*").
1833. „ SUND., Consp. Arachn., p. 16 (*ad max. part.*).
1833. **Steatoda** ID., ibid. (*ad partem*).
1850. **Ero** C. KOCH, Uebers. d. Arachn.-Syst., 5, p. 16 (*ad partem*).
1861. **Theridium** WESTR., Aran. Suec., p. 151 (*ad partem*).
1864. „ **[Theridion]** BLACKW., Spid. of Gr. Brit., II, p. 175 (*ad partem*).
1864. „ : sub-gen. **Theridium [Theridio]** SIM., H. N. d. Araignées, p. 167 (*ad max. part.*).
1866. **Steatoda** MENGE, Preuss. Spinn., I, p. 150.
1868. **Neottiura** ID., ibid., II, p. 162.
1868. **Euryopis** ID., ibid., p. 174 (*ad partem*).

Type: *Theridium sisyphium* (CLERCK).

If the great Latreilleian genus *Linyphia* forms a close and compact whole, from which only a few species need be detached and divided among more recently formed genera, this is by no means the case with *Theridium* WALCK. WALCKENAER himself, already in his Tableau des Aranéides, divided it into several "families", most of which form perfectly natural groups, and have been raised to the rank of independent genera by C. KOCH, who has been followed in this by OHLERT and others. Most of these two writers' genera I have thought it best to retain, though with some slight modifications; moreover I have adopted the genus *Euryopis* proposed by MENGE, and have added the new genus *Dipœna* NOB. (for *Atea melanogaster*: see preceding page). The name *Theridium* I have, according to the example of (C. KOCH and) SIMON, preserved for WALCKENAER'S "3ᵐᵉ Fam. les *Renflées*", which family appears to me to include the forms specially typical of his *Theridium*. By MENGE this genus has been called *Steatoda*, a name belonging to a quite different group (see following genus). His *Theridium* is our *Phyllonethis* (see p. 90). For *Th. bimaculatum* (LINN.), MENGE has formed the genus *Neottiura*, which I have thought it best to unite with *Theridium*. *Th. guttatum* REUSS, which has been looked upon as a *Theridium* by OHLERT, but for which MENGE has proposed a separate genus, *Crustulina*, ought, it appears to me, to be referred to the same genus as *Ar. bipunctata* LINN., i. e. to *Steatoda* (SUND.).

WESTRING and BLACKWALL preserve for *Theridium* about the same limits that it has in WALCKENAER, the latter even refers *Episinus* WALCK. to it.

In *Theridium* the claws are small and weak, the superior tarsal claws ordinarily provided with 5 or 6 teeth gradually increasing in length, of which the extreme ones are long, pointed and somewhat curved; the inferior is not much smaller than the superior claws, with one short, blunt, somewhat curved tooth, and frequently a little point behind it; the palpal claw has most generally 4—7 rather long pointed teeth.

Th. tepidariorum, *formosum* and *riparium* compose a little separate group, distinguished, as OHLERT has already remarked, also by some differences in the form of the palpal claw: that claw in these species is in fact strongly bent, with about 7 (in *Th. tepidariorum* 10) long, close-set, vertical teeth, which, together with the downward bent extremity of the claw, form a comb. The tarsal claws also are somewhat stouter than those of the other species of the genus.

Gen. 14. STEATODA (SUND.). 1833.

Deriv.: "*στεατώδης, sævum referens*": SUND. (*στέαρ*, tallow; *εἶδος*, appearance).

Syn.: 1805. **Theridium** WALCK., Tabl. d. Aran., p. 72 ("2^e^ Fam. Les Arrondis, *Rotundatæ*" *ad part.* + "5^e^ Fam. Les Triangulilabres, *Triangulilabræ*" *saltem ad part.*).
1833. **Steatoda** SUND., Consp. Arachn., p. 16 (*ad partem*).
† 1836. **Eucharia** C. KOCH, *in* HERR.-SCHÆFF., Deutschl. Ins., 134, 8—11.
1837. „ ID., Uebers. d. Arachn. Syst., 1, p. 7.
1839. **Phrurolithus** ID., Die Arachn., VI, (*ad part.:*) p. 114.
1856. **Steatoda** THOR., Rec. crit., p. 108 (*ad partem*).
1861. **Theridium** WESTR., Aran. Suec., p. 151 (*ad partem*).
1864. „ BLACKW., Spid. of Gr. Brit., II, p. 175 (*ad partem*).
1864. „ : sub-gen. **Steatoda** [**Steatodum**] *ad max. part.*, + **Phrurolithus** [**Phrurolithum**] *ad part.:* SIM., H. N. d. Araignées, p. 168.
1867. **Eucharia** OHL., Aran. d. Prov. Preuss, p. 32.
1868. **Crustulina** MENGE, Preuss. Spinn., II, p. 168.
1869. **Eucharia** ID., ibid., III, p. 260 (*ad partem*).

Type: *Steatoda castanea* (CLERCK).

C. KOCH, by an unaccountable mistake, has used the name *Steatoda* for *Theridium lineatum*, which, together with *Th. sisyphium* and *longimanum* (*tinctum* WALCK.), SUNDEVALL (loc. cit.) has expressly cited as examples of the species he allows to remain under the genus *Theridium*, after having therefrom separated *Steatoda*. According to SUNDEVALL'S characteristics of this genus, "*Th. 4-punctatum*" and "*Th. castaneum*" must be considered as its types, and these species are also the *first* entered by him as thereto belonging; *afterwards* he names "*Th. albo-maculatum*" (which we look upon as the type of a separate genus, *Lithyphantes* NOB. = *Phrurolithus* (C. KOCH) OHLERT), and *lastly* "*Th. lunatum*", which however approaches nearer to *Th. sisyphium* than to "*Th. 4-punctatum*", and is also by most recent arachnologists (lately by MENGE) referred to the same genus as the former of these two species. *Steatoda* (SUND.) ought then to be considered as very nearly synonymous with *Eucharia* (C. KOCH) OHL. — In Rec. crit. aran., p. 108 (1856), I adopted the genus *Steatoda* in the full extent it had received from SUNDEVALL, i. e. as comprising *St. bipunctata* and *castanea*, as also some species, which I then supposed ought to be united in the same genus with these, a compass which however, as has been said, must be considerably curtailed. MENGE has afterwards used the generic name *Steatoda* for *Th. sisyphium* etc., which according to what has here been shown, is not right; the species of *Steatoda* (SUND.) are by him called *Eucharia*. — Of C. KOCH'S *Phrurolithus*, at least *Ph. ornatus* (Die Arachn., loc. cit.) belongs to *Steatoda*, of his *Eucharia* again *E. atrica* to the Epeiroid genus *Zilla* (C. KOCH) WESTR. — The name *Steatoda* has the right of priority in preference to *Eucharia*, which moreover had already in 1816 been assigned by HÜBNER to a genus of *Lepidoptera*.

The claws of *Steatoda* are far more powerful than those of *Theridium*, but still tolerably long. The free end of the superior tarsal claw is somewhat thickened about the middle; the teeth are thick, not long, generally 7—8 in number; the inferior tarsal claw has *one* blunt tooth. The palpal claw has ordinarily about 6—7 teeth, gradually increasing in length, and pointing forwards. — Such is the case in *S. castanea* and *bipunctata*.

Gen. 15. LITHYPHANTES N.

Deriv.: λίθος, stone; ὑφάντης, weaver.

Syn.: 1805. **Theridium** WALCK., Tabl. d. Aran., p. 72 ("2e Fam., Les Arrondies, *Rotundatæ*" *ad partem*).
1833. **Steatoda** SUND., Consp. Arachn., p. 16 (*ad partem*).
1837. **Eucharia** C. KOCH, Uebers. d. Arachn.-Syst., 1, p. 7 (*ad partem*).
1839. **Phrurolithus** ID., Die Arachn., VI, (*ad part.:*) p. 100, 105—109.
1861. **Theridium** WESTR., Aran. Suec., p. 151 (*ad partem*).
1864. „ : subgen. **Phrurolithus** [**Phrurolithum**] SIM., H. d. Araignées, p. 168 (*ad partem*).
1867. **Phrurolithus** OHL., Aran. d. Prov. Preuss., p. 33.
1869. **Eucharia** MENGE, Preuss. Spinn., III, p. 260 (*ad partem*).

Type: *Lithyphantes corollatus* (LINN.).

If from C. KOCH's heterogeneous genus *Phrurolithus* we detach some not allied forms, as for instance *Ph. trifasciatus*, which is a *Singa*, *Ph. ornatus*, which seems to be the young of *Steatoda bipunctata*, as also *Ph. festivus* and *minimus*, which belong to the *Drassoidæ*, the remaining *Theridioidæ* form a perfectly natural group, which has accordingly been acknowledged by OHLERT as a separate genus, and by him characterized in a satisfactory manner. WESTRING has however as early as 1851 (Förteckn. etc., p. 46) reserved the name *Phrurolithus* to the above named *Drassoidæ* which KOCH had referred to this genus, so that OHLERTS *Phrurolithus* requires a new name. We have chosen the name *Lithyphantes*, as indicating the habits of the various species belonging to this genus. — By MENGE [1]) *Lith. corollatus* is referred to his *Euckaria*, i. e. *Steatoda* (SUND.) NOB.

In *L. corollatus* the superior tarsal claws are of about the same form as those of *Steatoda*, but somewhat stouter; they are pectinated, with about 8—10 strong teeth increasing rapidly in length from the base. The inferior claw has a thick, blunt tooth, behind which is another fine and more pointed. The female's palpal claw has about 4 coarse teeth, pointing forwards.

Gen. 16. LATHRODECTUS WALCK. 1805.

Deriv.: λάθρᾳ, secretly; δήκτης, biting (δάκνω, bite).

Syn.: 1805. **Lathrodectus** [**Latrodectus**] WALCK., Tabl. d. Aran., p. 81.
1806. **Theridium** [**Theridion**] LATR., Gen. Crust. et Ins., I, p. 98 (*ad partem*).
1836. **Meta** C. KOCH, Die Arachn., III, (*ad part.*:) p. 9, 10.
1864. **Lathrodectus** [**Latrodectus**] SIM., H. N. d. Araignées, p. 177.

Type: *Lathrodectus 13-guttatus* (ROSSI).

This genus is so nearly similar to *Lithyphantes*, as to differ from it almost only by a greater distance between the lateral eyes. If WALCKENAER'S genus *Theridium* is preserved undivided, the species that compose *Lathrodectus* must unquestionably be affiliated to it, as was done by LATREILLE and DUGÈS. — The name "*Latrodectus*" is evidently formed of λάθρᾳ and δήκτης, in the same manner as e. g. *Lathrobium* of λάθρᾳ and βιόω (to live), and ought therefore to be written *Lathrodectus*. The derivation "*λάτρον, merces, δεκτός, acceptus*" given in AGASSIZ' Nomenclator Zool., and that from "λατρεὺς, ouvrier" and δήκτης, which SIMON adopts, appear to me very improbable, as yielding no rational meaning for the name.

1) Preuss. Spinn., III, p. 264.

According to DUFOUR [1]) the claws of his *L. (Ther.) lugubris* — not to be confounded with *L. (Ther.) lugubris* MOTSCHOULSKY [2]) — are *destitute of teeth.* WALCKENAER says [3]) that the inferior claw is toothless in the genus *Lathrodectus* in general, and that in *L. 13-guttatus all* the claws on the *posterior* legs are so. All these statements, as VAN HASSELT [4]) suspected, depend upon some mistake, probably on the claw-teeth in the specimens examined having been broken off. In a *L. lugubris* (DUF.) from Spain I have found all the claws evidently pectinated. The superior tarsal claws are in this species short and strong, regularly and pretty much bent, with (on the first pair of legs) about 8 long, somewhat pointed comb-teeth, directed somewhat forwards; the inferior claw is considerably smaller, with two short, blunt teeth connected at the base, of which the foremost is much coarser but not longer than the other. On the 4th pair the teeth are not quite so many; the inferior claw there appears to me to show a rudiment of a third tooth (?).

Gen. 17. EURYOPIS (MENGE). 1868.

Deriv.: *εὐρύς*, wide, broad; *ὤψ*, face.

Syn.: 1836. **Micryphantes** C. KOCH, Die Arachn., III, (*ad part.:*) p. 67.
1847. **Argus** WALCK., H. N. d. Ins. Apt., IV, (*ad part.:*) p. 501.
1861. **Theridium** WESTR., Aran. Suec., p. 151 (*ad partem*).
1864. „ BLACKW., Spid. of Gr. Brit., II, p. 175 (*ad partem*).
1864. **Micryphantes** SIM., H. N. d. Araignées, p. 193 (*ad partem*).
1868. **Asagena** ID., ibid., p. 162 (*ad partem*).
1868. **Euryopis** MENGE, Preuss. Spinn., II, p. 174 (*ad partem*).

Type: *Euryopis flavo-maculata* (C. KOCH).

KOCH himself, in the passage above referred to, has remarked, that his *Micryph. flavo-maculatus* can hardly be suffered to remain within the genus *Micryphantes.* WESTRING, BLACKWALL and OHLERT accordingly detach it therefrom: they reckon this spider as a *Theridium.* MENGE, on the other hand, and, I think, with reason, has formed for it a distinct genus, *Euryopis.* To this genus he also assigns, though not without some doubt, *Ther. triste* HAHN, which seems to be by no means so happy an arrangement. We refer this species to *Theridium* sensu strictiori (*Steatoda* MENGE), although

1) Descr. de six Arachn. nouv., p. 356.
2) Note sur deux Araignées venim., p. 290.
3) Hist. Nat. d. Ins. Apt., I, p. 644—646.
4) Studien ov. de z. g. Curaçaosche Oranjespin, p. 65.

it must be owned, that it deviates pretty considerably from the typical species of that genus. On the other hand *Ther. lætum* WESTR. and *Ther. acuminatum* LUCAS [1]) belong to *Euryopis*, as I place the limits of that genus (vid. p. 77). Of the last-named species, which was discovered by LUCAS in *Algeria*, and which, as far as I am aware, has never yet been acknowledged as belonging to the fauna of Europe, I found several specimens, both ♂ and ♀, at *Livorno* (Leghorn) in the summer of 1853.

In *E. flavo-maculata* (♂) the superior tarsal claws (of the 1st pair of legs) are large, rather strong, but not broad towards their base, uniformly and almost semicircularly curved, with about 6 coarse, pointed comb-teeth, issuing from the side of the claw, from its base to the vicinity of its apex: their extremities form (together with that of the claw itself) an almost straight line; the inferior claw is small, coarse, with two small and very close-set, blunt teeth. The claw of the female's palpus is according to OHLERT [2]) strong, with 5 straight long teeth.

Gen. 18. ASAGENA SUND. 1833.

Deriv.: *ἀ priv.; σαγήνη*, net; "*reti carens*": SUND.

Syn.: 1801. **Phalangium** PANZ., Faun. Ins. Germ., (*ad part.:*) 78, 21.
1805. **Theridium** WALCK., Tabl. d. Aran., p. 72 ("6e Fam. Les Cachées, *Absconditæ*" *ad part.*).
1832. **Drassus** SUND., Sv. Spindl. Beskr., *in* Vet.-Akad. Handl. f. 1831, p. 132 (*ad partem*).
1833. **Asagena** ID., Consp. Arachn., p. 19.
1856. **Steatoda** THOR., Rec. crit., p. 108 (*ad partem*).
1861. **Theridium** WESTR., Aran. Suec., p. 151 (*ad partem*).
1864. „ BLACKW., Spid. of Gr. Brit., II, p. 175 (*ad partem*).
1864. **Asagena** SIM., H. N. d. Araignées, p. 162 (*ad partem*).
1867. „ OHL., Aran. d. Prov. Preuss., p. 33, 41.
1869. „ MENGE, Preuss. Spinn., III, p. 256.

Type: *Asagena phalerata* (PANZ.).

SUNDEVALL included this genus in his *Drassides* (*Tubitelariæ* NOB.), among which analogous forms occur, e. g. the genus *Titanœca* NOB. among the *Amaurobiinæ*. By C. KOCH it was first curiously enough aggregated to his "*Agelenides*" (Uebers. d. Arachn.-Syst., 1, p. 13), but afterwards rightly to the "Theridides". — WESTRING, BLACKWALL and others suffer it to re-

1) Explor. de l'Algérie, Arachn., p. 268, Pl. 17, fig. 10.
2) Klauenbild. d. Preuss. Spinn., p. 9.

main among the species of WALCKENAER'S *Theridium*. — SIMON classes as *Asagenæ* several spiders, which according to SUNDEVALL'S definition of this genus can in no wise belong to it.

The claws of the typical species are strong. The superior tarsal claws are rather strongly curved, broader towards their base, pectinated, with (on the 1st pair of legs) about 10 long, straight, parallel teeth from the base to near the extremity of the claw, which thus form a dense comb; the inferior claw has one very thick and blunt tooth and a fine point behind it. The female's strong, much curved palpus-claw is in the same manner as the superior claws of the tarsus closely pectinated, with about 12 very long teeth pointing somewhat forward.

* Gen. 19. PHOLCOMMA N.

Deriv.: *Pholcus*, generic name of spider (φολκός, squint-eyed); ὄμμα, eye.

Syn.: 1862. Theridion CAMBR., Descr. of ten new spec. of Brit. spid., (*ad part.:*) p. 7962.

Type: *Pholcomma projectum* (CAMBR.).

I have not seen the species, on which I have founded this new genus, but the excellent description given by CAMBRIDGE (loc. cit.) leaves no doubt of its differing more from *Theridium* sensu strict. than any of the species that have been removed from that and referred to newer genera. CAMBRIDGE says himself: "The extreme dissimilarity in size between the two "centre eyes of the front row, and the rest, and their position, is a striking "characteristic of the species, and would almost warrant its separation from "the genus Theridion". According to CAMBRIDGE, the two centre eyes of the front row are very minute, almost contiguous; on each side of these is a group of three almost contiguous eyes, in the form of an equilateral triangle. The eyes of these two groups are disproportionably large compared with the size of the spider. The male has a projecting ridge round the abdomen. "By the position of the eyes this species seems to be allied to "the genus Pholcus, though in general form and appearance it is much more "like the true Theridia" (CAMBR.).

Fam. II. SCYTODOIDÆ.

Syn.: 1864. "Scytodiformes" SIM., H. N. d. Araignées, p. 43.

As regards the proper place for the very peculiar spiders, that we bring together under this name, opinions have been, and still are, very much divided. The types of the two sub-families, *Pholcinæ* and *Scytodinæ*, into which we divide them, *Pholcus Pluchii* (SCOP.) and *Scytodes thoracica* LATR., were referred by LATREILLE 1804 [1]), together with the spiders, for which WALCKENAER in the following year formed the genus *Theridium*, to one and the same genus: *Scytodes* LATR., and even subsequently, after LATREILLE had acknowledged the genera *Pholcus* and *Theridium*, which had been formed by WALCKENAER, he continued [2]) to place *Scytodes* and *Pholcus* beside each other in his family *Inæquitelæ*. — WALCKENAER also at first gave them the same systematic position: *Scytodes* and *Pholcus* in his Tableau d. Aran. (p. 79, 80) immediately follow *Linyphia* and *Theridium;* but when he began to make the number of the eyes a basis for his classification of "les araignées" [3]), he was obliged to separate *Scytodes* and *Pholcus* from each other: *Scytodes* (together with *Rachus*) is referred to a separate group, "*Cellulicoles*" or "*Capteuses*", which is placed between "*Tubicoles*" *(Dysderidæ)* and "*Coureuses*" *(Citigradæ)* within the division "*Vagabondes*", whereas *Pholcus* (like *Artema*) has a place in the group "*Filitèles*" among "*Errantes*" [4]). Both genera were removed far from the "*Rétitèles*", which correspond to our *Theridioidæ*. WALCKENAER seems however to have perceived that that method of classifying these spiders was quite artificial, for he himself says: "Le genre *Scytode* appartient encore plus particulièrement aux *Théridions* (i. e. than do *Dysdera* and *Segestria* to *Clubiona* and *Tegenaria*) . . . . et le genre *Rack* est un *Pholque* dont les yeux intermédiaires sont oblitérés" [5]).

DUGÈS, who, like LATREILLE, acknowledged the close relationship between *Scytodes* and *Pholcus*, united these genera, together with *Filistata*, *Uroctea (Clotho)*, *Enyo*, *Laches (Lachesis)* [6]) and *Hersilia* into one family, which he calls "*Scythodés*", also "*Micrognathes*" [7]), on account of these spiders' mandibles, which are usually small and united at the base. These genera in fact show no small affinity with the *Scytodoidæ*, and this appears to be especially the case with *Enyo*, *Uroctea* and *Filistata.* The Scytodoidæ agree with all these genera in having maxillæ closely encircling the lip; they particularly resemble *Enyo* (and *Hersilia*) in their fine extremities, with the claw-joint of the tarsi distinctly separate; *Uroctea* in their mandibles

---

1) Nouv. Dict. d'Hist. Nat., XXIV, p. 134.
2) For inst. in Cours d'Entom., p. 125.
3) Mém. s. une nouv. Classif. d. Aran., p. 438.
4) Hist. Nat. d. Ins. Apt., I, p. 202; IV, p. 524, 525.
5) Ibid., IV, p. 522.
6) Concerning these names, vid. sup., p. 36, 37.
7) Observ. s. les Aran., p. 106.

united towards the base; *Filistata* not only in this character, but also by having the lip and sternum coalesced. The *Pholcinæ* at least are particularly related to the *Theridioidæ* in the construction of their webs: both WESTRING [1]) and BLACKWALL [2]) accordingly refer *Pholcus* to their *Theridiidæ;* whereas the *Scytodinæ* show a certain relationship with the *Dysderoidæ* (the nearest relations to *Filistata*) in the form of the male's organ of copulation, to which may be added the identity in the number of eyes, which is probably what principally induced BLACKWALL to place *Scytodes* in immediate juxtaposition with that family [3]). Even C. KOCH assigned *Scytodes* to the "*Dysderides*" [4]), whereas he first [5]) referred the *Pholcinæ* to his *Drassides*, and when he had for *Pholcus* formed the family *Pholcides* [6]), he placed it next to the *Drassides*. DOLESCHALL refers *Pholcus* to his *Tubicolæ* [7]), which correspond most nearly with our *Agalenoidæ*, to which family also AUSSERER [8]), as well as CANESTRINI and PAVESI [9]) affiliate that genus — for what reason, I am at a loss to understand. *Loxosceles* is by LOWE referred to the *Laterigradæ* [10]). The nearest relations of the *Scytodoidæ* are however *Filistatoidæ* and *Enyoidæ* (which last, on account of their elongated inferior spinners and free lip, we consider ought to form a separate family), and especially the *Theridioidæ*. That they, together with the two last named families, belong to the sub-order *Retitelariæ*, is evidenced by their whole appearance, especially their long, fine extremities, armed with *three* (in *Loxosceles* only, two) *fine, slender claws.* The pattern on the abdomen is often bright and lively, and depends in the Scytodoidæ, as in the Retitelariæ generally, on the pigment of the skin itself, not on the usually thin covering of hair. As tolerably constant distinguishing features of this family we may also mention the sloping, more or less projecting, long clypeus, and the presence of a spine or tooth at the extremity of the mandible, opposite to the claw, indicating an approach to the two-fingered mandibles of the *Opiliones* or *Phalangia.* (Conf. DUGÈS, loc. cit.). In the spiders belonging to this family (of the genera *Scytodes*,

1) Aran. Suec., p. 296.
2) Spid. of Gr. Brit., II, p. 207.
3) Loc. cit., p. 379.
4) Uebers. d. Arachn.-Syst., 1, p. 21.
5) Ibid., p. 20.
6) Ibid., 5, p. 31.
7) Syst. Verzeichn. d. in Oesterr. vork. Spinnen, p. 14.
8) Die Arachn. Tirols, I, p. 151.
9) Aran. Ital., p. 65.
10) Descr. of two spec. of spid., natives of Madeira, p. 321.

Loxosceles and *Pholcus*) known to me, the female's palpal claw is either more or less *rudimentary* or *entirely absent*.

SIMON has lately united the spiders in question in one family, "*Scytodiformes*", which appears to me perfectly natural and justifiable. (Conf. p. 33). The position he has given this family, which with him is the first, and is immediately followed by the "*Mygaliformes*", is however not the most appropriate, as may be seen from what has already been said. Like SIMON we divide the *Scytodoidæ* into two sub-families ("tribus": SIMON): 1. *Pholcinæ* ("*Phalangoidiens*" or "*Pholciens*" SIM.), 2. *Scytodinæ* ("*Scytodiens*" SIM.). We distinguish these sub-families and the few European genera which belong to them, as follows:

I. Oculi aut 8, aut 6, et tum tres in utroque latere frontis. (Palpi marium valde incrassati, clava complicata). . . . . I. PHOLCINÆ.

1. Oculi 8. Pedes omnium longissimi. . . . . 1. *Pholcus*.
2. Oculi 6. . . . . 2. *Spermophora*.

II. Oculi 6, in tria paria dispositi, duo in utroque latere frontis. (Palpi marium tenues, clava parum complicata). . . . . II. SCYTODINÆ.

3. Cephalothorax postice alte convexus. Mandibulæ parvæ, debiles. Ungues tarsorum trini [1]) . . . . 3. *Scytodes*.
4. Cephalothorax plus minus depressus. Mandibulæ fortiores. Ungues tarsorum bini. . . . . 4. *Loxosceles*.

Sub-fam. I. PHOLCINÆ.

Syn.: 1850. **Pholcides** C. KOCH, Uebers. d. Arachn.-Syst., 5, p. 31.

Gen. 1. PHOLCUS WALCK. 1805.

Deriv.: φολκός, squint-eyed.

Syn.: 1804. **Scytodes** LATR., *in* Nouv. Dict. d'Hist. Nat., XXIV, p. 134 (*ad part.*).
1805. **Pholcus** WALCK., Tabl. d. Aran., p. 80 (*ad part.*).
1861. „ WESTR., Aran. Suec., p. 296.
1864. „ BLACKW., Spid. of Gr. Brit., II, p. 207.
1865. „ SIM., H. N. d. Araignées, p. 54.

Type: *Pholcus Pluchii* (SCOP.).

---

1) Not *two* only, as is erroneously stated in my paper: Om hanen af Scytodes thoracicus (On the male of Sc. thor.), p. 199.

In *Ph. Pluchii* the superior tarsal claws are large, weak, of equable curvature, pectinated, with about 8 long, pointed, somewhat curved teeth proceeding from the side of the claw; the inferior claw is tolerably large, long and fine, bent to a hook, curved somewhat outwards at the extremity, with 1 (2?) strong, pointed tooth bent inwards at the tip. The palpal claw in the female is rudimentary, and consists of a little conical, somewhat curved process, surrounded by fine, pointed bristles, of which the two outermost are considerably stronger than the rest.

*Gen. 2. SPERMOPHORA HENTZ. 1841.

Deriv.: σπερμοφόρος, seminiferous (σπέρμα, seed; φέρω, bear).

Syn.: 1836. **Pholcus** DUGÈS, Observ. s. les Aran., p. 160 (*ad partem*).
1841. **Spermophora** HENTZ, Descr. of an Amer. Spid. etc., p. 116.
1847. **Rachus** WALCK., H. N. d. Ins. Apt., IV, p. 459.
1850. **Oophora** HENTZ, Aran. of the United States, *in* Bost. Journ. of Nat. Hist., VI, p. 285.
1864. **Rachus** SIM., H. N. d. Araignées, p. 52.

Type: *Spermophora meridionalis* HENTZ.

In the passage above referred to, DUGÈS gave in 1836 a very scanty description of a remarkable spider found in the south of France, which he called *Pholcus senoculatus*, but which differed from other species of the genus *Pholcus* by *being destitute of both the centre eyes*. This spider was also found in Algeria, and excellently described and figured in 1847 [1]) by LUCAS, who called it *Pholcus quadri-punctatus*. For this 6-eyed spider WALCKENAER (loc. cit.) in 1847 formed a new genus, *Rachus*. But already in 1841 HENTZ had proposed the name *Spermophora* for an approximate form (from Alabama), which name he subsequently changed to *Oophora* (vid. *Syn.*). The female of that species, *Spermophora meridionalis* HENTZ, carries with her the loosely conglutinated mass of eggs, holding it with her mandibles (as is also the case with the species composing the genus *Pholcus*), and this is no doubt what has led HENTZ to give it the said generic names. HENTZ himself says of *Spermophora* or *Oophora*: "This sub-genus is very closely related to *Pholcus*. Nay, had it 8 eyes instead of 6, and were its legs much longer, it could not be separated from that sub-genus" [2]). The whole

1) Explor. de l'Algér., Arachn., p. 239, Pl. XV, fig. 2.
2) Aran. of the United States, loc. cit., p. 285—286.

appearance of the animal, the position of the eyes and the structure of the mandibles, are the same in *Spermophora* as in *Rachus*, as may be seen by a glance at the figures given by LUCAS [1]), DUGÈS [2]) and HENTZ [3]). Only the legs appear to be somewhat shorter in proportion in the American than in the European and North-African species. Even their industry and the form of their webs are the same. Generically they cannot possibly be separated, and, as the name *Spermophora* has the right of priority, it must be preferred to *Rachus* (and *Oophora*). — In a paper that has lately appeared [4]), SIMON unites this genus with *Pholcus*, in spite of the difference in the number of eyes, a proceeding which I can by no means approve.

Sub-fam. II. SCYTODINÆ.

Syn.: 1864. **Scytodidæ** BLACKW., Spid. of Gr. Brit., II, p. 379.

Gen. 3. SCYTODES LATR. (1804).

Deriv.: *σκυτώδης*, resembling leather (*σκῦτος*, leather, hide; *εἶδος*, appearance).

Syn.: 1804. **Scytodes** LATR., *in* Nouv. Dict. d'Hist. Nat., XXIV, p. 134 (*ad partem*).
1805. „ WALCK., Tabl. d. Aran., p. 79.
1864. „ BLACKW., Spid. of Gr. Brit., II, p. 379.
1864. „ [**Scytoda**] SIM., H. N. d. Araignées, p. 45.

Type: *Scytodes thoracica* LATR.

This genus, which, as we have above mentioned, originally included also WALCKENAR'S *Pholcus* and *Theridium*, was reduced by WALCKENAER loc. cit. so as only to comprise the species that typified the genus, *S. thoracica* LATR. WALCKENAER afterwards enlarged it again so, that according to his characteristics, it corresponded with the whole of our sub-family *Scytodinæ*. We take it in the compass first assigned to it by WALCKENAER, and also adopted by SIMON, namely, as answering to the "1re Fam., les Gibbeuses, *Gibbosæ*" of *Scytodes* WALCK. in H. N. d. Ins. Apt., I, p. 270.

The superior tarsal claws are large, weak, equably curved, with about 6 or 7 long, strong saw-teeth, proceeding from one side of the

1) Loc. cit.
2) CUVIER, Règne anim., 3e Éd., Arachn., Atlas, Pl. 9, fig. 7.
3) Aran. of the United States, loc. cit., Pl. X, fig. 5.
4) Monogr. d. espèces europ. du genre Pholcus, p. 119.

claw; the inferior claw is very small, and without teeth (in the next genus it disappears altogether). The female's palpi are destitute of a claw, but have instead *three coarse bristles*, slightly curved towards the extremity, and *thickened like a button at tip*, which internally disclose a canal terminating in a blind, rounded enlargement in the button. These bristles are surrounded by numerous pointed bristles of the ordinary form. This all applies to *S. thoracica*. In an exotic species (from Caffraria) I have counted 5 such-like bristles dilated in the form of a button.

Gen. 4. LOXOSCELES HEIN. et LOWE. 1831.

Deriv.: λοξός, oblique; σκέλος, leg.

Syn.: 1820. **Scytodes** DUFOUR, Descr. de cinq Arachn. nouv., p. 202 (*ad partem*).
1831. **Loxosceles** LOWE, Descr. of two spec. of Aran., p. 321.
1833. **Omosites** WALCK., Mém. s. une nouv. Classif. d. Aran., p. 438.
1837. **Scytodes** ID., H. N. d. Ins. Apt., I, p. 270 (*ad part.:* "2° Fam. Les Déprimées, *Depressæ*").
1864. **Omosites** [**Omosita**] SIM., H. N. d. Araignées, p. 50.

Type: *Loxosceles citigrada* HEIN. et LOWE.

The genus *Loxosceles*, which was formed by (HEINEKEN and) LOWE in 1831 at the above mentioned place, is identical with *Omosites*, proposed by WALCKENAER in 1833 for *Scytodes rufescens* DUF. (loc. cit.), though WALCKENAER himself afterwards united it with *Scytodes*, as a "family" belonging to that genus. I however agree with SIMON in considering *Omosites* or *Loxosceles* as a group sufficiently characterized to deserve being preserved as an independent genus. In its appearance it bears a certain resemblance to some *Philodrominæ*, and was therefore by LOWE considered as belonging to the *Laterigradæ* 1). The species of *Loxosceles*, which I have had the opportunity of examining, differ from all other *Retitelariæ*, with which I am acquainted, in having only *two* claws on the tarsi. These claws are long and slender, strongly and regularly curved almost into a semicircle. In an Egyptian species I have found them provided with about 12 very pointed comb-teeth, the points of which lie in an almost straight line; in a specimen of *L. rufescens* (DUF.), from Spain, kindly sent to me by Mr. SIMON, I have found only 8 such teeth. On the other pairs of legs the

1) "Citissime currit. Quietus pedes omnes oblique in libella horizontali extendit. Telam prædæ nec jacit nec ullam nisi fila quædam struit", says LOWE loc. cit. of *Lox. citigrada*.

number of teeth is less. The female has no palpal claw, but a little conical process instead. The claw-joint of the tarsi is shorter and slenderer than in *Scytodes* and *Pholcus*.

Fam. III. ENYOIDÆ.

The species of this little family have been generally placed in close connexion with the following family, the *Urocteoidæ*, and have, together with them, sometimes been considered as *Retitelariæ*, and sometimes as *Tubitelariæ*. SUNDEVALL however included *Enyo* among his *Terididides*, while he united *Uroctea* with his *Drassides* [1]). Together with *Uroctea*, they were referred to the *Retitelariæ* or *Inæquitelæ* by e. g. SAVIGNY and AUDOUIN [2]), SIMON [3]) and C. KOCH [4]), who however had at first [5]) given both *Enyo* and *Uroctea* a place among his *Drassides;* by LATREILLE [6]), LUCAS [7]) and others both *Enyo* and *Uroctea* are placed among genera belonging to our *Tubitelariæ*. WALCKENAER, who at first [8]) referred these two genera to his "*Filitèles*", a group consisting exclusively of *Retitelariæ*, afterwards [9]) united all the forms known to him of *Enyoidæ* and *Urocteoidæ* in the genus *Clotho*, which subsequently took its place in the group "*Niditèles*" [10]), which answers nearest to our *Drassoidæ;* but he soon [11]) detached from it one of the three families ("Uroctées", "Enyo" and "Zodarions"), into which he had divided the genus, namely "*les Zodarions*", and aggregated it to the Theridioid genus *Argus* (= *Erigone* + *Walckenaera*). *Enyo* and *Uroctea* were by DUGÈS [12]) united with the *Scytodoidæ* and some others in his family "*Scythodés*" or "*Micrognathes*", as we have already (p. 99) mentioned. SIMON forms for them a separate "tribus", "*Clothéiens*", of the family "*Théridiformes*" (loc. cit.).

Although the Enyoidæ show a more or less striking resemblance with almost every one of the various groups of spiders, to which they have thus been referred, nevertheless they ought not in my opinion to be united with any of them. Among the *Tubitelariæ* it is only the *Urocteoidæ* and *Filistatoidæ*, with which they can be shown to have any intimate connexion,

1) Consp. Arachn., p. 17, 18.
2) Descr. de l'Égypte, (2 Édit.) XXII, p. 347—252.
3) Hist. Nat. d. Araignées, p. 152.
4) Uebers. d. Arachn.-Syst., 5, p. 23, 24.
5) Ibid., 1, p. 19, 20.
6) Gen. Crust. et Ins., IV, p. 370.
7) Explor. de l'Algérie, Arachn., p. 230.
8) Mém. s. une nouv. Classif. d. Aran., p. 438; Hist. Nat. d. Ins. Apt., I, p. 202.
9) Hist. Nat. d. Ins. Apt., I, p. 635.
10) Ibid., II, p. 512; IV, p. 526.
11) Ibid., II, p. 347.
12) Observ. s. les Aran., p. 160.

and I think it is only the relationship, in which they have been placed to the Urocteoidæ, that has caused them to be foisted, as a sort of appendage to these latter, into the sub-order Tubitelariæ. The characteristic features (the structure of the mandibles), which they have in common with the *Filistatoidæ* and *Urocteoidæ*, belong equally to the *Scytodoidæ*. Their entire appearance, especially the long, fine extremities, indicate beyond all doubt their place to be among the *Retitelariæ*, and of these the *Scytodoidæ* must be considered as nearest akin to them. As in the Scytodoidæ, the tarsus is augmented with a little *separate claw-joint*, and even in the structure of the mouth they seem to approach nearest to the Scytodoidæ, though the lip is separated by a suture from the sternum, as in the *Theridioidæ*. With the *Urocteoidæ*, in spite of the considerable difference in their general appearance, they show several striking points of contact, not only in the similar position of the eyes, and in the mandibles being connected towards the base and having a very small claw (as is also the case with most *Scytodoidæ* and *Filistatoidæ*), but even in the structure of the female's palpi, which are *thickened towards the extremity*, and *armed with a powerful, pectinated claw*. The eyes are 8 in number, arranged in two transverse rows, of which the posterior row is strongly curved forwards, so much so that the eyes may also be said to form three rows.

But the Enyoidæ differ from the Urocteoidæ and all the Retitelariæ, and indeed, as far as I know, from *all* other spiders, in the *structure of their spinners*. Seen in profile, these organs display a considerable resemblance to those of the Urocteoidæ, for we first remark a pair of long spinners, which appear to consist of a short basal joint, from which the remaining part of the spinner issues in the form of a compressed longer joint, somewhat tapering towards the extremity. But whereas in the *Urocteoidæ* the *superior* (posterior) pair of spinners are the longest, in the *Enyoidæ* the *inferior* pair are incomparably larger than the others. Moreover in the Enyoidæ — at least in the species, *Enyo græca* C. KOCH, which I have had the opportunity of examining — *the basal part of the inferior spinners is common to them both:* seen from beneath it is almost inversely heart-shaped, rounded off in front and cut transversely behind, with an incision in the edge, on both sides of which the two real spinners are inserted. Thus seen, they lie in tolerably close juxtaposition, are about as long as the basal piece, but taken together narrower than this, about double as long as broad at the base, and tapering a little towards the extremity. The basal piece must be considered as a part of the abdomen, as it has not divided itself into two separate basal *joints* for the spinners. The spinner itself is directed slightly upward, as in *Uroctea*, and appears to consist of two joints, of which the second is very short, and terminates in a flat surface, bearing the rather few and very short spinning-tubes, which open each through a slightly curved, cylindrical spinning-bristle. — As regards the superior and intermediary spinners, they are very small and difficult to observe. I believe however that I have found them somewhat above the larger spinners, which are situated at the extremity of the abdomen, in the shape of four pale cylindrical nipples, with a few spinning-tubes in their apices.

To this family we reckon two European genera, *Zodarium* and *Enyo*. Of these genera, *Zodarium* corresponds to the "Race *Zodarionides*" of WALCKENAER'S *Argus*, *Enyo* to the "Famille *Enyo*" of *Clotho* WALCK. By SAVIGNY and AUDOUIN, C. KOCH, SIMON and others they are considered — perhaps rightly — as constituting but one genus, *Enyo*. We distinguish them in the following manner:

1. Series oculorum anticorum procurva. . . . . 1. *Zodarium*.
2. Series oculorum anticorum sub-recta. . . . . 2. *Enyo*.

Among exotic genera, *Laches* NOB. (*Lachesis* SAV. et AUD.) and *Storena* WALCK. perhaps belong to the Enyoidæ: the former genus has, according to AUDOUIN [1]), the latter, according to CAMBRIDGE [2]), the inferior spinners longer than the superior, and three claws on the tarsi.

Gen. 1. ZODARIUM WALCK. 1847.

Deriv.: ζωδάριον, little animal.

Syn.: 1825–27. Enyo SAV. et AUD., Descr. de l'Égypte, (Éd. 2:) XXII, p. 349 (*ad partem*).
† 1837. Lucia C. KOCH, Uebers. d. Arachn.-Syst., 1, p. 19 (*ad partem*) [3]).
† 1837. Clotho WALCK., H. N. d. Ins. Apt., I, p. 635 (*ad part.:* "3e Fam., Les Zodarions, *Zodariones*").
† 1841. Argus ID., ibid., II, p. 344 (*ad part.:* "1e Fam., 2e Race, Les Zodarionides").
1847. Zodarium [Zodarion] ID., ibid., IV, p. 563.
1864. Enyo SIM., H. N. d. Araignées, p. 159 (*ad partem*).

Type: *Zodarium longipes* (SAV. et AUD.).

In the "Description de l'Égypte" loc. cit., under Genus *Enyo*, we find: "M. WALCKENAER vient d'établir ce genre sous le nom de *Zodarion*. Il lui trouve plusieurs points de ressemblance avec les théridions, et il le place entre ceux-ci et les drasses." It would seem from this, that the name *Zodarium* is older than *Enyo;* AUDOUIN does not however indicate the source,

1) Descr. de l'Égypte, (Édit. 2:) XXII, p. 309.
2) Descr. and sketches of some new species of Aran. etc., p. 2 et seq.
3) *Lucia* SWAINS. [Lepidopt.] 1833.

from which he took his information, and in any work of WALCKENAER published previous to 1837, I have not found "*Zodarion*" mentioned. As far as I am aware, WALCKENAER speaks for the first time of any spider of this genus in his Hist. Nat. d. Ins. Apt., Vol. I; and he seems not to have considered his "Zodarions" or "Zodarionides" as forming a distinct genus till in 1847. (Conf. Syn.).

I have at Nizza met with a female specimen of a *Zodarium*, which seems to be identical with *Enyo græca* C. KOCH. Perhaps it is also the same as *E. longipes* SAV. et AUD., and as *E. occitanica* DUGÈS; but if such be the case, the species must vary considerably in colour.

In *Z. græcum* the free claw-joint is large and easily seen, much as in *Scytodes;* the superior tarsal claws are weak, strongly curved, saw-toothed, with about 6 large teeth proceeding from the outer side of the claw; the inferior claw is very small, without teeth. The female's palpal claw is strongly and uniformly curved, and from the base nearly to the tip armed with about 10 straight, parallel, vertical comb-teeth, the points of which are situated in a slightly curved, nearly straight line: the teeth are accordingly longest in the middle of the claw and shorter towards its extremity and base. They are not inserted in the middle line of the claw, but on one side. The claw is surrounded by numerous hairy, pointed bristles.

* Gen. 2. ENYO (SAV. et AUD.). 1825—1827.

Deriv.: *Ἐνυώ*, mythol. proper name.

Syn.: 1825—27. **Enyo** SAV. et AUD., Descr. d. l'Égypte, (Ed. 2:) XXII, p. 349 (*ad partem*).
† 1837. **Lucia** C. KOCH, Uebers. d. Arachn.-Syst., I, p. 19 (*ad partem*).
† 1837. **Clotho** WALCK., H. N. d. Ins. Apt., I, p. 635 (*ad part.:* "2e Fam. Les Enyo").
1864. **Enyo** SIM., H. N. d. Araignées, p. 159 (*ad partem*).

Type: *Enyo nitida* SAV. et AUD.

I have seen no species of this genus. *Enyo germanica* C. KOCH (which is supposed by WALCKENAER to be the same as *E. nitida* SAV. et AUD.), and *E. italica* CANESTR. are the only European species of the genus as yet known.

E. amaranthina LUC., which LUCAS only provisionally united with *Enyo* [1]), and for which WALCKENAER formed the family "les *Incertaines*" of his genus *Clotho* [2]), appears to differ in highly important features both from *Zodarium* and *Enyo*, and ought in my opinion to be made the type of a separate genus.

1) Explor. d. l'Algérie, Arachn., p. 232. 2) Hist. Nat. d. Ins. Apt., IV, p. 454.

Sub-ordo III. TUBITELARIÆ.

Syn.: 1817. **"Tubitèles"** LATR., *in* CUV., Règne Anim., III, p. 81.
1823. **Textores** SUND., Gen. Aran. Suec., p. 10.
1825. **Tubitelæ** LATR., Fam. Nat. du Règne Anim., p. 314.
1833. **Drassides** SUND., Consp. Arachn., p. 17.
1833. **Araneæ Tubitelariæ** PERTY, Delect. Anim. Art. Bras., p. 192.

The best way of briefly characterizing the *Tubitelariæ* is perhaps the following: all known spiders, which cannot be classed under any of the other sub-orders, belong to this! — Their ordinary form and appearance are too well known to need describing here; but within this polymorphous group we meet with transition-forms to many different families, not only of *Retitelariæ* and *Territelariæ*, but also of *Laterigradæ*, *Citigradæ* and *Saltigradæ* — indeed of all the other sub-orders, except the *Orbitelariæ*. It is probably impossible to mention any sure characteristic, that at once distinguishes these spiders from all the other sub-orders, with which they are thus related: I have therefore instead of this endeavoured, in the case of each of these latter, to indicate such marks of distinction as appear to me decisive of the limits between them and the Tubitelariæ, and I refer to what is said on this subject under the heads of these sub-orders as well as under the different families of the Tubitelariæ.

The *Tubitelariæ*, as we already know, correspond to LATREILLE'S *Tubitelæ*, but certain of the genera included by us in this division have been otherwise classified by other authors. *Uroctea* is often assigned to the *Inæquitelæ* or *Retitelariæ*, *Filistata* again to the *Territelariæ*, *Zora* to the *Citigradæ*, etc. To this we shall return in treating of the different families. — The Tubitelariæ seem, as we have also had occasion to observe, to be the lowest sub-order among spiders, that in fact, from which the others have mediately or immediately been developed. It may be divided into several families, which do not however all seem to be very sharply defined. To the usually received three European families, *Agalenoidæ*, *Drassoidæ* and *Dysderoidæ*, we add for the European fauna three more, *Urocteoidæ*, *Hersilioidæ* and *Filistatoidæ*, which 6 families we characterize as follows:

I. Stigma tubi trachealis utrinque pone stigma sacci trachealis (pulmonalis) in latere ventris non adest. Oculi sæpissime 8.

A. Tarsi articulo unguifero aucti. Mamillæ superiores reliquis multo longiores, articulis trinis aut binis: subtus tubulis textoriis præditæ. Series oculorum 8 ambæ recurvæ. Tarsorum ungues trini. . . . II. *Hersilioidæ*.

B. Tarsi articulo unguifero distincto carentes.

a. Pars cephalica impressionibus lateralibus a parte thoracica sæpissime distincta. Mamillæ superiores inferioribus plerumque multo longiores.

α. Cephalothorax brevis, sub-reniformis vel inverse cordatus, parte cephalica parva. Mamillæ superiores reliquis multo longiores, articulis binis: 2do longo, compresso. Mandibulæ parvæ, debiles. Maxillæ in labium valde inclinatæ. Oculi 8. Tarsorum ungues trini. . . . . I. *Urocteoidæ.*

β. Cephalothorax oblongus, parte cephalica majore, sæpissime elevata, convexa. Mamillæ superiores reliquis plerumque longiores et tum subtus tubulis textoriis præditæ. Oculi 8, rarissime (in gen. *Haditis*) nulli. Tarsorum ungues trini (excepto in gen. *Agræca*). . . . . III. *Agalenoidæ.*

b. Pars cephalica a parte thoracica non distincta. Mamillæ superiores inferioribus non vel parum longiores.

α. Mandibulæ inter se liberæ, ungue mediocri vel longo. Labium non cum sterno coalitum. Oculi 8, rarissime (in gen. *Thysa*) 6. Ungues tarsorum bini. . . . . IV. *Drassoidæ.*

β. Mandibulæ versus basin inter se unitæ. Labium cum sterno coalitum. Oculi 8. Ungues tarsorum trini. . . . . . VI. *Filistatoidæ.*

II. Stigmata 4, bina in utroque latere ad basin ventris: anteriora saccorum, posteriora tuborum trachealium. Oculi 6, rarissime (in gen. *Stalita*) nulli. Ungues tarsorum trini aut bini. . . . . V. *Dysderoidæ.*

Fam. I. UROCTEOIDÆ.

The few spiders belonging to this family appear to me to stand just upon the boundary-line between Tubitelariæ and Retitelariæ, and might with almost equal reason be attributed to either of these sub-orders. By their ordinarily short extremities, and in a certain degree also by their general appearance, the Urocteoidæ exhibit an approach to the more short-legged forms among the *Theridioidæ*, e. g. *Asagena* and *Euryopis*. The small mandibles united towards the base show their relationship with the *Scytodoidæ* and *Filistata*. With the last-named genus and the *Enyoidæ* they agree in the structure of the female's palpal claw, and with the Enyoidæ also in the position of the eyes (the 8 eyes form two transverse rows, curved forwards); but they differ from them in the absence of a separate claw-joint on the tarsi, in their entire general appearance, and especially in their spinners. (Conf. p. 106). We class them among the *Tubitelariæ* principally because the *superior* (posterior) *spinners are considerably longer than the others*, and are, at least in *Uroctea*, *along the underside of the elongated 2nd joint provided with spinning-tubes*, thereby plainly showing the relationship of these spiders to the *Hersilioidæ* and *Agalenoidæ*.

That they cannot however be united with either of these two families, follows from certain peculiarities in their organisation. The cephalothorax is short, reniform or inversely heart-shaped. The first joint of the superior spinners is very short, whereas the second is long, *compressed, and almost lancet-formed. The anus is surrounded by a double crown of a peculiar kind of bristles*, which, as far as I am aware, has not been found in any other spider, and whose functions are unknown [1]). Respecting the different views, that have previously to the present time been maintained on the subject of the systematic position of the Urocteoidæ, we refer to what has been stated above, p. 105.

Beside *Uroctea* DUF. or *Clotho* (WALCK.), I include in this group only the genus *Œcobius* LUC. [2]). That the 6-eyed genus *Sicarius* WALCK. (*Thomisoides* NIC.), which SIMON [3]) refers to his "*Clothéiens*", i. e. our *Urocteoidæ* and *Enyoidæ*, should belong to that group, seems to me highly improbable; I imagine that it ought to be referred to the *Thomisoidæ*, with which also according to GAY and NICOLET it is most nearly related [4]).

Uroctea and *Œcobius* are easily distinguished in the following manner:

1. Oculi omnes rotundati, convexi. Cephalothorax sub-reniformis. Pedes robusti. Mamillæ superiores subtus tubulis textoriis vestiti. . . 1. *Uroctea.*
2. Oculi intermedii postici sub-trianguli, deplanati. Cephalothorax inverse subcordatus. Pedes graciliores. . . . . 2. *Œcobius.*

Gen. 1. UROCTEA DUF. (1820).

Deriv.: *οὐρά*, tail; *κτείς*, comb.

Syn.: † 1809. Clotho WALCK., *in* LATR., Gen. Crust. et Ins., IV, p. 370.
1820. Uroctea DUF., Descr. de cinq Arachn. nouv., p. 198.

---

1) DUFOUR, who did not succeed in observing any spinning-tubes on the spinners of *Uroctea*, and accordingly supposed that these organs were not the true spinning apparatus, believed that "les véritables filières" were to be found between the circles of bristles, and that the bristles themselves "servent de peigne ou de carde pour enchevêtrer les fils dont l'araignée fabrique sa demeure." (Descr. de cinq Arachn. nouv., p. 200).

2) Explor. de l'Algérie, Arachn., p. 232.

3) Hist. Nat. d. Araignées, p. 156.

4) GAY, Hist. fis. e. pol. de Chile, Zool., III, p. 351.

1837. Clotho WALCK., H. N. d. Ins. Apt., I, p. 635 (*ad part.:* "1° Fam. Les Uroctées, *Urocteæ*").
1864. „ SIM., H. N. d. Araignées, p. 152.

Type: *Uroctea Durandii* (WALCK.).

Before this genus of spiders received the name of *Clotho*, that name had already (in 1808) been appropriated by FAUJAS DE ST. FONDS to a genus of shells (= *Saxicava* FLEUR.), and had therefore here to be replaced by the more recent, synonymous denomination *Uroctea*, given by L. DUFOUR. (Conf. p. 9, note 2).

In *U. Durandii* the tarsal claws are coarse, strongly curved, broad at the base, and have from the base to a little beyond the middle about 10—15 long stout comb-teeth, the points of which lie in an almost straight line. The inferior claw is comparatively small, with *one* tolerably long tooth near the base. The female's palpal claw is very strong, curved almost into a half-circle, with about 10 strong blunt teeth, gradually, but slightly increasing in length when reckoned from the base, where they are very short.

The second joint of the superior spinners forms in this species (the only one of the genus, with which I am acquainted) an angle with the short basal joint, and is directed obliquely upward; it is of considerable length and strongly compressed from the sides, almost lancet-formed, curved upwards and inwards, rounded at the extremity, without any trace of a separate lamina there; the spinning-tubes form a narrow, close band beginning at the apex of the spinner and continued throughout the entire length of its inferior surface; they are cylindrical, small, and very numerous. The anterior or inferior spinners are short, with a plainly visible but short 2nd joint. The intermediate spinners are very small.

Gen. 2. ŒCOBIUS LUC. 1845.

Deriv.: οἰκόβιος, living in houses (οἶκος, house; βιόω, live).

Syn.: 1845. Œcobius LUC., Explor. d. l'Algérie, Arachn., p. 101.
1847. „ WALCK., H. N. d. Ins. Apt., IV, p. 386.
1864. „ SIM., H. N. d. Araignées, p. 157.

Type: *Œcobius domesticus* LUC.

This genus was created by LUCAS loc. cit. for two small spiders from Algeria, *Œc. domesticus* and *Œc. annulipes* LUC., and were reckoned by him among the genera of spiders that have but *six* eyes. SIMON, who has discovered both these species in Spain, and has thus enriched the European Fauna with this interesting genus, has had the kindness to send me a specimen of each of them, whereby I have been enabled to observe, that this genus, as well as *Uroctea*, has 8 eyes, and not only 6, as LUCAS and all others who mention it, have stated. The posterior central eyes, which were supposed to be missing, have however quite a different appearance from the rest. They are posited much nearer to the lateral eyes than to each other, and are of an oblong triangular form, with the longest side turned towards the lateral eyes; they diverge rapidly backwards with their points, whereas the obliquely cut off base-sides diverge forwards. They also show a more or less evident transverse depression passing from the longest side to the opposite almost right angle. They are very flat, and clear as glass, and very much resemble the similarly flat and clear, oval, or almost triangular posterior central eyes of certain *Drassoidæ* (e. g. of the genera *Drassus* and *Gnaphosa*).

The anus is surrounded by a double ring of bristles, exactly as in the case of *Uroctea*. In *Œc. annulipes* the bristles of the outer ring are curved almost in the form of an ∽. I have not been able to discover any row of spinning-tubes on the underside of the superior spinners. The legs are finer and slenderer in proportion than those of *Uroctea*, especially in *Œc. domesticus;* but in other respects the species of *Œcobius* are in the highest degree similar to *Uroctea*, and SIMON very justly remarks: "Un observateur peu exercé prendrait les écobes pour de très-jeunes clothos, tant l'aspect de toutes ces araignées est semblable" [1]). It having been now shown that the number of eyes is the same in both, the mutual agreement between these two genera is still more striking.

The superior tarsal claws are slender, uniformly and much curved: in *Œc. annulipes* I have found those of the 1st pair of legs armed with about 10 comb-teeth, not long, but increasing in length from the base. On the 4th pair the claws are still more slender than on the 1st, with about 8 teeth. The inferior claw has 3 teeth, the palpal claw about 12. All this applies to *Œc. annulipes* ♀.

BLACKWALL has, under the name of *Œcobius navus*, described a spider from Madeira, which has 6 eyes, infra-mammillary organ and calamistrum, 2 claws on the tarsi, and three-jointed (?) spinners with spinning-tubes on the underside. None of these characteristics however belong to the genus *Œcobius* LUC., and it is clear that the species described by BLACKWALL

1) Hist. Nat. d. Araignées, p. 158.

does not even belong to the family before us. BLACKWALL has proposed a separate family for it, which he calls *Œcobiidæ* [1]), and which we also have adopted; but as it requires a new name, we call the family *Omanoidæ* and the species in question *Omanus* [2]) *navus*. (Conf. p. 44 above).

Fam. II. HERSILIOIDÆ.

The genus *Hersilia*, which was formed (1825—27) by SAVIGNY and AUDOUIN in "Descr. d. l'Égypte" (T. XXII, p. 317 of the 2nd Edit.), is, as the reader, from the few details above given (p. 109) under the head of the family *Hersilioidæ*, is probably already aware, so peculiar, as scarcely to admit of being united with any of the other families that are referred to the Tubitelariæ. By SAVIGNY and AUDOUIN *Hersilia* was placed between the genera *Arachne* (*Nyssus* WALCK.) and *Erigone*. WALCKENAER places this genus between *Ctenus* and *Sphasus*, and refers it to his "*Coureuses*" or LATREILLES *Citigradæ*, on account of a certain similarity, in the position of the eyes, to *Dolomedes* and *Sphasus* [3]). It has the same systematic position in the works of LUCAS — who nevertheless has remarked that, in his opinion, it belongs to another "section" or ought to form a new one [4]) — as also in SIMON, who formed for it a "tribus", "*Herséliens*" in the family "*Lycosiformes*" [5]). DUGÈS [6]) reckoned *Hersilia* to his "*Micrognathes*" or "*Scythodés*", SUNDEVALL to his *Drassides*, i. e. our *Tubitelariæ* [7]), C. KOCH to the family *Agelenides* among these [8]). It seems to me that the *Hersilioidæ* stand nearest to the *Agalenoidæ*, with which they agree in the structure of the spinning organs (especially in the spinning-tubes appearing not only on the extremity but along the underside of the superior spinners), and the *Urocteoidæ*, which they also resemble in the structure of the parts of the mouth: in that respect they also approximate to the other spiders

---

1) *Œcobius navus* was first described in 1859, in "Descr. of newly disc. spid. capt. by J. Y. Johnson" (Ann. and Mag. of Nat. Hist., 3 Ser., IV, p. 258); the family *Œcobiidæ* was formed in 1862, in "Descr. of newly disc. spid. from the Isl. of Madeira (ibid., 3 Ser., IX, p. 382).

2) *Omanus*, mythol. prop. name.

3) Mém. s. une nouv. Classif. d. Aran., p. 438; — Hist. Nat. d. Ins. Apt., I, p. 202 and 372.

4) Observ. sur les Aran. du genre Hersilia, p. 4.

5) Hist. Nat. d. Araignées, p. 343. 6) Observ. sur les Aran., p. 160.

7) Consp. Arachn., p. 22.

8) Uebers. d. Arachn.-Syst., 1, p. 14; ibid., 5, p. 25.

distinguished by DUGÈS as "*Micrognathes*", accordingly to the *Filistatoidæ*, *Scytodoidæ* and *Enyoidæ*, of which the two last-named families have a separate claw-joint on the legs, like the *Hersilioidæ*. With the *Citigradæ* this family seems to me to be far less nearly related.

Only one species of this remarkable family has, as far as I am aware, as yet been met with in Europe, viz. *Hersilia oraniensis* LUC., which SIMON has found in Spain, and of which he had the kindness to send me a specimen (a young ♂) under the name of "*Hersiliola oraniensis*". As this species differs considerably from the typical species of the genus *Hersilia*, we form for it a new genus, with the name proposed by SIMON, *Hersiliola*, and we define it thus:

1. Tarsorum articulus unguiferus ipso tarso multo brevior. Mamillarum superiorum articuli bini: primus et secundus sub-æquales. Pedes 3tii paris reliquis non valde breviores. . . . . 1. *Hersiliola.*

Gen. 1. HERSILIOLA N.

Deriv.: Dimin. of *Hersilia*, histor. prop. name.

Syn.: 1845. Hersilia LUC., Explor. de l'Algérie, Arachn., p. 127 (*ad partem*).
1847. „ WALCK., H. N. d. Ins. Apt., IV, p. 404 (*ad part.:* "2e Fam. Les Orthopodes, *Orthopodes*").
1864. „ SIM., H. N. d. Araignées, p. 343 (*ad partem*).

Type: *Hersiliola oraniensis* (LUCAS).

In *Hersilia caudata* SAV. et AUD., the typical species for *Hersilia*, the claw-joint of the tarsus is *as long* as the tarsus itself, the superior spinners extraordinarily long, 3-jointed, and the 3rd pair of legs in an unusual degree shorter than the other legs; even the armature of teeth on the claws is quite different from that of *Hersiliola oraniensis* [1]). LUCAS has himself explained the properties that distinguish *H. oraniensis* from the other known species of *Hersilia;* he formed for it a separate division of that genus, characterised by the shorter spinners and comparatively long 3rd pair of legs, but considered these characteristics as not of sufficient importance to justify the formation of a new genus [2]).

In *Hersiliola oraniensis* the superior spinners are about double as long as the inferior, and consist of two cylindrical joints about twice as

---

1) Conf. Descr. de l'Égypte, (Éd. 2:) T. XXII, p. 317 et seq.
2) Explor. de l'Algérie, Arachn., p. 127.

long as they are broad, and of about equal length, the second joint being somewhat conically terminated. This second joint on the underside exhibits a row of (about 6) unusually long and stout spinning-tubes, about as long as the medium diameter of the joint, and terminating with a short, almost cylindrical, truncated spinning-bristle. A similar large spinning-tube is situated just under the extremity of the first joint. Moreover the end of that joint and the whole underside of the 2nd joint are occupied by a number of shorter and much finer spinning-tubes, which also terminate in a somewhat short, fine spinning-bristle. At the apex of the spinner a few spinning-tubes of different sizes are observable. The inferior spinners are as thick as the superior, but scarcely half as long, somewhat tapering; their second joint is extremely short, the apex thickly covered with small spinning-tubes [1]). The intermediate spinners are somewhat shorter, and of much less diameter than the inferior, cylindrical, with a few spinning-tubes at the apex.

The superior tarsal claws are rather weak, but large, of uniform curvature, with about 10 or 12 close-set comb-teeth, gradually increasing in length towards the extremity of the claw; the uttermost teeth are somewhat sinuated (i. e. curved a little in the form of an ∽) and divergent; the inferior claw is small, and has only *one* rather coarse and somewhat curved tooth. The claw-joint is plainly visible, a little slenderer than the tarsus, rather longer than it is broad. Inside this joint, in my specimen (a ♂ jun.) lies a new outfit of claws, ready to take the place of the old ones, which fall away when the spider changes its integument. This circumstance I have also noticed in younger specimens of species destitute of a separate claw-joint, e. g. in an *Histopona*, and it would seem therefore to be a general law, *that previous to every moult new claws are formed within the tarsus itself.* Ohlerts' conjecture, that the old claws are retained, and only their skin changed [2]), is not reconcileable with these observations, and must accordingly be considered as erroneous.

In *Hersiliola oraniensis* (and perhaps in other species of the same family) it is a remarkable fact, *that the palpus also of the male is armed with a pectinated claw.* This is at least the case in the ♂ jun. of this species in my possession. Only one similar case was previously known, that namely of *Dolomedes fimbriatus*, in which Ohlert has observed a pectinated claw at the extremity of the male's as well as the female's palpus [3]).

---

1) These tubes are cylindrical, narrow, and apparently destitute of a spinning bristle at the tip (?).

2) Klauenbild. d. Preuss. Spinn., p. 2. 3) Ibid., p. 12.

Fam. III. AGALENOIDÆ.

Syn.: 1837. **Agelenides** C. KOCH, Uebers. d. Arachn.-Syst., 1, p. 13 (*ad max. part.*).
1852. **Tubicolæ** DOLESCH., Syst. Verzeichn. etc., p. 14 (*ad max. part.*).

The Agalenoidæ were detached as a separate family from LATREILLE'S *Tubitelæ* or SUNDEVALL'S *Drassides* by C. KOCH 1837 (loc. cit.), and that family has since been acknowledged by BLACKWALL, OHLERT, and others. In WALCKENAER it also forms a group, "les *Tapitèles*", answering to one of our families. WESTRING on the other hand preserves SUNDEVALL'S *Drassides* undivided, and accordingly assigns the *Agalenoidæ* to that family. SIMON has, as aforesaid (p. 33), united most of the spiders belonging to this family, together with *Linyphia* and others, into a "tribus", "*les Linyphiens*" of the family "les *Théridiformes*" — a way of classing them, which, in my opinion, is quite inadmissible. Species of the genera *Dictyna* and *Titanæca* have formerly, before their relationship with *Amaurobius* was detected, been reckoned as *Theridioidæ*, by e. g. WALCKENAER (who also has described a couple of species of *Dictyna* under the head of his *Drassus*), and SUNDEVALL, and even still by SIMON and OHLERT [1]). *Hyptiotes* on the contrary, which builds a regular, geometrical net in the form of a circular sector, and is nearly related to *Uloborus* (vid. sup. p. 69 et seq.), has been erroneously referred by AUSSERER [2]) and CANESTRINI [3]) to the *Agalenoidæ*, to which these authors, as well as DOLESCHALL [4]), also assign *Pholcus* (and *Rachus* or *Spermophora*), which I believe to be equally unnatural. The genus *Textrix*, which exhibits sundry remarkable analogies with the *Lycosoidæ*, has sometimes, e. g. by LUCAS [5]), been placed in juxtaposition with genera belonging to this latter family, which also, through the medium of *Dolomedes*, nearly approaches the *Agalenoidæ*. But it nevertheless appears to be generally admitted that the *Drassoidæ* are the nearest relations of this last-named family: the transition from the *Agalenoidæ* to the *Drassoidæ* is in fact so gradual, that the demarcation can only be made in a tolerably arbitrary and artificial manner; several genera, situated just upon the boundary-line between the two families, have therefore been referred sometimes to the one, sometimes to the other, and sometimes they have been formed into a separate family. Thus according to C. KOCH the

1) Aran. d. Prov. Preuss., p. 33.
2) Die Arachn. Tirols, I, p. 14.
3) Aran. Ital., p. 65.
4) Syst. Verzeichn. etc., p. 14.
5) Explor. de l'Algérie, Arachn., p. 121: Gen. *Lycosoides* LUCAS ad partem = *Textrix* SUND.

species forming the genera *Amaurobius* and *Cœlotes* belong to the *Drassoidæ*, to which also Ohlert refers the first-named of these genera, whereas by Blackwall and L. Koch *Cœlotes* is assigned to the *Agalenoidæ*, and *Amaurobius* (*Ciniflo* Blackw.) is made the type of a separate family, *Cinifionidæ* Blackw. or *Amaurobiidæ* L. Koch. These spiders are classified in like manner by Canestrini and Pavesi [1]). *Agrœca* and *Liocranum*, which C. Koch includes in his *Agelenides*, belong according to L. Koch to the *Drassoidæ*. Blackwall refers the species of *Agrœca* to the former, those of *Liocranum* to the latter family; and so forth. — If attention be fixed exclusively on the *number (3) of the tarsal claws*, *Agrœca* must be detached from the *Agalenoidæ*, though in its whole appearance closely allied to that family, but having only *2* claws on the tarsus; if again, with Ohlert, we assume *elongated superior spinners* as the indispensable characteristic of the *Agalenoidæ*, then we are obliged to exclude not only *Agrœca*, but also *Argyroneta*, *Cybæus* and the *Amaurobiinæ*. L. Koch, in his excellent works on the Amaurobiinæ and Drassoidæ, detaches, in company with Blackwall, as we have already seen, the *Amaurobiinæ* as a separate family on account of the presence of the *infra-mammillary organ and calamistrum;* he appears to consider *two-jointed superior spinners and three tarsal claws* as essentially necessary characteristics of the *Agalenoidæ*, and is therefore in doubt to what family to refer *Cybæus* [2]), which, like the Agalenoidæ, has no infra-mammillary organ or calamistrum, but has only *one-jointed* superior spinners, and on account of its 3 tarsal claws cannot be referred to the *Drassoidæ.* For my part I prefer, in determining the boundary between *Drassoidæ* and *Agalenoidæ*, to lay, in cases of doubt, the principal stress on the presence of a *distinctly marked pars cephalica* in these latter in contradistinction from the former. Not only *Cybæus*, but also *Cœlotes* appears to me much more nearly related to *Amaurobius* than to the *typical* Agalenoidæ, and I am therefore obliged to consider the presence of the infra-mammillary organ and the calamistrum, which distinguishes the Amaurobiinæ (but which also occurs in genera of the most widely differing families), as a feature of tolerably trifling importance [3]), and which barely allows the forming of a separate *sub-family* for the genera of Agalenoidæ, which are provided with these organs. As regards the superior spinners, their length varies so considerably within

1) Aran. Ital., p. 61—63.

2) Die Arachn.-gattungen Amaurobius, Cœlotes u. Cybæus, p. 4.

3) Menge does not seem to lay any weight on the organs in question: at least he includes the genera *Dictyna* and *Lethia* (= *Ciniflo* Blackw. ad part.) in his family *Therididæ* (Preuss. Spinn., III, p. 244, 249).

the limits of this family (compare, for example, *Hadites tegenarioides* and *Tegenaria cinerea (cicurea)*), that it does not appear to me unjustifiable to refer to it even genera, in which their 2nd joint is so short, that it can only have spinning-tubes quite at the *extremity* (*Amaurobiinæ*, *Argyroneta*, *Agrœca*), or in which it has been reduced to a mere *flat lamina* bearing the spinning-tubes, as appears to me to be the case in the genus *Cybæus*. That the *inferior tarsal claw* should sometimes be absent within a family, in which it is generally met with, is exemplified not only in the *Dysderoidæ* and *Scytodoidæ* (*Loxosceles*) but also in the *Eresoidæ* (*Palpimanus*), and I therefore consider that I ought to aggregate *Agrœca* to the Agalenoidæ and not to the Drassoidæ, although it has but two claws, for in its general appearance it seems to me to approximate much more nearly to the former than to the latter.

It cannot however be denied that the family Agalenoidæ, as I have understood its compass, comprises tolerably heterogeneous elements, and I therefore break up the European forms belonging to it into 3 sub-families, *Amaurobiinæ*, *Agaleninæ* and *Argyronetinæ*. C. KOCH also divided his "Agelenides" (nearly answering to the two last-mentioned sub-families) into 3 such groups: "Eigentliche Trichter-spinnen", "Wanderspinnen" and "Wasserspinnen". The middlemost of these is an unnatural section, containing species, which ought to be distributed among the *Agalenoidæ* (: *Philoica* C. KOCH) and the *Drassoidæ* (: *Anyphæna*). — We distinguish the sub-families and genera of the European fauna that belong to family Agalenoidæ according to the following scheme:

§ Nulla stigmata trachealia in medio ventris paullo pone plicam genitalem.

† Organum infra-mamillare et calamistrum adsunt. . . . I. AMAUROBIINÆ.

A. Maxillæ in labium sub-triangulum inclinatæ. Pedes omnes aculeis carentes.

1. Oculi laterales inter se sub-contingentes; antici eorum a mediis anticis longius (diametro saltem oculi) distantes. . . . . 1. *Dictyna*.

2. Oculi laterales et omnes oculi seriei anterioris inter se valde et æque appropinquantes. . . . . 2. *Argenna*.

B. Maxillæ sub-parallelæ.

a. Oculi laterales disjuncti.

1. Pedes saltem 6 posteriores aculeis carentes. Labium semi-ovale . . . . 3. *Titanœca*.

2. Pedes omnes aculeati. Labium ad basin constrictum, apice truncatum vel sub-emarginatum. Oculi seriei 1mæ sub-æquales; medii postici paullo longius a lateralibus posticis quam inter se distantes. . . . . 5. *Amaurobius*.

b. Oculi laterales sub-contingentes, medii antici reliquis multo minores: medii postici inter se parum plus diametro oculi, at paullo longius quam a lateralibus posticis distantes. Pedes pilosi et setosi, non aculeati. . . . . 4. *Lethia.*

†† Organum infra-mamillare et calamistrum desunt. . . . . . II. AGALENINÆ.

* Mamillæ superiores reliquis longiores, articulis binis: 2do acuminato, in latere inferiore, non in apice tantum, tubulis textoriis instructo [1]).

A. Oculi 8.

a. Mandibulæ ad basin geniculato-convexæ. Mamillarum superiorum articulus 2^{dus} 1mo paullo brevior vel ejus fere longitudine. 7. *Cœlotes.*

b. Mandibulæ dorso recto vel leviter modo convexæ, non ad basin geniculatæ.

α. Series oculorum posticorum, desuper visa, plus minus recurva vel sub-recta: simulque est mamillarum superiorum articulus 2^{dus} vix vel non brevior, plerumque longior quam 1^{mus}.

1. Series oculorum anticorum sub-recta vel recurva, posticorum, ex quibus medii lateralibus multo majores sunt, desuper visa fortiter recurva. Cephalothorax antice carinato-elevatus, fronte prominenti. . . . . 13. *Textrix.*
2. Series oculorum anticorum procurva vel sub-recta; series posticorum, inter se parum inæqualium, desuper visa sub-recurva vel recta. . . . . 12. *Histopona.*

β. Series oculorum posticorum, desuper visa, procurva vel saltem recta.

a. Mamillæ superiores et inferiores in trapezium postice paullo latius vel in aream sub-rectangulam dispositæ.

I. Series oculorum anticorum fortiter procurva.

1. Series oculorum posticorum, desuper visa, procurva. Mamillarum superiorum articulus 2^{dus} 1mo longior. . . . . 11. *Agalena.*
2. Series oculorum posticorum, desuper visa, sub-recta. Mamillarum superiorum articulus 2^{dus} 1mo saltem dimidio brevior. . . . . 9. *Cryphœca.*

II. Series oculorum anticorum sub-recta vel paullo procurva. Mamillarum superiorum articulus 2^{dus} 1mo sæpissime multo brevior. . . . . 8. *Tegenaria.*

---

1) According to BLACKWALL and some other writers, the superior spinners of these spiders consist of *three* joints; but as I have not been able to discover that the elevation, to which the joint considered by these authors as the 2nd, is articulated, is separated by any articulation from the abdomen, I cannot consider it as a separate joint.

b. Mamillæ longæ, superiores inter se valde remotæ, cum inferioribus in lineam transversam recurvam dispositæ, et iis fere dimidio longiores. Series oculorum anticorum sub-recta vel procurva. . . . . 9. *Hahnia.*

B. Oculi nulli. Mamillæ superiores valde longæ, articulo 2do æque fere longo atque 1mo. . . . . 14. *Hadites.*

** Mamillæ superiores inferioribus non vel parum longiores, in ipso apice tantum tubulis textoriis præditæ.

1. Mamillæ superiores articulis distinctis binis. Ungues tarsorum bini. . . . . 15. *Agrœca.*

2. Mamillæ superiores articulo 2do exserto nullo. Ungues tarsorum trini. . . . . 6. *Cybœus.*

Pone plicam genitalem alia plica, stigmata trachealia duo in medio ejus sita continens, ad basin ventris adest. Pedes posteriores præsertim subtus (et in lateribus) pilis longis natatoriis vestiti. . . . . III. ARGYRONETINÆ.

1. Mamillæ superiores et inferiores eadem fere longitudine, articulo 2do brevi. Series oculorum antica fere recta, postica desuper visa paullo recurva. . . . . 16. *Argyroneta.*

Dolomedes agalenoides WALCK. [1]) probably forms a separate genus of this family. *Apostenus* WESTR., the species of which BLACKWALL [2]) appears to refer to *Agalena*, and which genus also AUSSERER [3]) reckons to that family, we aggregate to the *Drassoidæ*, as also *Anyphæna* SUND., which by C. KOCH had been united with the Agalenoidæ [4]).

Sub-fam. I. AMAUROBIINÆ.

This sub-family corresponds to BLACKWALL'S *Ciniflonidæ*, when we detach therefrom the genera, which belong to other sub-orders, and agree with *Amaurobius* or *Ciniflo* BLACKW. only in having an infra-mammillary organ and calamistrum [5]). AUSSERER places these spiders, as also we do,

1) Hist. Nat. d. Ins. Apt., II, p. 454. 2) Spid. of Gr. Brit., I, p. 151—162.
3) Die Arachn. Tirols, I, p. 151. 4) Uebers. d. Arachn.-Syst., 5, p. 26.
5) In MENGE'S Preuss. Spinn., Abth. III, which I received after the five first sheets of the present work were printed, several important observations on the infra-mammillary and the respiratory organs of spiders are communicated. MENGE thinks (loc. cit. p. 244) that the infra-mammillary-organ answers to the small conical process (*colulus* MENGE), which in other spiders is seen immediately under or in front of the spinners, and that both may be considered as a separate terminal part (*hypopygium*) of the coalesced abdominal segments (?). In at least one species of the genus *Dictyna*, *D. albo-maculata* MENGE, two tracheal tubes have their

in the family *Agalenoidæ* [1]). That in the form and armature of the *claws* they agree with the typical Agalenoidæ, has already been pointed out by OHLERT [2]). Even in the cases, when the inferior tarsal claw has but *two* teeth (there are usually more), these teeth are distinguished by their form: they are long, curved, generally very pointed, and the palpal claw of the female is *at the same time* armed with several powerful teeth. The spinning-tubes are very small and short, sometimes, as in *Dictyna*, difficult to perceive. — The European species known to me may be divided into five genera: *Dictyna*, *Argenna*, *Titanœca*, *Lethia* and *Amaurobius*.

Gen. 1. DICTYNA SUND. 1833.

Deriv.: *Δίκτυνα*, mythol. proper name (of *Diana*).

Syn.: 1805. **Theridium** WALCK., Tabl. d. Aran., p. 72 (*ad part.:* "7ᵉ Fam. Les Minimes, *Minimæ*").
1805. **Drassus** ID., ibid., p. 45 (*ad part.:* 3ᵉ Fam. Les phytophiles apparentes, *Phytophilæ conspicuæ*").
1833. **Dictyna** SUND., Consp. Arachn., p. 16.
1833. **Clubiona** BLACKW., Charact. of some undescr. gen. and spec. of Aran., p. 437 (*ad partem*).
1834. **Drassus** ID., Res. in Zool., p. 337 (*ad part.; sec.* BLACKW., Spid. of Gr. Brit.).
[1840. **Operaria**..., *in* Proceed. of the Linn. Soc., I, p. 66.]
1841. **Ergatis** BLACKW., The differ. in the numb. of eyes etc., p. 608.
1847. **Argus** WALCK., H. N. d. Ins. Apt., IV, p. 500: ("Fam. des Ergatides, *Ergatides*", *ad max. part.*).
1861. **Dictyna** WESTR., Aran. Suec., p. 382.
1861. **Ergatis** BLACKW., Spid. of Gr. Brit., I, p. 146.
1864. **Dictyna** SIM., H. N. d. Araignées, p. 186.
1869. „ MENGE, Preuss. Spinn., III, p. 244.

Type: *Dictyna arundinacea* (LINN.).

It is BLACKWALL that we have to thank for having assigned to this genus, which had usually before been reckoned among the *Retitelariæ*, its proper place in the vicinity of *Amaurobius*, and for having united with it

stigmata in the infra-mammillary organ, which MENGE here (loc. cit. p. 248) even calls the *tracheal area* ("Luftröhrenfeld"). Conf. our note p. 30. But in other species of *Dictyna*, *D. arundinacea* or *benigna* for inst., the tracheæ do not open in the infra-mammillary organ, but just behind the rima genitalis, according to MENGE. — The ordinary air-sacs are said (l. c., p. 248) to be rudimentary in *D. albo-maculata*.

1) AUSSERER, Die Arachn. Tirols, 1, p. 150.

2) Klauenbild. d. Preuss. Spinn., p. 9, 18.

those species of *Drassus* WALCK., which WALCKENAER referred to the "5e fam. les *Phytophiles*" of that genus 1). It is however still by SIMON, OHLERT and MENGE referred to the *Theridioidæ*, from which the appearance of its claws is sufficient to distinguish it; the inferior tarsal claw is in fact armed with *several* (4—6) long, curved teeth, which is never the case in the sub-orders *Orbitelariæ* and *Retitelariæ*.

As regards the name *Operaria* see below (p. 128) under the head of Gen. 5. *Cœlotes*.

Gen. 2. ARGENNA N.

Deriv.: Ἀργεννός, mythol. proper name.

The spider for which we have formed this genus, and which we have called *A. Mengei* 2), seems to occupy an intermediate position between *Dictyna*, *Amaurobius* and *Hahnia*, and on a hasty inspection reminds an observer strongly of the last mentioned genus. I have but two dried specimens of it, a ♂ and a ♀ (the first much injured), which I found many years ago here in the vicinity of Upsala. It is distinguished *by the eyes of the anterior row being situated very close together, not more distant than are the lateral eyes from each other.* In the form of the maxillæ and lip, as well as in the unarmed legs, this spider is nearly related to *Dictyna*, but the form of the cephalothorax and the mandibles is much the same as in *Amaurobius.*

The breadth of the large, arched, thin-haired pars cephalica is nearly $= \frac{2}{3}$ of the maximum breadth of the cephalothorax (in ♀; it is somewhat less in ♂). The eyes are of nearly equal magnitude, the anterior central eyes a trifle smaller than the others. The anterior row of eyes is straight, the posterior, when seen from before, curved downwards, when seen from above, slightly curved forwards. The distance of the anterior series from the border of the clypeus is a little greater than an eye's diameter. The 4 central eyes describe a trapezoid broader behind; the distance between the

1) Hist. Nat. d. Ins. Apt., I, p. 630.

2) *Argenna Mengei.* — Rufescenti-fusca, pedibus extus plus minus distincte fusco-annulatis, abdomine fusco vel nigro, sericeo-pubescenti, in dorso maculis parvis testaceis picto: primum 4, fere in quadratum dispositis, quarum duæ posteriores, majores, versus medium dorsi sitæ sunt, tum pone eas pluribus, minutis, in tres series, versus anum convergentes, dispositis.

Longit. c:a 2—2½ millim. (♂♀).

Ad Upsaliam rarissime inventa.

posterior central eyes is about an eye's diameter, and a little less than the distance between them and the posterior lateral eyes. The eyes of the anterior row, like the lateral eyes, are so near each other as almost to be contiguous. The mandibles are strong, and, seen from the side, almost pear-formed; when seen from in front, slightly tapering at the extremity, convex and somewhat projecting at the base, the length about double the breadth. The maxillæ are dilated at the base, sinuated a little inwards at the extremity, and somewhat inclined towards the lip, which is large, almost triangular, and rounded at the apex. The last joint of the female's palpus is cylindrical, not gradually tapering. Legs short, of almost equal length, hairy, but without spines. The abdomen is short, inversely ovate; the spinners are tolerably far apart (almost as in *Cryphœca*): the superior somewhat longer and thicker than the inferior, distinctly two-jointed, with the second joint slenderer and much shorter than the first. The superior tarsal claws are much curved, with about 9 very long, parallel comb-teeth of about equal length; the inferior claw is small, with *two* long, fine, curved teeth. The palpal claw has at least 3 teeth.

Gen. 3. TITANŒCA. N.

Deriv.: *τίτανος*, lime-stone; *οἰκέω*, inhabit.

Syn.: ? 1805. **Theridium** WALCK., Tabl. d. Aran., p. 72 ("6e Fam. Les Cachées, *Abscondatæ*", *ad part.*).
1831. „ HAHN, Die Arachn., I, (*ad part.*:) p. 84.
1837. **Asagena** C. KOCH, Uebers. d. Arachn.-Syst., 1, p. 13 (*ad partem*).
1850. **Lathrodectus** ID., ibid., 5, p. 23 (*ad partem*).
?1864. **Theridium**: sub-gen. **Eucharia** SIM., H. N. d. Araignées, p. 165 (*ad partem*).
1867. **Amaurobius** AUSS., Die Arachn. Tirols, I, p. 150 et 162 (*ad partem*).

Type: *Titanœca quadri-guttata* (HAHN).

I have found it necessary to create this new genus for HAHN'S *Theridium 4-guttatum* (= *Amaurobius Kochii* AUSS.; *Ther. obscurum* WALCK.?), which is widely different from both *Theridium* and *Lathrodectus*. It has in fact infra-mammillary organ and calamistrum, and is, in the form of the cephalothorax, mandibles, and parallel maxillæ, intimately related to the genus *Amaurobius*. Even the position of the eyes is almost identically the same as in that genus. In its broad, heart-formed sternum, in the absence of spines on the (6 posterior) extremities, and in its colour, it much resembles certain genera among the *Theridioidæ*, especially *Asagena*. The lateral eyes are however still more distant than in that genus, which is probably the reason why C. KOCH, after having first assigned it to *Asagena*, subsequently transferred it *Lathrodectus*.

The anterior row of eyes is very slightly bent forward, almost straight. The mandibles are a little thicker than the femora, perpendicular, their back straight, only a little convex towards the base (in ♀). The maxillæ are almost double as long as the lip, parallel, without impression, slightly rounded on the outside, straight on the inside, the apex rounded exteriorly. The relative lengths of the legs is 1, 4, 2, 3. The calamistrum is remarkably strongly developed: its bristles proceed from short, almost cylindrical nipples, directed obliquely backwards, which form a row following the superior border of the compressed metatarsus. The inferior spinners are somewhat thicker and longer than the superior, two-jointed, with very short 2nd joint. The palpal claw is armed with about 10 strong, closely set comb-teeth, pointing much forwards, and situated along almost the whole length of the claw: the superior tarsal claws, which are remarkably powerful, have about 9 stout comb-teeth directed somewhat forwards, and their free extremity is somewhat thickened in the middle; the inferior claw is small, but stout, with *three* pointed, curved teeth gradually increasing in length.

I have found several examples of this species at Kissingen in Bavaria, but only females and young males, under stones in dry chalky declivities. In these the 1st pair of legs have but *one* spine near the extremity of the thighs: according to AUSSERER (loc. cit. p. 163) the adult ♂ has 8 pairs of short, knife-formed spines on the underside of the tibiæ of the first pair of legs, and the mandibles are excavated inwards and in front, as in *Dictyna*, but less distinctly. The 6 other legs are without spines.

T. 4-guttata appears then to stand about half-way between *Dictyna* and *Amaurobius*, which latter it more resembles in its habits. Both AUSSERER and L. KOCH [1]) have already expressed the opinion, that it ought to form an independent genus, distinct from *Amaurobius*.

Gen. 4. LETHIA MENGE. 1869.

Deriv.: "*λήθια, occulta*": MENGE (*λήθω = λανθάνω*, to be hidden).

Syn.: 1855. Ciniflo BLACKW., (*ad part.:*) Descr. of two newly disc. spec. of Aran., p. 120.
1861. „ ID., Spid. of Gr. Brit., I, p. 139 (*ad partem*).
1869. Lethia MENGE, Preuss. Spinn., III, p. 249 (*saltem ad part.*).

Type: *Lethia humilis* (BLACKW.).

1) Die Arachn.-gatt. Amaur., Cœl. u. Cybæus, p. 31.

Ciniflo humilis BLACKW. (Spid. of Gr. Brit., I, p. 145, Pl. IX, fig. 2), of which species I have myself taken a female at Pyrmont in Germany, and received English specimens from the Rev. Mr. CAMBRIDGE, differs too much in the relative size and position of the eyes etc. from the genus *Amaurobius* (C. KOCH) NOB. (*Ciniflo* BLACKW. *ad max. part.*), to be allowed to remain in that genus. This spider has recently been described by MENGE (loc. cit.) under the name of *Lethia varia*. — MENGE reckons *Lethia* to his *Therididæ*.

The superior tarsal claws of *L. humilis* ♀ are rather stout, much curved, strongly pectinated, with (on the 1st pair of legs) about 8—10 long straight, coarse, parallel and very close-set teeth directed a little forward; the inferior claw has two very long, curved, pointed teeth and a very small point behind them. The claw of the palpus is tolerably weak, uniformly and much curved, and armed with about 4 rather long and pointed teeth pointing forward and gradually increasing in length.

Gen. 5. AMAUROBIUS (C. KOCH). 1837.

Deriv.: ἀμαυρόβιος, living in the dark (ἀμαυρός, dark; βιόω, live).

Syn.: 1805. Clubiona WALCK., Tabl. d. Aran., p. 41 (*ad part.:* "4e Fam. Les Parques, *Parcæ*").
1837. Amaurobius C. KOCH, Uebers. d. Arachn.-Syst., 1, p. 15 (*ad max. part.*).
1841. Ciniflo BLACKW., The differ. in the numb. of eyes etc., p. 607.
1861. „ ID., Spid. of Gr. Brit., I, p. 139 (*ad max. part.*).
1861. Amaurobius WESTR., Aran. Suec., p. 373.
1864. „ SIM., H. N. d. Araignées, p. 138 (*ad max. part.*).
1868. „ L. KOCH, Die Arachn.-gatt. Amaur., Cœl. u. Cyb., p. 4.

Type: *Amaurobius fenestralis* (STRØM) (= *Ar. atrox* DE GEER).

Instead of the name given by C. KOCH to this genus, BLACKWALL makes use of the newer name *Ciniflo* BLACKW., on the ground, that KOCH has united under the name of *Amaurobius* species, that can never be allowed to remain together under the same generic name, nay, that in BLACKWALL'S opinion belong to quite different families [1]). That this reason cannot be admitted, is easily seen: one would thus for consistency's sake be obliged to cashier a great many good and universally accredited generic names, e. g. both *Theridium* and *Drassus*, because WALCKENAER referred to them species, which belong to the genus *Dictyna*, and consequently to another fa-

1) BLACKW., Spid. of Gr. Brit., I, p. 171.

mily than either *Theridium* or *Drassus*. In these and similar cases it is quite sufficient to detach from the old genus such species as one considers not to belong to it, and to assign to them a new generic name, as also BLACKWALL did, when he formed the genus *Cœlotes* of species detached from C. KOCH'S *Amaurobius*.

In *Amaurobius* the claws are very nearly similar in form to those of the typical *Agaleninæ*, coarse and strong, with many and long comb-teeth; on the inferior tarsal claw the teeth are sometimes 3, sometimes only 2 in number, but always long, pointed and curved.

Sub-fam. II. AGALENINÆ.

In this sub-family we combine the *typical* Agalenoidæ, characterized by having spinning-tubes distributed along the *underside* of the superior spinners [1]), and also a couple of genera standing just on the points of transition, the one to the *Drassoidæ*, and the other to the *Amaurobiinæ*, viz. *Agræca* and *Cybæus*. We accordingly begin with the last named.

Gen. 6. CYBÆUS L. KOCH. 1868.

Deriv.: *cybæus*, (a ship of burden;) thick and bellied (as such a ship).

Syn.: 1839. **Amaurobius** C. KOCH, Die Arachn., VI, (*ad part.*:) p. 43.
1864. „ SIM., H. N. d. Araignées, p. 168 (*ad partem*).
1868. **Cybæus** L. KOCH, Die Arachn.-gatt. Amaur., Cœl. u. Cyb., p. 46.

Type: *Cybæus tetricus* (C. KOCH).

On the systematic position of this interesting genus, vid. p. 118 et seq. In *C. angustiarum* L. KOCH, the female's palpal claw is slender, slightly curved, with a long extremity, and armed towards the base with about 4 pointed, rather short comb-teeth pointing forwards. Of the tarsal claws (of the 1st pair) the superior have about 9, the inferior only 2 teeth. On the 4th pair the claws are longer and slenderer, with very long extremities, and about 7 teeth, of which the outmost are rapidly divergent; the teeth of the

1) BLACKWALL seems to be the first who (in 1833) observed these spinning-tubes and showed the erroneousness of the commonly received opinion, that the long superior spinners in the Theraphosoidæ and Agalenoidæ were not spinning-organs, but a sort of palpi (anal palpi, "filières tentacules"). Vid. BLACKW., Spid. of Gr. Brit., I, p. 154.

inferior claw are short and pointed. In *C. tetricus* the powerful superior tarsal claws have about 12 long, closely set comb-teeth, the inferior 3. — Of both the above named species specimens have been kindly presented to me by Dr. L. KOCH.

Gen. 7. CŒLOTES BLACKW. 1841.

Deriv.: *κοιλόω*, hollow, excavate.

Syn.: 1820. **Drassus** DUF., Observ. gén. sur l. Arachn., p. 9 (356) (*ad partem*).
1830. „ WALCK., Faune Franç., Arachn., p. 169 (*ad part.:* "IV. Les Spéophiles, *Speophilæ*").
1833. **Clubiona** BLACKW., Charact. of some undescr. gen. and spec. of Aran., p. 436 (*ad partem*).
† 1834. **Aranea** REUSS, Zool. Misc., Arachn., p. 210 (216) (*ad partem*).
1837. **Amaurobius** C. KOCH, Uebers. d. Arachn.-Syst., 1, p. 15 (*ad partem*)
[1840. **Cavator** ..., Proceed. of the Linn. Soc., I, p. 66.]
1841. **Cœlotes** BLACKW., The differ. in the numb. of eyes etc., p. 618.
1861. „ ID., Spid. of Gr. Brit., I, p. 169.
1864. **Amaurobius** SIM., H. N. d. Araignées, p. 138 (*ad partem*).
1868. **Cœlotes** L. KOCH, Die Arachn.-gatt. Amaur., Cœl. u. Cyb., p. 32.

Type: *Cœlotes saxatilis* BLACKW.

In a short notice of BLACKWALL's above cited work, "The difference in the number of eyes with which Spiders are provided," etc., in the Proceedings of the Linn. Soc. for Apr. 21, 1840, we read as follows:

"In the first tribe [Octonoculini] he proposes three new genera, two "of them belonging to a family, which he characterizes under the name of "*Ciniflonidæ:* these genera he also characterizes under the names of *Ciniflo*, "founded on *Clubiona atrox* of LATREILLE, and *Operaria*, comprising the "*Theridion benignum* WALCK., *Drassus exiguus* BLACKW. and *Drassus viridissi-*"*mus* WALCK. The third genus characterized by Mr. BLACKWALL, is referred "by him to the family of *Agelenidæ*, under the name of *Cavator:* it is founded "on the *Clubiona saxatilis* BLACKW."

From this we may conclude that in BLACKWALL'S above-mentioned paper, before it was printed, his genus *Ergatis*, or *Dictyna* SUND., was called *Operaria*, and *Cœlotes*, *Cavator*. Although I certainly do not think that an author has the right of arbitrarily changing a name, when it has once been published, yet in this case *Cœlotes* seems to me preferable to *Cavator*, and so much the more so, as no author of the notice that occurs in the "Proceedings" is named, and the two denominations in question, there introduced, are not used by BLACKWALL in the work in which they are said to have been proposed, so that one has not even a right to cite BLACKWALL as authority for them. It is best to consider them as "nulles et non avenues".

As may be seen from the synonyms, the species of *Cœlotes* have been referred to very different genera, and it was not till BLACKWALL had shown that their spinning-organs are of exactly the same structure as those of the typical *Agalenoidæ*, that they received a secure position in the vicinity of these spiders.

In the typical species the palpal claw is strong, of tolerably uniform curvature, and armed with about 7 comb-teeth, gradually increasing in length, and directed slightly forwards; the superior tarsal claws are long, strong, and armed with about 13 similar long and powerful teeth. The inferior claw has only *two* long, pointed, teeth.

Gen. 8. TEGENARIA (LATR.) 1804.

Deriv.: uncertain. Perhaps from τέγος, roof, or τήγανον, pan (with reference to the form of the web)[1].

Syn.: 1804. **Tegenaria** LATR., *in* Nouv. Dict. d'Hist. Nat., XXIV, p. 134 (*ad partem*).
† 1806. **Aranea** ID, Gen. Crust. et Ins., I, p. 94 (*ad partem*).
1832. **Agelena** SUND., Sv. Spindl. Beskr., *in* Vet.-Akad. Handl. f. 1831, p. 125 (*ad part.*).
1837. **Tegenaria** C. KOCH, Uebers. d. Arachn.-Syst., 1, p. 13 (*ad max. part.*).
1837. **Philœca** [**Philoica**] ID., ibid.
1841. **Tegenaria** WALCK., H. N. d. Ins. Apt., II, p. 1 (*ad part.:* "1e Fam. Les Familières, *Familiariæ*", et "2e Fam. Les Agrestes, *Agrestæ*").
1861. „ WESTR., Aran. Suec., p. 303.
1861. „ BLACKW., Spid. of Gr. Brit., I, p. 163 (*ad max. part.*).
1864. „ SIM., H. N. d. Araignées, p. 201 (*ad max. part.*)

Type: *Tegenaria civilis* WALCK.

The generic name *Aranea* was in 1804 restricted by LATREILLE to the spiders now called *Epeira*, but in 1806 he adopted for them this latter name, which had been given them by WALCKENAER, and transferred the denomination *Aranea* to those that he had before called *Tegenaria*. In this signification it was adopted by several writers. But afterwards, especially since SUNDEVALL[2]) called attention to the fact, that the order of Spiders in its entire extent ought to be called *Araneæ*, the use of this word as a generic name has gradually been abandoned.

1) In AGASSIZ' Nomencl. Zool. it is derived from "Τέγεα, nom. prop." SIMON (and also STAVELEY) thinks that it comes from "τεγη, toit; αιρω, élever".
2) Svenska Spindlarnes Beskr., *in* Vet. Akad. Handl. för 1832, p. 372.

We take the genus *Tegenaria* in the compass assigned to it by WESTRING, i. e. we include in it also most of the forms, which C. KOCH and SIMON refer to the genus or sub-genus *Philœca [Philoica]*. As however this last genus may at some future time be resumed, it will not be deemed superfluous to indicate here in a few words its relation to *Tegenaria* properly so called, and to WESTRING'S genus *Agrœca*.

In consequence of C. KOCH'S contradictory and confusing definitions of his *Philœca*, it is utterly impossible to determine, which species ought properly to be united under that name, unless we accept the limitations *first* given by him of *Tegenaria* and *Philœca*. When these genera were first separated (1837, in Uebers. d. Arachn.-Syst., 1), KOCH expressly gave "*Araneus domesticus* CLERCK" as type of the genus *Philœca*, and at the same time set up as the type of *Tegenaria*, "*Aranea domestica* LINN.", by which KOCH, as one sees from e. g. Die Arachn., VIII, p. 37, rightly understood *Tegenaria civilis* WALCK. But in 1850, in the 5th Number of Uebers. d. Arachn.-Syst., this relation is reversed: there we find reckoned up under the head of *Tegenaria: T. domestica* (CLERCK), *T. intricata*, *T. campestris*, etc., whereas to *Philœca* are now referred *T. civilis*, *T. atrica* and near-related species, as also two spiders belonging to totally different genera, "*Ph. notata*" (*Liocranum domesticum* (REUSS)) and "*Ph. linotina*" (*Agrœca brunnea* (BLACKW.)). In "Die Arachniden" (vid. Vol. XVI, p. 49) KOCH reckons to *Philœca* only these two last-mentioned species, and a third, "*Ph. advena*", which appears to be a young specimen of *Tegenaria atrica*. If then the genus *Philœca* is to be retained as separate from *Tegenaria*, which however to me appears superfluous, we are obliged *by the law of priority* so to limit these genera, that *T. domestica* (CLERCK) may belong to *Philœca*, and *T. civilis* to *Tegenaria*, and not *vice versa*, as SIMON has done (loc. cit.). For *Phil. linotina* C. KOCH, which cannot be united with either of the above genera, WESTRING has very properly formed a new genus, *Agrœca*, and has with so much greater reason given it a new appellation, as that the name *Philœca* is quite unreasonable for that species, which never lives in houses.

Of WALCKENAER'S *Tegenaria* (loc. cit.) the first two (see Syn.), and possibly also the 3d family ("les Brévilabes, *Brevilabiæ*") belong to *Tegenaria* NOB.; the 4th fam., "les Caudées, *Caudatæ*", appears to agree with the genus *Histopona* NOB. The 5th, "les Tisseuses, *Textrices*", answers to SUNDEVALL'S and BLACKWALL'S *Textrix*.

The superior tarsal claws are long and powerful, armed with numerous comb-teeth, in *T. atrica*, for example, with 16—18 on the 1st and about 15 on the 4th pair of legs. In this species the inferior tarsal claw has 4 long, curved teeth, and the female's palpal claw about 10 teeth gradually increasing in length. In other species the number of teeth on the superior tarsal and on the palpal claw is somewhat less.

Gen. 9. CRYPHŒCA N.

Deriv.: *κρύφος*, hiding-place; *οἰκέω*, inhabit.

Syn.: 1834. **Tegenaria** C. KOCH, *in* HERR.-SCHÆFF., Deutschl. Ins. (*ad part.:*) 125, 26.
1845. **Hahnia** ID., Die Arachn., XII, (*ad part.:*) p. 158.
1847. **Tegenaria** WALCK., H. N. d. Ins. Apt., IV, p. 464 (*ad part.:* "6e Fam. Les Argusides, *Argusides*").
1850. **Amaurobius** MENGE, Verzeichn. Danz. Spinn., p. 63 (*ad partem*).
1861. **Hahnia** WESTR., Aran. Suec., p. 315 (*ad partem*).
1861. **Tegenaria** BLACKW., Spid. of Gr. Brit., I, p. 163 (*ad partem*).
1864. **Agelena**: sub-gen.: **Hahnia** SIM., H. N. d. Araignées, p. 212 (*ad partem*).
1869. **Hahnia** MENGE, Preuss. Spinn., III, p. 251 (*ad partem*).

Type: *Cryphœca silvicola* (C. KOCH).

This genus I have thought it necessary to form for C. KOCH'S *Hahnia silvicola*, which especially by the different arrangement of the spinners essentially differs from the typical species of the genus *Hahnia*. Even OHLERT [1]), although he refers it to *Hahnia* (as does also MENGE in his Preuss. Spinn.), thinks it ought preferably to form a separate genus. BLACKWALL refers it to *Tegenaria*, and unites the other species of KOCH'S *Hahnia* with *Agelena*. It appears to differ from *Hahnia* also in the armature of the palpal claw: whereas this claw in *Hahnia* has no tooth or only one that is scarcely visible, it has in *Cryphœca silvicola* 4 or 5 long, gradually increasing teeth pointing slightly forwards. The superior tarsal claws have about 10 long, closely set comb-teeth; the inferior claw has 3 (4?) long, curved, pointed teeth gradually increasing in length.

Gen. 10. HAHNIA (C. KOCH) 1841.

Deriv.: HAHN, proper name.

Syn.: 1841. **Hahnia** C. KOCH, Die Arachn., VIII, (*ad part.:*) p. 61, 63.
1841. **Agelena** BLACKW., The differ. in the numb. of eyes etc., (*ad part.:*) p. 619 etc.
1847. **Argus** WALCK., H. N. d. Ins. Apt., IV, (*ad part.:*) p. 465, 503, 506.
1861. **Hahnia** WESTR., Aran. Suec., p. 315 (*ad partem*).

1861. **Agelena** BLACKW., Spid. of Gr. Brit., I, p. 152 (*ad partem*).
1864. „ : sub-gen. **Hahnia** SIM., H. N. d. Araignées, p. 212 (*ad partem*).
1869. **Hahnia** MENGE, Preuss. Spinn., III, p. 251 (*ad partem*).

Type: *Hahnia montana* (BLACKW.) (= *H. pusilla* C. KOCH).

The species belonging to this genus are referred by BLACKWALL to *Agalena*, from which genus they are however easily distinguished by the different position of the eyes, etc. — On the superior, strongly curved tarsal claws I have in the typical species counted about 8 long, powerful, close-set, slightly divergent comb-teeth; on the inferior 3: the female's palpal claw is, according to OHLERT [1]), toothless or provided with one scarcely perceptible point below the middle. Also in *H. elegans* (BLACKW.) (*H. pratensis* C. KOCH) this claw is destitute of teeth, according to MENGE [2]). — By MENGE *Hahnia* (with *Cryphœca*) is now referred to the family *Therididæ*, from which it seems to me to be widely separated.

Gen. 11. AGALENA WALCK. 1805.

Deriv.: *α priv.*, and *γαλήνη*, calm, tranquillity [3]).

Syn.: 1805. **Agalena [Agelena]** WALCK., Tabl. d. Aran., p. 51.
1841. „ „ ID., H. N. d. Ins. Apt., II, p. 19 (*ad part.:* 1:e Fam. Les Labyrinthiques, *Labyrinthicæ*").
1861. „ „ WESTR.. Aran. Suec., p. 308.
1861. „ „ BLACKW., Spid. of Gr. Brit., I, p. 152 (*ad partem*).
1864. „ „ : sub-gen. *id.*, SIM., H. N. d. Araignées, p. 211.

Type: *Agalena labyrinthica* (CLERCK).

BLACKWALL assigns to this genus much wider limits than we can adopt, and even refers to it some species, to which the characteristics he gives of *Agalena* by no means correspond, and which we refer to *Hahnia*, *Apostenus* and *Agrœca*. Even his *Ag. Hyndmanni* is hardly an *Agalena*, but still less does it belong to any of the three last-named genera. Also *Ag. boopis* CAMBR. [4]) seems to me to be the type of a separate genus: its anterior row of eyes is straight, and the central eyes of the posterior row are very

1) Klauenbild. d. Preuss. Spinn., p. 11. 2) Preuss. Spinn., III, p. 254.

3) With respect to the animal's rapid and restless motions. To derive this name, as some have done (Vid. e. g. Dict. Univ. d'Hist. Nat. par D'ORBIGNY) from ἀγέλη, herd, has no other foundation than the accidental similitude of the letters in the two words.

4) Descr. of twenty-four new spec. of Spid., p. 11 (8571).

disproportionately large, protruding and wide apart, their outer brims extending nearly to the entire length of the anterior row, according to CAMBRIDGE loc. cit.

Of the two families into which WALCKENAER divided this genus, perhaps the 2nd, "les Nysses, *Nyssæ*", deserves to form a separate genus: *Nyssus* WALCK. 1805 [1]) = *Arachne* SAV. et AUD. [2]). According to WALCKENAER [3]), *Megamyrmecium [Megamyrmækion]* REUSS [4]) or *Dyction* WALCK. [5]) is identical with *Arachne* SAV. et AUD.

The derivation of the name given above, is that generally adopted, and the only one which affords a rationel meaning to it. I therefore write *Agalena*, not *Agelena*, as is usually the custom. It is an additional reason for writing *Agalena*, that WALCKENAER himself, when he used that word as a specific name (in "*Epeira agalena*") always wrote it thus.

The long, powerful, superior tarsal claws, in the typical species, have 10 or 12 comb-teeth, the inferior 3 or 4 long, curved, pointed teeth. On the palpal claw, which is more slender, I have counted 6 teeth rapidly increasing in length, and pointing more forwards. The superior tarsal claws of the 4th pair have about 14 teeth.

Gen. 12. HISTOPONA N.

Deriv.: ἱστός, web; πονέω, work.

Syn.: 1834. **Agelena** C. KOCH, *in* HERR.-SCHÆFF., Deutschl. Ins., (*ad part.:*) 125, 11.
1837. **Tegenaria** ID., Uebers. d. Arachn.-Syst., 1, p. 13 (*ad partem*).
1841. „ WALCK., H. N. d. Ins. Apt., II, p. 1 (*ad partem*).
1841. **Textrix** C. KOCH, Die Arachn., VIII, (*ad part.:*) p. 48.
1864. „ [**Teotrix**] SIM., H. N. d. Araignées, p. 219 (*ad partem*).

Type: *Histopona torpida* (C. KOCH).

The spider we have chosen as type for this genus has, as we see, been referred by C. KOCH first to *Agalena*, then to *Tegenaria*, and lastly to *Textrix*. To me it appears to stand about midway between the two last mentioned genera; it differs from *Textrix* in that the cephalothorax is less high and less compressed in front, with a forehead that is not prominent,

1) Tableau d. Aran., p. 52.
2) Descr. de l'Égypte, (2 Éd.:) XXII, p. 314.
3) Hist. Nat. d. Ins. Apt., II, p. 419.
4) Zool. Misc., Arachn., p. 211 (217).
5) Hist. Nat. d. Ins. Apt., I, p. 380.

and that the eyes, of which the posterior lateral ones are almost of equal size with the central, are, when seen from before, arranged in two rows uniformly and slightly curved forwards; the posterior row, seen from above, is just a little bent backwards. In another, apparently undescribed species from Nizza, the anterior row is straight, or, if bent, bent rather backward than forward; the posterior row, seen from above, is scarcely perceptibly curved backwards, and the second joint of the superior spinners is longer than the first. This genus differs from *Tegenaria* chiefly in the greatly elongated superior spinners, which are exactly like those of *Textrix*. C. KOCH'S *Textrix montana* 1) belongs beyond a doubt to *Histopona*, as do probably also the spiders, which WALCKENAER refers to the 4th Fam. "les Caudées, *Caudatæ*" 2) of his genus *Tegenaria*.

In *H. torpida* the claws are of the form usual in the Agaleninæ, powerful, pretty much curved, with a long, strong extremity, and about 10 somewhat diverging teeth, of which those at the base are much the smallest; the inferior claw has 3 teeth, of which the inmost is very small. The female's palpal claw is slender, pretty much curved, with about 7 gradually increasing, sharp teeth pointing forwards. In the above mentioned species from Nizza the superior tarsal claws are very closely pectinated, with about 14—17 teeth; the inferior claw has 3.

Gen. 13. TEXTRIX SUND. 1833.

Deriv.: *textrix*, female weaver.

Syn.: † 1831. **Aranea** DUF., Descr. et fig. de quelques Aran. nouv. ou mal conn., p. 358.
1832. **Agelena** SUND., Sv. Spindl. Beskr., *in* Vet.-Akad. Handl. f. 1831, p. 125 (*ad partem*).
1833. **Textrix** ID., Consp. Arachn., p. 19.
1833. „ BLACKW., Charact. of some undescr. gen. and spec. of aran., p. 108.
1841. **Tegenaria** WALCK., H. N. d. Ins. Apt., II, p. 1 (*ad part.:* "5e Fam. Les Tisseuses, *Textrices*").
1845. **Lycosoides** LUCAS, Explor. de l'Algérie, Arachn., p. 12 (*ad partem*).
1861. **Textrix** WESTR., Aran. Suec., p. 310.
1861. „ BLACKW., Spid. of Gr. Brit., I, p. 171.
1864. „ [**Teotrix**] SIM., H. N. d. Araignées, p. 219 (*ad partem*).

Type: *Textrix denticulata* (OLIV.) (= *T. lycosina* SUND.).

---

1) Die Arachn., VIII, p. 53, Tab. CCLXVII, f. 630.
2) Hist. Nat. d. Ins. Apt., II, p. 13.

As regards our limitation of this genus, we refer to what has been said of the preceding genus, or *Histopona*. — The superior tarsal claws in *T. denticulata* or *lycosina* are of the form usual in the family, but not particularly strong, pretty regularly curved, with about 10—12 comb-teeth gradually increasing in length, the outermost pointing forwards and diverging. The inferior tarsal claw has only *two* teeth. In a few other species (among which is *T. vestita* or *ferruginea* C. KOCH) I have also found only 2 teeth on that claw.

Gen. 14. HADITES KEYSERL. 1862.

Deriv.: ᾅδης, *Hades*.

Syn.: 1862. **Hadites** KEYSERL., Beschr. ein. neuen Spinne aus d. Höhlen v. Lesina, p. 3 (541).

Type: *Hadites Tegenarioides* KEYSERL.

Of this remarkable, blind spider, which has hitherto been found only in the subterraneous caverns of the isle of *Lesina*, Count KEYSERLING has kindly presented me with a female specimen. — The superior spinners are very long, two-jointed: the first joint is more than double as long as it is broad, and somewhat longer than the inferior, thicker spinners; the 2nd joint is not so thick as the first, but equally long, conically pointed at the extremity, covered on the underside with very long spinning-tubes; on the apex of the joint a similar, very coarse spinning-tube is situated. Such a tube is found also at the apex of the slender intermediate spinners. — The palpal claw is weak, pretty regularly and slightly curved, with about 8—10 gradually increasing, pointed comb-teeth directed forwards. The tarsal claws are of the form usual in the *Agalenince*, somewhat weak, with about 12 long, pointed comb-teeth directed forwards; the inferior claw is small, with 3 long, pointed teeth. — KEYSERLING has found only 7 or 8 teeth on the superior and 2 on the inferior tarsal claw (loc. cit., p. 5).

Gen. 15. AGRŒCA WESTR. 1861.

Deriv.: ἀγροῖκος, living in the country (ἀγρός, country; οἰκέω, inhabit).

Syn.: 1833. **Agelena** BLACKW., Charact. of some undescr. gen. and spec. of Aran. (*ad part.:*) p. 351.

1843. **Philœca** [Philoica] C. KOCH, Die Arachn. X, (*ad part.:*) p. 108.

1861. **Agrœca** WESTR., Aran. Suec., p. 311.

1861. **Agelena** BLACKW., Spid. of Gr. Brit., I, p. 152 (*ad partem*).

1864. **Tegenaria**: sub-gen. **Philœca** [**Philoica**] SIM., H. N. d. Araignées, p. 202 (*ad partem*).

1868. **Agrœca** L. KOCH, Die Arachn.-fam. d. Drassiden, p. 2.

Type: *Agrœca brunnea* (BLACKW.) (= *A. linotina* (C. KOCH)).

As regards the systematic position of this genus vid. sup. p. 118, 119. Concerning *Philœca* C. KOCH v. p. 129: Gen. 7. *Tegenaria*. — The species of this genus are referred by BLACKWALL and CAMBRIDGE to *Agalena*, from which they differ widely by the totally dissimilar structure of the spinners etc. — In *A. brunnea* the female's palpal claw is moderately curved, with 5 tolerably long comb-teeth gradually increasing in length, and pointing somewhat forward. The two tarsal claws on the 1st pair of legs are pretty powerful, with about 4—6 strong comb-teeth; on the 4th pair they are thin and slender, much weaker and longer than those of the 1st pair, springing at a right or slightly acute angle from a narrow, high basement, and armed with about 5 or 6 sparse teeth gradually increasing in length and pointing somewhat forwards. Thus the form of the claws on that pair differs from that which is usual among the Agalenoidæ, and indicates that *Agrœca* stands just upon the point of transition to the *Drassoidæ*. As in these latter, the tarsi have no inferior claw. The inferior spinners are a trifle longer and thicker than the superior; their 2nd joint is very short, scarcely perceptible, with rather few, short spinning-tubes at the apex.

Sub-fam. III. ARGYRONETINÆ.

Argyroneta aquatica seems to me to deserve to be taken as the type of a separate sub-family, as well on account of its peculiar habits, as of the structure of its respiratory organs. *Argyroneta* has in fact, as has been shown by GRUBE [1]) and MENGE [2]), two large tracheal tubes opening close to each other in a transversal groove, situated a little behind the ordinary genital- or "pulmonary" groove, in which the two tracheal sacs have their stigmata. These large air-tubes run through the petiolum into the cephalothorax, there sending out bundles of fine tracheæ into the legs, palpi and mandibles etc.: near the stigmata they give off two such bundles for the abdomen [3]). In certain species of *Dictyna*, *D. arundinacea* for instance,

1) Einige Resultate aus Unters. üb. die Anat. d. Spinnen, p. 300.

2) Ueber d. Lebensweise d. Arachn., p. 23.

3) MENGE, loc. cit.

the tracheæ have a similar distribution [1]). Also in *Anyphæna* (of the family *Drassoidæ*), L. KOCH [2]) has discovered a transversal groove under the abdomen, into which tracheæ probably debouch. With respect to the position of its stigmata, *Argyroneta* (as also *Dictyna arundinacea* etc.) is related to the *Dysderoidæ*, which have also 4 stigmata, of which two lead to tracheal tubes: but these stigmata lie, each behind the corresponding one of the stigmata of the tracheal sacs, at the sides of the abdomen, whereas in *Argyroneta* (and *Dictyna*) the two tracheal tubes terminate near the middle-line of the belly.

Gen. 16. ARGYRONETA LATR. 1804.

Deriv.: ἄργυρος, silver; νέω, spin.

Syn.: 1804. **Argyroneta** LATR., *in* Nouv. Dict. d'Hist. Nat., XXIV, p. 134.
1861. „ WESTR., Aran. Suec., p. 367.
1861. „ BLACKW., Spid. of Gr. Brit., I, p. 136.
1864. „ SIM., H. N. d. Araignées, p. 127.

Type: *Argyroneta aquatica* (CLERCK).

In this spider the superior tarsal claws are large and powerful, almost straight at the base, afterwards curved strongly and much downwards, with (on the 1st pair of legs) about 9—12 long, vertical, parallel comb-teeth, of which the 2 or 3 innermost are much smaller than the others. The inferior claw has 3—4 pointed teeth gradually increasing in length. On the remaining pairs of legs the number of teeth on the superior claws is somewhat less. The first half of the palpal claw shows about 6 somewhat diverging teeth, of which the innermost is much smaller than the rest.

Fam. IV. DRASSOIDÆ.

Syn.: 1833. **Drassides** SUND., Consp. Arachn., p. 17 (*ad partem*).
1852. **Cellicolæ** DOLESCH., Syst. Verzeichn. etc., p. 6 (*ad partem*).

In the arrangement of this family — which may be considered as including all not laterigrade (nor saltigrade) spiders, which are provided with only 2 stigmata and only 2 tarsal claws, and are destitute of a distinctly

1) MENGE, Preuss. Spinn., III, p. 246.
2) Die Arachn.-fam. d. Drassiden, p. 194.

marked pars cephalica, and whose 2nd pair of legs is not longer than the others — I have adopted the limitations of the genera given by L. Koch in his excellent work: Die Arachniden-familie der Drassiden. As aforesaid however, I exclude from this family *Agrœca*, which in my opinion ought to be classed among the *Agalenoidæ*, although it has not, like the other genera of that family, three, but only two claws at the extremity of the tarsus, as also *Storena* (Conf. p. 107). *Apostenus* is not received as a separate genus in L. Koch's work; neither is *Thysa* adduced in it, this last genus having been later made known to arachnologists.

The want of a distinctly marked pars cephalica, together with the presence of only 2 tarsal claws, distinguishes in doubtful cases the spiders belonging to this family from the Agalenoidæ, into which they gradually pass, through e. g. *Apostenus* in the one family and *Agrœca* in the other. From certain not distinctly laterigrade *Thomisoidæ* (*Misumena*), the Drassoidæ are easily distinguished by the relative length of the legs: the 2nd pair being *not* longer than the others. All European Drassoidæ have 8 eyes, except *Thysa*, which has but 6. As their eyes, of which the two central ones of the anterior row are never considerably larger than the rest, are, excepting in *Zora*, arranged in *two* transverse rows, they are thus easily distinguished from the *Attoidæ*. From certain other (exotic) *Saltigradæ* (*Otiothops*, *Myrmecium* etc.), which approach near the Drassoidæ in the position of the eyes, the Drassoidæ are probably best distinguished by their cephalothorax being less high and broad anteriorly. *Zora* in the position of the eyes approaches the *Lycosoidæ*, but not only the number and form of the claws, but also the double row of long, moveable spines under the tibiæ and metatarsi of the anterior legs, indicate for that spider a place in the vicinity of *Apostenus* among the Drassoidæ.

The structure of the tarsal claws is very various. While in the *Agalenoidæ* they are gradually tapering, more equably curved, they are generally in the *Drassoidæ* of a more uniform breadth and straight at the base, and bent downwards only towards the extremity. Yet the genera, that in other respects approach the Agalenoidæ, as *Liocranum*, also have claws more like theirs than those of the *typical* Drassoidæ. Even among these we find that large and strong species, e. g. *Drassus 4-punctatus*, *Gnaphosa lucifuga*, have claws gradually tapering from the base. The claw-teeth are in general less numerous and stouter than those of the Agalenoidæ. In the genera, which, at least in the structure of the claws, form the transition to the *Thomisoidæ* (*Clubiona*, *Chiracanthium*), these organs are very elongated and closely pectinated with many teeth. The palpal claw is often entirely toothless; frequently it has a few, rarely many teeth.

This family chiefly corresponds to WALCKENAER'S "*Nidicoles*" [1]), which group however originally [2]) included also the *Dysderoidæ* and *Micrommata* (*Sparassus*), and in which he at last erroneously placed e. g. *Enyo* and *Lathrodectus* [3]). From SUNDEVALL LATREILLE'S *Tubitelæ* received the name of *Drassides*, which many arachnologists, e. g. WESTRING, continue to give them. *Agalenoidæ* and *Dysderoidæ* were however soon (1837) detached from them as separate families by C. KOCH. BLACKWALL'S *Drassidæ* are pretty nearly identical with our Drassoidæ, as also SIMON'S "*Drassiens*", which however also include genera which we refer to the *Agalenoidæ*.

The European genera we include in this family may be distinguished in the following manner (Conf. L. KOCH, Die Arachn.-fam. d. Drassiden, p. 2):

§ Oculi 8.

† Maxillæ convexæ, non impressæ.

* Series oculorum postica, desuper visa, recurva.

A. Pedes aculeati.

1. Oculi laterales inter se longius quam medii antici a mediis posticis distantes; series oculorum 4 posteriorum adeo recurva, ut oculi potius tres quam duas series designent. . . . . 1. *Zora*.

2. Oculi laterales inter se non longius remoti, quam medii antici a mediis posticis: series oculorum postica leviter tantum recurva. . . . . 2. *Apostenus*.

(?) *B*. Pedes non aculeati. . . . . 3. *Trachelas*.

** Series oculorum postica procurva vel recta.

A. Abdomen subtus pone plicam genitalem aliam plicam transversam præbet. . . . . 5. *Anyphæna*.

B. Abdomen plica pone plicam genitalem caret.

a. Mandibulæ ad basin inermes.

α. Pedes 4[ti] paris reliquis longiores.

1. Labium ad summum dimidiam maxillarum longitudinem æquat. 4. *Liocranum*.

2. Labium $\frac{1}{3}$ brevius quam maxilla. . . . . . . 6. *Clubiona*.

β. Pedes 1[mi] paris reliquis longiores. . . . . 7. *Chiracanthium*.

b. Mandibulæ ad basin aculeo armatæ. . . . . 8. *Phrurolithus*.

†† Maxillæ in medio impressæ.

* Cephalothorax linea media impressa caret. . . . . 9. *Micaria*.

** Cephalothorax linea media impressa præditus.

---

1) Hist. Nat. d. Ins. Apt., I, p. 202. 2) Tabl. d. Aran., p. 1.
3) Hist. Nat. d. Ins. Apt., II, p. 512; IV, p. 526.

A. Series oculorum postica, desuper visa, plus minus procurva, et evidenter longior quam series antica. . . . . 10. *Drassus.*

B. Series oculorum postica, desuper visa, recta vel recurva.

1. Series oculorum postica sub-recta, non recurva, parum longior quam series antica. Margo posterior sulci unguicularis mandibularum intus inermis vel dentibus tantum parvis armatus. 11. *Melanophora.*
2. Oculi laterales inter se evidenter longius distantes quam medii antici a mediis posticis; series oculorum posticorum sæpissime evidenter recurva. Margo posterior sulci unguicularis mandibulæ intus in laminam denticulatam (rarissime in dentem tantum fortem) productus. . . . . 12. *Gnaphosa.*

§§ Oculi 6. . . . . 13. *Thysa.*

Gen. 1. ZORA (C. KOCH). 1848.

Deriv.: probably ζωρός, strong, fiery.

Syn.: 1820. Dolomedes DUF., Descr. de cinq Arachn. nouv. (*ad part.:*) p. 204.
†1833. Lycæna SUND., Sv. Spindl. Beskr., *in* Vet.-Akad. Handl. f. 1832, p. 265.
1833. „ [Lycodia] ID., Consp. Arachn., p. 22.
†1833. Hecaërge BLACKW., Charact. of some undescr. gen. and spec. of Aran., p. 193.
?1847. Lycosoides LUCAS, Explor. de l'Algérie, Arachn., p. 12 (*ad partem*).
1848. Zora C. KOCH, Die Arachn., XIV, p. 91 (*ad partem*).
1851. Lycodia WESTR., Förteckn. etc., p. 46.
1861. Hecaërge BLACKW., Spid. of Gr. Brit., I, p. 41.
1861. Zora WESTR., Aran. Suec., p. 324.
1864. „ SIM., H. N. d. Araignées, p. 371 (*ad partem*).
1866. „ L. KOCH, Die Arachn.-fam. d. Drassiden, p. 2 (*ad partem*).

Type: *Zora lycæna* (WALCK.) [1]).

The species of this genus were by WALCKENAER and others united with *Dolomedes* among the *Lycosoidæ;* they really constitute a transition from the Drassoidæ to that family, from which however, as is justly remarked by WESTRING, they are excluded by their low and weak cephalothorax, the peculiar spines with which the legs are armed, their habits etc., as also by the number and form of the claws, to which OHLERT has drawn attention.

---

1) DUFOUR has already in 1820 (loc. cit.) given the specific name *spinimanus* to another spider belonging to the genus *Zora*, and accordingly the typical species, *Z. spinimana* (SUND.), must be denominated by the specific name next following in order of time (*lycæna* WALCK.), under which it has been described. The older names produced by WALCKENAER and SIMON as synonyms, *Dolomedes errans* DUF. and *D. hippomane* SAV. et AUD., assuredly do not belong to *Zora spinimana* (SUND.).

Nevertheless they have been considered as Lycosoidæ by most writers, e. g. by C. KOCH, BLACKWALL, SIMON. — *Zora ocreata* C. KOCH [1]) probably does not belong to this genus.

The name *Lycodia* SUND. (Consp. Arachn.) is either a slip of the pen or a misprint for *Lycœna*, as is evident partly from the passage cited in the Consp. Arachn.: "*Lycodia* Act. Holm. 1832" — in Act. Holm. (Vet.-Akad. Handl.) 1832 the word is *Lycœna*, not *Lycodia* — partly from SUNDEVALL'S own express declaration in "Årsber. om nyare zool. arb. o. uppt. 1837—40", p. 340. It is on this account that the name *Zora* is to be preferred to *Lycodia*. The names *Lycœna* and *Hecaërge* were already applied to genera of butterflies before they were given to the spiders now under consideration [2]). Conf. WESTRING, Aran. Suec., p. 325.

In the typical species the tarsal claws are weak, slender, and of almost uniform breadth (but somewhat stronger at the place which bears the teeth), issuing from a high base, uniformly and much curved. The inner claw has about 4 or 5 saw-teeth, gradually increasing in length, of which the outermost are pointed, those nearest the base blunt and very short. The outer claw has but from 2 to 4 teeth [3]), the innermost tooth being situated under the middle of the claw. The hair-tuft under the claws is rather small, the hairs shorter than the claws and dilated towards the extremity. The female's palpal claw is very small, uniformly and pretty much curved, with 3 or 4 short, triangular teeth gradually increasing in length.

Gen. 2. APOSTENUS WESTR. 1851.

Deriv.: ἀποστενόω, to make narrow (στενός, narrow).

Syn.: ?1841. Agelena BLACKW., The differ. in the numb. of eyes etc., (*ad part.:*) p. 624.
?1847. Argus WALCK., H. N. d. Ins. Apt., IV, p. 504 (*ad part.:* "Fam. des Agélénides, *Agelenides*").
1851. Apostenus WESTR., Förteckn. etc., p. 46.
1861. „ ID., Aran. Suec., p. 322.
?1861. Agelena BLACKW., Spid. of Gr. Brit., I, p. 152 (*ad partem*).
?1861. Drassus CAMBR., Descr. of ten new spec. of spid. lately disc. in Engl. (*ad part.:*), p. 3 (430).
1866. Zora L. KOCH, Die Arachn.-fam. d. Drassiden, p. 2 (*ad partem*).

Type: *Apostenus fuscus* WESTR.

---

1) Die Arachn., XIV, p. 105.

2) *Lycœna* FABR. [Lepidopt.] 1808. — *Hecaërge* OCHSENH. [Lepidopt.] 1816.

3) According to OHLERT (Klauenbild. d. Preuss. Spinn., p. 17), the teeth of the tarsal claws are more numerous, 7 and 4 respectively.

This genus, which is not received by L. KOCH in his "Die Arachn.-fam. d. Drassiden", appears to me to form a transition from the Drassoidæ, on the one side to the *Agalenoidæ*, and on the other through *Zora* to the *Lycosoidæ*. By BLACKWALL a couple of species belonging, as far as I can see, to this genus, are referred to *Agalena*, namely his *A. celans* and *gracilipes* [1]). — *A. fuscus* WESTR. I have found at Kissingen in Bavaria, and afterwards also at Söderköping in Sweden; another species is described by AUSSERER [2]) under the name of *A. saxatilis*.

The tarsal claws of *A. fuscus* are weak, much curved, and provided nearer the base with 4—5 divergent, rather long comb-teeth gradually increasing in length. On a conical process beneath them are only *two*, colossal hairs (the claw-tuft), in the form of flat thin slices, narrow at the base, gradually dilated, and cut off obliquely at the broad extremity. They are much longer than the claws. The palpal claw, which is weak like those of the tarsi, has 2 or 3 teeth pointing forward, near its base.

To *Apostenus* or some nearly related genus belongs perhaps *Aranea spinicrus* DUF. [3]), which however is by WALCKENAER referred to the genus *Sparassus* or *Micrommata* [4]), as also *Drassus sub-niger* CAMBR. loc. cit.

*Gen. 3. TRACHELAS L. KOCH. 1866.

Deriv.: τραχηλᾶς, thick-necked.

Syn.: 1866. Trachelas L. KOCH, Die Arachn.-fam. d. Drassiden, p. 2.

Type: ?

This South-European genus is known to me only by the few words with which it is characterized by L. KOCH in the above cited passage. I am not even sure that I have assigned it a right place in my schema, for L. KOCH does not say that the posterior row of eyes, seen from above, is curved backwards, but only: "die hintere Augenreihe durch Tieferstehen der Mittelaugen gebogen." CANESTRINI and PAVESI [5]) refer *Trachelas* to the *Theridioidæ*, not to the *Drassoidæ*.

---

1) Spid. of Gr. Brit., I, p. 161, 162, Pl. X, fig. 103, 104. — *A. celans* BLACKW. is by CANESTRINI and PAVESI (Aran. Ital., p. 37) referred to the genus *Liocranum* of L. KOCH.

2) Die Arachn. Tirols, I, p. 163.

3) Descr. et fig. de quelques Aran. nouv. ou mal connues, p. 361, Pl. X, fig. 3.

4) Hist. Nat. d. Ins. Apt., I, p. 586.

5) Araneidi Italiani, p. 46.

Gen. 4. LIOCRANUM L. KOCH. 1866.

Deriv.: λεῖος, smooth; κρᾶνον, head, skull.

Syn.: 1834. **Tegenaria** C. KOCH, *in* HERR.-SCHÆFF, Deutschl. Ins., 124, (*ad part.:*) 14, 15.
1834. **Clubiona** REUSS, Zool. Misc., Arachn., (*ad part.:*) p. 208 (214).
1841. **Philœca** [**Philoica**] C. KOCH, Die Arachn., VIII, (*ad part.:*) p. 55.
1861. **Clubiona** BLACKW., Spid. of Gr. Brit., I, p. 121 (*ad partem*).
?1861. **Drassus** CAMBR., Descr. of ten new spec. of spid. lately disc. in Engl., (*ad part.:*) p. 1 (428).
1866. **Liocranum** L. KOCH, Die Arachn.-fam. d. Drassiden, p. 2.

Type: *Liocranum domesticum* (REUSS).

The typical species of this genus, formed by L. KOCH, belongs also to the Fauna of Sweden: I found a few half-grown specimens under stones at Söderköping in the summer of 1862. BLACKWALL still refers it to *Clubiona*. — The tarsal claws are pretty strong, short, with about 5 divergent teeth on the 1st pair of legs. On the 4th pair the claws are somewhat longer and weaker, also with 5 teeth. There is no claw-tuft. The palpal claw is pretty much curved, with about 3 teeth.

Drassus prælongipes CAMBR. loc. cit. appears to belong to this genus.

Gen. 5. ANYPHÆNA SUND. 1833.

Deriv.: ἀνυφαίνω, unravel a web.

Syn.: 1805. **Clubiona** WALCK., Tabl. d. Aran., p. 41 (*ad part.:* "2e Fam. Les Hamadryades, *Hamadryades*").
1832. **Agelena** SUND., Sv. Spindl. Beskr., *in* Vet.-Akad. Handl. f. 1831, p. 125 (*ad partem*).
1833. **Anyphæna** ID., Consp. Arachn., p. 20.
1861. „ WESTR., Aran. Suec., p. 370.
1861. **Clubiona** BLACKW., Spid. of Gr. Brit., I, p. 121 (*ad partem*).
1864. „ SIM., H. N. d. Araignées, p. 131 (*ad partem*).
1866. **Anyphæna** L. KOCH, Die Arachn.-fam. d. Drassiden, p. 2, 194.

Type: *Anyphæna accentuata* (WALCK.).

The genus *Anyphæna*, still by BLACKWALL included in *Clubiona*, was formed by SUNDEVALL in 1833 (loc. cit.) for WALCKENAER'S *Clubiona accentuata*. To this genus C. KOCH subsequently, in 1837 [1]), referred, to-

---

1) Uebers. d. Arachn.-Syst., 1, p. 18.

gether with this or the typical species, also *Clubiona nutrix* WALCK., which he however some time afterwards detached from *Anyphæna* and united with a couple of other species into a new genus, *Chiracanthium*. To take, as SIMON has done, the generic name of *Anyphæna* for just these species, which SUNDEVALL never referred to that genus, is of course an error.

L. KOCH has discovered (vid. loc. cit., p. 194) that *Anyphæna* is distinguished by both sexes having on the underside of the abdomen, sometimes in the middle of the belly, sometimes a little fore or aft of that point, a small transversal groove or fold of the skin. I imagine that in this groove there are one or two tracheal stigmata, as is the case with e. gr. *Argyroneta aquatica*, which has a similar groove under the anterior part of the belly. (Vid. sup. p. 136).

The tarsal claws of *A. accentuata* are rather small, strong, with about 14—20 long, closely set comb-teeth on the inner claw and only about half that number on the outer. The claw-tuft is formed of uncommonly broad, flattened, platelike hairs, which are dilated outwards, cut almost transversely at the extremity, and somewhat longer than the claw.

Gen. 6. CLUBIONA (LATR.). 1804.

Deriv. unknown [1]).

Syn.:	1804.	Clubiona	LATR., in Nouv. Dict. d'Hist. Nat., XXIV, p. 134 (*ad partem*).
	1805.	„	WALCK., Tabl. d. Aran., p. 41 (*ad part.: saltem* "1° Fam. Les Dryades, *Dryades*").
	1861.	„	WESTR., Aran. Suec., p. 388.
	1861.	„	BLACKW., Spid. of Gr. Brit., I, p. 121 (*ad partem*).
	1864.	„	SIM., H. N. d. Araignées, p. 131 (*ad max. partem*).
	1866.	„	L. KOCH, Die Arachn.-fam. d. Drassiden, p. 2, 291.

Type: *Clubiona holosericea* (DE GEER).

This genus is still preserved almost in its original compass, as more accurately limited by WALCKENAER, by, for example, BLACKWALL, who however detaches from it the species, which, in consequence of their having an infra-mammillary organ and calamistrum, he refers to *Ciniflo (Amaurobius)*.

The tarsal claws of these spiders are rather long, almost straight, curved only at the extremity, closely pectinated with long, strong teeth,

1) The usually received etymology, κλέος, fame; βιόω, live, seems highly improbable. — Perhaps the name is formed of κλωβίον, a bird-trap (with reference to the sack-like tube which these spiders inhabit).

about double as many on the interior as on the exterior claw. The chief part of the claw forms almost a right angle with its base. The claw-tuft varies greatly; in the larger species it is strongly developed, especially on the 4th pair of legs, where the hairs are long and thin and more numerous than on the 1st pair, and almost similar to the tuft-hairs of *Chiracanthium*. The claws themselves are also on that pair considerably longer than on the 1st pair. The palpal claw is small and without teeth.

Gen. 7. CHIRACANTHIUM C. Koch. 1839.

Deriv.: *χείρ*, hand; *ἀκάνθιον*, little thorn (*ἄκανθα*, thorn).

Syn.: 1805. Clubiona Walck., Tabl. d. Aran., p. 41 ("3e Fam. Les Nymphes, *Nymphæ*" *ad part.*).
1834. Drassus Reuss, Zool. Misc., Arachn., (*ad part.:*) p. 204 (210).
1837. Anyphæna C. Koch, Uebers. d. Arachn.-Syst., 1, p. 18 (*ad partem*).
1839. Chiracanthium [Cheiracanthium] id., Die Arachn., VI, p. 9.
1861. „ „ Westr.. Aran. Suec., p. 377.
1861. Clubiona Blackw., Spid. of Gr. Brit., I, p. 121 (*ad partem*).
1864. Anyphæna Sim., H. N. d. Araignées, p. 144 (*saltem ad part.*).
1866. Chiracanthium [Cheiracanthium] L. Koch, Die Arachn.-fam. d. Drassiden, p. 2, 231.

Type: *Chiracanthium nutrix* (Walck.).

Blackwall refers the species of this genus to *Clubiona*. By C. Koch it was singularly enough reckoned among the "*Theridides*" (Uebers. d. Arachn.-Syst., 5, p. 22). Concerning *Anyphæna* Simon, vid. sup., p. 144 under head of Gen. 5. *Anyphæna*.

The tarsal claws are of the same form as among the *Philodrominæ* in the next sub-order: they are small, long, slender, almost straight, except at the extremity, where they are bent into a hook. They spring at a right angle from a slender, high base: the inner claw has on the underside about 15 coarse, short, vertical, conical, somewhat sparse comb-teeth, that gradually increase in length; on the outer claw the number of teeth is somewhat less. The claw-brush is very thick and longer than the claw itself; its hairs are long and slender, slightly broader just at the extremity (as in *Micrommata*). The palpal claw is toothless, as in *Clubiona*.

Gen. 8. PHRUROLITHUS (C. Koch). 1839.

Deriv.: *φρουρέω*, guard; *λίθος*, stone.

Syn.: 1839. Phrurolithus C. Koch, Die Arachn., VI (*ad part.:*) p. 110—112.

1851. **Phrurolithus** WESTR., Förteckn. etc., p. 46.
1861. „ ID., Aran. Suec., p. 326.
1861. **Drassus** BLACKW., Spid. of Gr. Brit., I, p. 104 (*ad partem*).
1864. **Theridium** [**Theridio**]: sub-gen. **Phrurolithus** [**Phrurolithum**] SIM., H. N. d. Araignées, p. 168 (*ad partem*).
1866. **Phrurolithus** L. KOCH, Die Arachn.-fam. d. Drassiden, p. 2, 224.

Type: *Phrurolithus festivus* C. KOCH.

Under the name of *Phrurolithus*, C. KOCH united a number of spiders of various families, chiefly *Theridioidæ* and *Drassoidæ*. For the species among C. KOCH'S *Phrurolithi*, that are *Drassoidæ*, WESTRING in 1851 adopted this generic name, and has been followed in this by L. KOCH (Vid. Syn.). OHLERT'S *Phrurolithus*, embracing the greater part of the *Theridioidæ* included by C. KOCH in that genus, I have called *Lithyphantes*. Vid. sup., p. 94.

The tarsal claws of *Ph. festivus* are very small, rather short, much and pretty regularly curved (on the 4th pair of legs longer and weaker, straight at the base, much curved at the extremity), without teeth. There is a claw-tuft, but it consists only of a few much dilated hairs. The female's palpal claw is small, weak, and toothless.

Gen. 9. MICARIA WESTR. 1851.

Deriv.: *micare*, shine.

Syn.: 1805. **Drassus** WALCK., Tabl. d. Aran., p. 45 (*ad partem*).
1832. **Clubiona** SUND., Sv. Spindl. Beskr., *in* Vet.-Akad. Handl. f. 1831, p. 138 (*ad partem*).
? 1832. **Herpyllus** HENTZ, On North Amer. Spid., p. 120 (*ad partem*).
† 1835. **Macaria** C. KOCH, *in* HERR.-SCHÆFF, Deutschl. Ins., 129, 14—16.
1851. **Micaria** WESTR., Förteckn. etc., p. 46.
1861. „ ID., Aran. Suec., p. 330.
1861. **Drassus** BLACKW., Spid. of Gr. Brit., I, p. 104 (*ad partem*).
1864. **Macaria** SIM., H. N. d. Araignées, p. 112.
1866. **Micaria** L. KOCH, Die Arachn.-fam. d. Drassiden, p. 2, 52.

Type: *Micaria fulgens* (WALCK.).

The name *Micaria*, under which C. KOCH had introduced this genus, being already occupied [1]), it was in 1851 changed by WESTRING to *Micaria*.

---

1) *Macaria* CURT. [Lepidopt.] 1826.

The tarsal claws are small, straight at the base, but towards the extremity curved almost to a semicircle, with few teeth. In *M. pulicaria* the teeth are only 2 in number, very short and blunt; in *M. fulgens* they are 3, longer, but sparse, thick, and very obtuse. The hairs of the claw-tuft are few, dilated, rounded at the extremity; the whole underside of the tarsus is thinly covered with suchlike hairs.

Under this genus SIMON [1]) takes up as synonyms *Corinna* C. KOCH [2]) and *Drassina* GRUBE [3]), both of which appear to me to be very remote from it. *Drassina* is stated to have *three* claws on the tarsi, and, if this be really so, cannot even belong to the family Drassoidæ. *Corinna* would seem to stand on the point of transition from the family *Drassoidæ* to the *Myrmecioidæ*, to which last the genus is referred by C. KOCH: to me it appears rather to belong to the former family. L. KOCH however has not received it among the Drassoidæ.

Gen. 10. DRASSUS WALCK. (1805).

Deriv.: *δράσσομαι*, seize, catch.

Syn.: 1805. **Drassus** WALCK., Tabl. d. Aran., p. 45 (*ad partem*).
1805. **Clubiona** ID., ibid., p. 41 (*ad part.:* "5° Fam. Les Furies, *Furiæ*").
1832. **Herpyllus** HENTZ, On North Amer. Spid., p. 102 (*ad partem*).
1834. **Filistata** REUSS, Zool. Misc., Arachn., p. 197 (213) (*ad partem*).
1837. **Drassus** C. KOCH, Uebers. d. Arachn.-Syst., 1, p. 18.
1851. **Drassodes** WESTR., Förteckn. etc., p. 48.
1861. „ ID., Aran. Suec., p. 360.
1861. **Drassus** ID., ibid., p. 337.
1861. „ BLACKW., Spid. of Gr. Brit., I, p. 104 (*ad partem*).
1864. „ SIM., H. N. d. Araignées, p. 123.
1866. „ KOCH, Die Arachn.-fam. d. Drassiden, p. 2, 76.

Type: *Drassus quadri-punctatus* (LINN.).

In common with L. KOCH, we unite WESTRING'S *Drassodes* with his *Drassus* in one genus, since, as L. KOCH has shown, it is not at present possible to determine any sure line of separation between them, however different in their general appearance the more typical species of these two groups may be. It must not however be forgotten, that the form of the *cocoons* in WESTRING'S *Drassus* and *Drassodes* is quite different, although of course that circumstance alone cannot be considered as possessing any de-

1) Hist. Nat. d. Araignées, p. 539.
2) Die Arachn., IX, p. 17 et seq.
3) Beschr. neuer im Amurlande u. in Ostsibirien gesammelter Aran., p. 15.

cisive importance. — In BLACKWALL the genus *Drassus* has a far wider compass, and comprises also the groups *Phrurolithus*, *Micaria*, *Melanophora* and *Gnaphosa*, which we have considered as separate and independent genera. WALCKENAER, as is well known, also referred to this genus many other and widely separate forms, among which are some species of the genera *Cœlotes* and *Dictyna*, to which BLACKWALL first assigned their true places in the system.

The genus *Herpyllus* HENTZ seems very nearly to correspond to *Drassus* WALCK., and comprises not only species of the genus now before us, and of *Gnaphosa*, but probably also of several others, *Micaria* and *Melanophora* among the rest. Conf. HENTZ, Aran. of the United States, *in* Boston Journ. of Nat. Hist., V, p. 454—461, Pl. XXIV, fig. 2—20.

In the genus *Drassus* the tarsal claws are powerful, straight at the base, generally somewhat long, especially on the posterior legs, armed below with 5 or 6 strong comb-teeth; the palpal claw has about 3 teeth at some distance apart. Such is the case with e. g. *D. lapidicola*. In *D. quadri-punctatus* the claws are still more powerful, but in other respects very similar. The hairs of the claw-tufts are in general dilated towards the extremity, flattened, and mostly short; they are often continued on the tarsus throughout its underside, especially on the first pair of legs. In *D. braccatus* (vid. infra) I have counted 3 thick, almost parallel comb-teeth and a little point before them on the tarsal claws, 2 or 3 teeth on the palpal claws.

In the vicinity of Söderköping I have met with a particularly fine species of *Drassus* [1]) (no doubt identical with *D. braccatus* L. KOCH, though the cephalothorax in that species is said to be black, whereas in my specimens it is reddish brown), which in some respects appears to form the transition to *Gnaphosa*. In size and colour it is very like *Gn. variana*, but the position of the eyes is exactly the same as in *Drassus*. The posterior edge of the claw-furrow of the mandibles forms in this species with the corresponding longitudinal inner edge of the mandible a *strong*, *right-*

1) ***Drassus braccatus*** L. KOCH cephalothorace rufescenti-fusco, palpis et partibus oris infuscatis, pedibus rufescenti-testaceis, femoribus 4 anterioribus nigricantibus; abdomine fuligineo, maculis 6 albicantibus in dorso: 2 ad basin, minoribus, rotundatis; reliquis 4 fere in medio, transversis, obliquis, in rectangulum vel trapezium postice angustius et paullo latius quam longius dispositis.

Long. ♀ 7—8, ♂ c:a 5 millim.

Femora supra in medio aculeis 2, 1^{mi}, 2^{di} et 4^{ti} paris præterea uno ad apicem in latere interiore, 3^{tii} paris 2 ad apicem; pedes cetero supra non aculeati; tibiæ et tarsi pedum 4 posteriorum subtus et in lateribus aculeati.

Sub lapidibus ad Söderköping rarissime inventus.

angled corner. The cephalothorax is broad in front, almost as in *D. troglodytes;* the maxillæ are almost parallel, scarcely at all inclined towards the lip, of considerable length, narrower in the middle, slightly rounded, nearly cut transversely, at the broad extremity.

Gen. 11. MELANOPHORA C. KOCH. 1833.

Deriv.: *μέλας*, black; *φέρω*, bear.

Syn.: 1805. **Drassus** Tabl. d. Aran., p. 45 (*ad partem*).
?1832. **Herpyllus** HENTZ, On North Amer. Spid., p. 120 (*ad partem*).
1833. **Melanophora** C. KOCH, *in* HERR.-SCHÆFF., Deutschl. Ins., 120, 20—23.
1834. **Filistata** REUSS, Zool. Misc., Arachn., p. 197 (213) (*ad partem*).
1837. **Melanophora** C. KOCH, Uebers. d. Arachn.-Syst., 1, p. 17.
1861. „ WESTR., Aran. Suec., p. 354 (*ad partem*).
1861. **Drassus** BLACKW., Spid. of Gr. Brit., I, p. 104 (*ad partem*).
1864. **Melanophora** SIM., H. N. d. Araignées, p. 116.
1866. „ L. KOCH, Die Arachn.-fam. d. Drassiden, p. 2, 142.

Type: *Melanophora atra* (LATR.).

We adopt this genus, which however might perhaps without harm be suppressed and united with *Gnaphosa*, in the extent assigned to it by L. KOCH, which appears to coincide with its original limits assigned by C. KOCH. — The structure of the tarsal claws is the same as in the genus *Gnaphosa*.

Gen. 12. GNAPHOSA (LATR.) 1804.

Deriv.: *γνάπτω*, scratch, tear.

Syn.: 1804. **Gnaphosa** LATR., *in* Nouv. Dict. d'Hist. Nat., XXIV, p. 134 (*ad partem*).
1805. **Drassus** WALCK., Tabl. d. Aran., p. 45 (*ad part.:* "1° Fam. Les Lithophiles *Lithophilæ*", etc.).
1832. **Herpyllus** HENTZ, On North Amer. Spid., p. 120 (*ad partem*).
1834. **Filistata** REUSS, Zool. Misc., Arachn., p. 197 (213) (*ad partem*).
1837. **Pythonissa** C. KOCH, Uebers. d. Arachn.-Syst., 1, p. 16.
1861. „ WESTR., Aran. Suec., p. 350.
1861. **Melanophora** ID., ibid., p. 354 (*ad partem*).
1861. **Drassus** BLACKW., Spid. of Gr. Brit., I, p. 104 (*ad partem*).
1864. **Pythonissa** SIM., H. N. d. Araignées, p. 120.
1866. „ L. KOCH, Die Arachn.-fam. d. Drassiden, p. 2, 6.
1868. **Gnaphosa** THOR., *in* EISEN *et* STUXBERG, Om Gotska Sandön, p. 379.

Type: *Gnaphosa lucifuga* (WALCK.).

This genus, for which LATREILLE in his Genera Crust. et Ins. (I, p. 125) still uses the name *Gnaphosa* — a name which he afterwards changed for the more recent Walckenaerian denomination *Drassus* — has in LATREILLE for its type *Gnaphosa melanogaster* LATR. (*Aranea lucifuga* WALCK. 1802), and it answers, in the more restricted meaning in which we now, in right of priority, restore it to science, the genus *Pythonissa* C. KOCH, for which without doubt the same species is typical. As regards the more accurate determination of the limits of *Gnaphosa* or *Pythonissa*, we follow L. KOCH, and accordingly refer to this genus *P. variana* C. KOCH, which, as well by the position of the eyes, as by the presence of the little lamina, into which the posterior edge of the mandible's claw-furrow is drawn out, shows itself to belong to this genus and not to *Melanophora*, to which WESTRING refers it. The cocoon of this spider is however of an altogether different form from that of the other species both of *Gnaphosa* and *Melanophora* known to me: it is not plano-convex and of a firmer substance, resembling paper, but loose and lenticular, as in e. g. *Drassus lapidicola*. We also consider *Pyth. maculata* C. KOCH (*Ar. nocturna* LINN.) as a *Gnaphosa*, though standing on the limit between that genus and *Melanophora*, to which WESTRING refers it. The claws in this spider are very dissimilar to those of the other species both of *Gnaphosa* and *Melanophora*, which I have examined.

The tarsal claws are in *Gnaphosa* usually small, but coarse and powerful, of about the same form as those of *Drassus:* of uniform breadth or slightly tapering and straight nearest to the base, much bent towards the extremity, which is long and strong. On the underside they have only a few comb-teeth (in *G. lucifuga* e. g. about 5—6). On the 4th pair the claws are weaker and more uniformly curved. The palpal claw is tolerably strong, with some few (in *G. lucifuga* about 5) coarse comb-teeth. Deviations from this however occur: in *G. exornata* for example, the tarsal claws of which have 5—7 rather long and close-set comb-teeth, the palpal claw is long and slender, slightly and uniformly curved, with about 15 fine, long, very closely set comb-teeth. *G. nocturna* deviates still more: in this species the palpal claw is toothless; the tarsal claws are weaker, more equably curved, and armed from the base nearly to the extremity with about 5 or 6 conical teeth, proceeding from the side of the claw; the free extremity of the claw is very short.

Remarks. LATREILLE is the first, who, after WALCKENAER had in 1802, in his Faune Parisienne, separated *Mygale* from the great Linnean genus *Aranea*, divided the remaining spiders into several smaller groups distinguished by generical names. (See Nouv. Dict. d'Hist. Nat., XXIV, p. 133—136). These groups are: *Eriodon*, *Dysdera*, *Segestria*, *Argyroneta*, *Gnaphosa*, *Clubiona*, *Tegenaria*, *Scytodes*, *Linyphia*, *Aranea*, *Heteropoda*, *Misumena*, *Micrommata*, *Oxyopes*, *Dolomedes*, *Lycosa* and *Salticus*. Although he did not himself immediately, but only some time afterwards [1]), expressly call these groups "genres", it is beyond a doubt that they ought to be considered as genera formed by LATREILLE, and their names accordingly to have right of priority before subsequently proposed, synonymous denominations. This is also usually admitted as regards most of them, those in fact which were retained by WALCKENAER in his Tableau des Aranéides (1805). As to the groups which received new names from WALCKENAER, LATREILLE in his subsequent works retained the appellations he had given to *a part* of them (*Eriodon*, *Micrommata*, *Oxyopes* and *Salticus*), whence also some arachnologists have adopted these names, whereas others have made use of the corresponding Walckenaerian denominations; but the names *Gnaphosa*, *Aranea*, *Heteropoda* and *Misumena* LATREILLE himself in time abandoned, and adopted the corresponding Walckenaerian synonyms, whereby these names have gradually fallen into oblivion. In the mean time, as no rational cause can be assigned, why these names should not be retained, as well as those, which belong to the two first named categories, I have adopted *all* LATREILLE'S generic names, with the single exception of *Aranea*, *Araneæ* being the general name for the entire order of spiders.

* Gen. 13. THYSA KEMP.

Deriv.: probably *Θύσαι*, a name of the female bacchanals (*θύω*, to rage).

Syn.: 1867. **Thysa** KEMP., Thysa pythonissæformis, p. 607 (1).

Type: *Thysa pythonissæformis* KEMP.

The remarkable spider, for which this genus has been formed, and of which only one specimen, a female, has been found (at Erlau in Hungary), is known to me only through KEMPELEN'S description and figures (loc. cit.). According to him it is related to *Pythonissa (Gnaphosa)*, but has only *six* eyes. If we imagine to ourselves a *Gnaphosa* without the anterior central eyes, and with the posterior row curved strongly backwards, we have much about the same position of the eyes as in *Thysa*. But this animal

---

1) In his Cours d'Entomologie, p. 501, he says: "... je perfectionnai ma distribution et *j'y établis la plupart des genres admis aujourdhui. (Nouv. Dict. d'Hist. Nat.).*"

differs also in other respects from *Gnaphosa*, as e. g. it is stated that "the head is considerably elevated above the thorax, especially anteriorly". The systematic position of this animal cannot yet be considered as definitely determined: it is only provisionally that we place it in this family and next after *Gnaphosa*.

Fam. V. DYSDEROIDÆ.

Syn.: 1837. **Dysderides** C. KOCH, Uebers. d. Arachn.-Syst., 1, p. 20.
1852. **Cellicolæ** DOLESCH., Syst. Verzeich. etc., p. 6 (*ad partem*).

The spiders belonging to this family are without difficulty distinguished from all others, except the *Territelariæ*, by their having *two stigmata, the one immediately behind the other, on each side of the belly near its base.* In other spiders provided with tracheal tubes as well as two air-sacs, the former usually debouch near the spinners, rarely *(Argyroneta, Dictyna, Anyphæna?)* in the middle line of the belly. The Dysderoidæ differ from the *Territelariæ* principally in having the mandibular claw, when at rest, bent inwards or obliquely inwards and backwards, not directly backwards and in that the two posterior stigmata lead to tracheal *tubes*, not to tracheal or air-*sacs* ("pulmonary" sacs). They are, in general, further distinguished by remarkably short tarsi and long patellæ, and have, as far as is known, never eight, but only *six* (or *two*, if the exotic genus *Nops* MAC LEAY belong to them) or *no* eyes. *Stalita Schiœdtei* NOB. (*S. tænaria* KEYSERL.) has, curiously enough, small *rudiments* of 6 eyes. The tarsal claws vary greatly in form and armature: the inferior claw is wanting about as frequently as it is present; the palpal claw is always small and without teeth.

DUFOUR, who first discovered that *Dysdera* had 4 stigmata, and who believed that these all led to tracheal or so-called pulmonary sacs [1]), as in the case of the *Territelariæ* (*Tetrapneumones* LATR.), united that genus with these last mentioned or "les araignées quadripulmonaires" [2]), and was in this followed by LATREILLE [3]). SUNDEVALL [4]) and WALCKENAER [5]) however powerfully opposed a so one-sided over-estimation of an anatomical pecu-

1) That the "pulmonary sacs" or "lungs" of spiders and of other arachnoidea are only peculiarly modified tracheæ, has been shown by LEUCKART (Ueb. d. Bau u. d. Bedeut. d. sog. Lungen bei d. Arachn., p. 246 et seq.).

2) Observ. sur quelques Arachn. quadripulm., p. 26 etc.

3) Fam. Nat. du Règne Anim., p. 312; Cours d'Éntom., p. 512.

4) Svenska Spindl. Beskr., *in* Vet.-Akad. Handl. f. 1829, p. 192 (1830).

5) Mém. sur une nouv. Classific. d. Aran., p. 436. (1833).

liarity, which moreover, as DUGÈS [1]) shortly after showed, had not even been correctly understood, since the posterior stigmata do not, like the anterior, lead to tracheal *sacs*, but to a pair of tracheal *tubes*. The Dysderoidæ are now generally admitted to stand in nearer relationship to SUNDEVALL'S *Drassides* than to his *Mygalides*. WALCKENAER [2]) was, I believe, the first who considered them as a separate group comparable with our families; they were by him called "*Tubicoles*": the name *Dysderides* they received from C. KOCH (loc. cit.). This family is also adopted by BLACKWALL. WESTRING includes it in his *Drassidæ*, and SIMON in his "*Drassiformes*" (as a separate "tribus", "*Ségestriens* ou *Pulmo-trachéens*"), i. e. in our *Tubitelariæ*.

The genus *Nops*, which SIMON refers to his "*Drassiens*" (ad max. part. = *Drassoidæ* NOB.) belongs probably to the *Dysderoidæ:* Conf. MAC LEAY, On some new forms of Arachn., p. 2 et seq. In that paper (p. 4) we read of another spider, which MAC LEAY also refers to the *Dysderoidæ:* "I possess specimens of a translucid West Indian spider closely allied to *Filistata*, and having Mygalidous eyes situated on the balloon-shaped cephalothorax of a *Nops*. In these specimens the antennæ [mandibles], maxillæ etc. are so rudimentary and inconspicuous, as would almost make us doubt that the species can be an animal of prey, did we not find it make an irregular web in the corners and crevices of houses. I call it *Hemerachne* [3]) *tenuipes*". — This spider would seem to belong rather to the *Scytodoidæ* than to the *Dysderoidæ* or *Filistatoidæ*.

The following genera belong to the European Fauna:

§ Oculi 6 perfecte explicati.

* Series oculorum antica ex 4, postica ex 2 oculis constans; oculi non omnes valde appropinquantes.

1. Maxillæ longæ, rectæ, sub-parallelæ. Ungues tarsorum trini. 1. *Segestria*.

2. Maxillæ breves, latæ, basi gibbosæ, in labium paullo inclinatæ. 2. *Schænobates*.

** Series oculorum antica ex 2, postica ex 4 oculis constans.

A. Oculi laterales seriei posticæ, sub-rectæ, longius ab oculis mediis ejusdem seriei disjuncti; oculi duo antici inter se longe remoti. Ungues tarsorum trini. . . . . 3. *Ariadne*.

B. Oculi saltem seriei posticæ inter se valde appropinquantes.

---

1) Sur les organes de la Respir. dans les Aran. Segestria et Dysdera, p. XII, XIV. (1835).

2) Mém. sur une nouv. Classif. d. Aran., p. 438.

3) This name (which is from ἥμερος, tame, and ἀράχνη, spider) ought of course to be written *Hemerarachne*.

a. **Oculi duo anteriores, reliquis plerumque manifeste majores, plus minus longe disjuncti. Series oculorum postica, desuper visa, procurva.**

1. Mandibulæ sub-porrectæ, ungue longo et valido. Ungues tarsorum bini. . . . . 5. *Dysdera.*

2. Mandibulæ verticales, ungue brevi. Ungues tarsorum trini. 6. *Harpactes.*

b. **Oculi omnes inter se valde appropinquantes, in tria paria dispositi, 2 utrinque, 2, reliquis majores, in medio. Tarsi articulo libero ungui-fero aucti. Ungues tarsorum bini. . . . . 7. *Oonops.***

§§ Oculi aut 6 valde imperfecti, aut nulli. Ungues tarsorum trini. . 4. *Stalita.*

Gen. 1. SEGESTRIA LATR. 1804.

Deriv.: *segestre*, a coarse coverlet.

Syn.:	1804.	**Segestria**	LATR., *in* Nouv. Dict. d'Hist. Nat., XXIV, p. 134.
	1861.	„	WESTR., Aran. Suec., p. 298.
	1864.	„	BLACKW., Spid. of Gr. Brit., II, p. 373.
	1864.	„	SIM., H. N. d. Araignées, p. 98.

Type: *Segestria senoculata* (LINN.).

The superior tarsal claws are powerful, somewhat long, with pretty many long, almost parallel, vertical, comb-teeth, of which the outermost are somewhat sinuated and divergent; in front of these the free extremity of the claw is somewhat swelled at the root. The inferior claw is small but stout, with one long, fine, curved tooth. On the 4th pair the teeth of the superior claws are somewhat fewer in number (about 7 in *S. senoculata*, which on the claws of the 1st pair has about 9). The palpal claw is weak, slightly curved, toothless.

* Gen. 2. SCHŒNOBATES BLACKW. 1850.

Deriv.: σχοινοβάτης, rope-dancer (σχοῖνος, rope; βαίνω, go).

Syn.:	1850.	**Schœnobates [Schænobates]**	BLACKW., Descr. of some newly disc. spec. etc., *in* Ann. and Mag. of Nat. Hist., 2 Ser., IV, p. 343.
	1864.	„ „	ID., Spid. of Gr. Brit., II, p. 375.

Type: *Schœnobates Walkeri* (BLACKW.).

Of this genus only one species, and of that only one specimen has been found. It is only on BLACKWALL'S authority that I have taken it up in this family.

Gen. 3. ARIADNE SAV. et AUD. 1825—27.

Deriv.: *Ἀριάδνη*, *Ariadne*, mythol. proper name.

Syn.: 1825—27. **Ariadne** [**Ariadna**] SAV. et AUD., Descr. de l'Égypte, (Éd. 2:) XXII, p. 308.
1837. **Dysdera** WALCK., H. N. d. Ins. Apt., I, p. 261 (*ad part.:* "3e Fam. Les Ariadnes, *Ariadna*").
1864. „ SIM., H. N. d. Araignées, p. 105 (*ad partem*).

Type: *Ariadne insidiatrix* (FORSK.).

In everything, except the position of the eyes, *Ariadne* comes much nearer to *Segestria* than to *Dysdera*, to which last genus it has been aggregated by WALCKENAER and several others. LATREILLE [1]) and C. KOCH [2]) however recognize it as an independent genus. Like *Segestria*, *Ariadne* is remarkable for *keeping the 3 first pairs of legs stretched forwards*, and only the 4th pair backwards. — *A. insidiatrix*, of which I have specimens which I caught in Rome, where that species is common, is in habits and industry quite similar to *Segestria Florentina* and *Filistata testacea*. — I am not aware that any species of this genus has previously been adduced as belonging to the fauna of Europe.

The orthography *Ariadne* is surely preferable to *Ariadna*, as being the ordinary Latin form of the word. — Of *Ariadne* DOLESCH. vid. p. 63.

The superior tarsal claws in *A. insidiatrix* are stout and powerful, pretty much and rather uniformly curved, gradually diminishing in breadth from the base, with 7—8 coarse, somewhat divergent comb-teeth; the inferior claw is small but powerful, with *one* little tooth. The superior tarsal claws on the 4th pair have but about 4 teeth. The female's palpal claw is small and toothless.

Gen. 4. STALITA SCHIÖDTE. 1847.

Deriv.: *στηλίτης*, belonging to pillars (*στήλη*, *Dorice στάλα*, pillar).

Syn.: 1847. **Stalita** SCHIÖDTE, Forelöbig Beretn. om d. underjord. Fauna, p. 80.
1849. „ ID., Bidr. t. d. underjord. Fauna, p. 22.

Type: *Stalita tænaria* SCHIÖDTE.

Through the kindness of Prof. SCHIÖDTE I have had the opportunity of comparing a male specimen of the typical species, the true *S. tænaria*,

1) Cours d'Entom., p. 514. 2) Die Arachn., X, p. 90.

with the spider described by KEYSERLING [1]) as *St. tænaria*, which, as SCHIÖDTE suspected, is quite a different species from the genuine *St. tænaria* so accurately described by this latter author. This is in fact easily seen since the appearance of SCHIÖDTE'S paper: On the genus Stalita [2]), in which special attention has been paid to the points in which the last mentioned spider differs from KEYSERLING'S description. Of KEYSERLING'S species I possess a full-grown ♀, $8^{mm.}$ long, exclusive of the mandibles, which are of $2^{mm.}$ length; it agrees in every essential particular with the description given by KEYSERLING. The length of the cephalothorax is $5^{mm.}$, and the breadth full $3^{mm.}$, the breadth of the pars cephalica little more than $2^{mm.}$. The length of the pars cephalica is a little greater than its breadth, and it is tapering behind. The mandibles are thinly covered with hairs on the whole of the dorsal surface, but more thickly hairy at the extremity, along the claw-furrow. The posterior edge of the claw-furrow has two teeth. The last joint of the palpus is longer and slenderer than the preceding joint. The patellæ are destitute of spines. The superior tarsal claws are long, slender, and much curved, with about 13 long, closely set comb-teeth; the toothless part of the claw is very long and much bent downwards. The inferior claw is long, slender and abruptly inflected downwards, and without teeth. All this refers to the 1st pair of legs. On the 4th pair the claws are still longer and slenderer, with about 6 divergent teeth near the base. The palpal claw is small and toothless. The abdomen is $4\frac{3}{4}^{mm.}$ long and $2\frac{1}{2}^{mm.}$ broad, with thin fine hairs. The posterior stigmata are as broad as the anterior.

A particular interest is attached to this spider, (which I call *S. Schiœdtei*), from the circumstance of its having *six rudimentary eyes!* In position these eyes agree nearest with those of *Ariadne* (which genus also, like *Stalita*, has 3 claws on the tarsi). They are small like points, about equal in size, and rather lighter in colour than the cephalothorax, and therefore easily visible with a good common magnifying lens, and occupy an area the breadth of which is about a third of that of the head, and which is about three times as broad as it is long. They are arranged in two rows very near the margin of the clypeus, 4 eyes in the posterior, and 2 in the anterior row. The posterior row is straight and considerably longer than the anterior. The two posterior central eyes are somewhat nearer to each other than to the lateral eyes. The distance between the two lateral eyes is about two eye-diameters, and perhaps somewhat greater than the

1) Beschr. einer neuen Spinne aus d. Höhlen v. Lesina, p. 2 (540).

2) Om slägten Stalita, p. 4—5 (74—75).

distance between the two anterior eyes and the very low clypeus, and equal to about $\frac{1}{3}$ of the distance between the two anterior eyes. — The specimen of *S. Schiœdtei* here described was kindly presented to me by Count KEYSERLING.

As to *S. tœnaria*, which shows no traces of eyes, I need but refer to SCHIÖDTE'S description of that species (locis cit.).

Gen. 5. DYSDERA LATR. (1804).

Deriv.: *δύσδηρις*, hard to contend with (*δυς-*, ill-; *δῆρις*, contention) [1]).

Syn.:	1804.	**Dysdera**	LATR., *in* Nouv. Dict. d'Hist. Nat., XXIV, p. 134 (*ad partem*).
	1837.	„	WALCK., H. N. d. Ins. Apt., I, p. 261 (*ad part.:* "1e Fam. Les Agones, *Agonæ*").
	1864.	„	BLACKW., Spid. of Gr. Brit., II, p. 369 (*ad partem*).
	1864.	„	SIM., H. N. d. Araignées, p. 105 (*ad partem*).

Type: *Dysdera punctoria* (VILL.). (*D. erythrina* WALCK.).

The tarsal claws in this genus are only *two* in number, and a claw-tuft is met with under them, whereas in *Ariadne* and *Harpactes*, which are usually united with *Dysdera*, there are 3 claws, and no claw-tuft (as is the case with all spiders that have 3 claws). The superior tarsal claws are slender, somewhat sinuated at the base, outwards curved strongly and almost into a semicircle, with several (in *D. punctoria* about 10, in *D. punctata* C. KOCH about 5) long saw-teeth, issuing from the side of the claw from about its middle to near the extremity, which is thus rather short. The claw-tuft is thickly set, and consists of linear hairs, slightly dilated at the extremity only. The palpal claw is small and toothless.

Gen. 6. HARPACTES TEMPLETON. 1834.

Deriv.: *ἁρπακτής*, robber (*ἁρπάζω*, rob).

Syn.:	1834.	**Harpactes**	TEMPL., On the Spid. of the gen. Dysdera, p. 401.
	1837.	**Dysdera**	WALCK., H. N. d. Ins. Apt., I, p. 261 (*ad part.:* "2e Fam. Les Agores, *Agoræ*").
	1861.	„	WESTR., Aran. Suec., p. 301.
	1864.	„	BLACKW., Spid. of Gr. Brit., II, p. 369 (*ad partem*).
	1864.	„	SIM., H. N. d. Araignées, p. 105 (*ad partem*).

Type: *Harpactes Hombergii* (SCOP.).

1) In AGASSIZ' Nomencl. Zool. it is derived from *δυς-*, and "*δέρη*, collum."

Not only the presence of a third tarsal claw, but also peculiarities in the structure of the parts of the mouth, and a longer, slenderer form of the body, distinguish this genus from *Dysdera*, to which it is otherwise very similar, and with which it is commonly united. — The superior tarsal claws of the typical species are slender, curved nearly to a semicircle, and provided with about 6 long, parallel, vertical comb-teeth. The inferior claw is toothless.

The genus *Pylarus* HENTZ [1]) is near related to *Harpactes*.

Gen. 7. OONOPS TEMPL. 1834.

Deriv.: ὠόν, egg; ὤψ, eye.

Syn.: 1834. **Oonops** TEMPL., On the Spid. of the gen. Dysdera, p. 404.
1837. **Deletrix** BLACKW., Charact. of a new gen. etc., p. 100.
1847. **Dysdera** WALCK., H. N. d. Ins. Apt., IV, p. 382 (*ad part.:* "4e Fam. Les Albionides, *Albionidæ*").
1864. „ SIM., H. N. d. Araignées, p. (105,) 455 (*ad partem*).
1864. **Oonops** BLACKW., Spid. of Gr. Brit., II, p. 377.

Type: *Oonops pulcher* TEMPL.

The typical species of this interesting genus, of which the Rev. O. P. CAMBRIDGE has kindly sent me specimens, is found not only in Great Britain and Ireland, but also in Italy, according to CANESTRINI and PAVESI [2]).

The two tarsal claws of *O. pulcher* are weak and slender, uniformly and rather slightly bent, with 5 or 6 tolerably coarse, pointed comb-teeth directed somewhat forwards. In stead of a claw, the female's palpus is at the extremity provided with a strong conical process (in a young specimen). By the presence of a small separate claw-joint this spider forms a transition to the *Scytodoidæ*; I place it among the *Dysderoidæ* principally on the authority of BLACKWALL, for I have not myself been able to see more than two stigmata in the somewhat damaged specimens I possess.

Fam. VI. FILISTATOIDÆ.

Syn.: 1867. **Filistatidæ** AUSS., Die Arachn. Tirols, I, p. 140.

This family comprises only the genus *Filistata*, which was referred by WALCKENAER to "les *Théraphoses*" or our *Territelariæ*, although it has

1) Aran. of the United States, *in* Boston Journ. of Nat. Hist., IV, p. 225.
2) Araneidi Italiani, p. 27.

6 spinners, the mandibular claw directed inwards, not backwards, and only *two* air-sacs, so that it is destitute of all the characteristics that usually distinguish the spiders belonging to the sub-order *Territelariæ*. Even LATREILLE, who first [1]) referred it to his "*Tubitèles*", assigned it in his later works, in consequence of the erroneous assumption that it had 4 "pulmonary" sacs, to his *Tetrapneumones* or the Territelariæ. C. KOCH [2]) gives it the same systematic position. DUGÈS referred it to his "*Micrognathes*" or "*Scythodés*" [3]), a group, that comprises spiders of widely separated families, but which agree with each other in the structure of the mandibles. (Conf. p. 99). SIMON, who rightly insists upon the relationship of the Filistatoidæ with the *Drassoidæ* and other *Tubitelariæ*, forms for them a separate "tribus", "*Filistatiens* ou *Mygalo-drasses*", within the family "*Drassiformes*" [4]). Lastly, in AUSSERER [5]), as also in CANESTRINI and PAVESI [6]), we find the *family Filistatidæ* placed between *Mygalidæ* and *Dysderidæ*.

It is strictly speaking only by the position of the eyes, that *Filistata* agrees more with the *Territelariæ* than with the *Tubitelariæ*, and it seems chiefly to have been this agreement that induced WALCKENAER and C. KOCH to refer *Filistata* to the first-named sub-order. Mandibles directed somewhat forwards and united at the base [7]), form a feature occurring in many other genera which have never been referred to the Territelariæ, and especially among the *Scytodoidæ*, which we unreservedly consider as the nearest relatives of the *Filistatoidæ*. The parts of the mouth exhibit the same structure in both these families, and also in *Filistata* the weak mandibles, armed with a very small claw, remind an observer of the two-fingered claw of the *Opiliones*, by their having a spine or tooth at their extremity opposite the claw. Both families appear to have been developed from a common root: the *Scytodoidæ* form the beginning of the series of genera, which constitute the sub-orders *Retitelariæ* and *Orbitelariæ*, while from the *Filistatoidæ* and forms nearly related to them the other sub-orders have probably descended.

The general appearance of the Filistatoidæ is very peculiar and unlike that of other spiders: it reminds one most of certain *Scytodoidæ (Loxosceles)* and *Theraphosoidæ*, but also of some *Tubitelariæ*, e. g. *Uroctea*. Their

1) CUV., Règne Anim., III, p. 83. (1817).
2) Uebers. d. Arach.-Syst., 1, p. 35; ibid., 5, p. 76.
3) Observ. sur les Aran., p. 106.
4) Hist. Nat. d. Araignées, p. 95.
5) Die Arachn. Tirols, I, p. 140.
6) Aran. Ital., p. 23.
7) In *F. capitata* HENTZ, they are however *not* united at the base, according to HENTZ, Aran. of United States, *in* Bost. Journ. of Nat. Hist., IV, p. 228.

generally strong extemities, as also their habits and the structure of their webs at once separating them from the *Retitelariæ*, they cannot be referred to any other sub-order than the Tubitelariæ. If by a certain outward appearance, by the structure of the mandibles, and by the form and armature of the female's palpi, they exhibit affinities with the *Urocteoidæ*, they, on the other hand, as LUCAS [1]) has remarked, and as I have myself in Southern Europe observed, agree with *Segestria*, and especially with *S. Florentina*, in their habits and economy: the tubular web has just the same appearance, and is met with in the same localities (especially in the holes and crevices of old walls), as that of the last mentioned spider. Also HENTZ remarks concerning this genus, that "by its habits it is closely related to *Pylarus* and to *Segestria*" [2]).

Gen. 1. FILISTATA LATR. 1810.

Deriv. uncertain: *filum*, thread; *stare*, stand. Or perhaps *filum* and *ἵστημι*, set, place (*ἱστός*, warp, web).

Syn.: 1810. Filistata LATR., Consid. gén. sur les Crust., les Arachn. et les Ins., p. 121.
1839. Teratodes C. KOCH, Die Arachn., V, p. 6.
1864. Filistata SIM., H. N. d. Araignées, p. 95.

Type: *Filistata testacea* LATR.

In the typical species the superior tarsal claws are very strong, long, much and uniformly curved, with about 12 long, strong, almost parallel, almost equally long comb-teeth; the free extremity of the claw is not long, a little swelled at the root below. The inferior claw is very small, but strong, with two very long, strong teeth, sitting close together. The palpal claw of ♀ is long, of almost uniform substance, much and regularly curved, armed from the base throughout about two thirds of its length with about 16 rather short, strong, parallel comb-teeth slightly increasing in length outwards, the points of which form a much curved line following the direction of the claw.

Sub-ordo IV. TERRITELARIÆ.

Syn.: Vid. infra sub Fam. *Theraphosoidæ*.

As an, in cases of doubt, decisive characteristic of the spiders belonging to this sub-order, we consider the to them peculiar direction of the

1) Observ. sur le genre Eriodon, p. 312.
2) Aran. of United States, *in* Bost. Journ. of Nat. Hist., IV, p. 227.

mandibular claw: it moves, as is known, in a vertical plane very nearly parallel to the longer axis of the body, and, when at rest, is directed *backwards;* in all other spiders on the contrary, it moves in a plane almost at right angles to the longer axis of the body, and lies with its point turned *inwards*, sometimes obliquely inwards and backwards. The mandibles themselves are generally more projecting and larger than in other spiders, and can only be opened to an inconsiderable amount. The Territelariæ have generally *four* air-sacs; the spinners are with few exceptions only four in number: the superior are usually considerably longer than the inferior, and consist of three, sometimes (at least in the genera *Diplura* C. KOCH and *Eriodon* LATR. or *Missulena* WALCK.) of four joints. The tarsal claws are mostly two, sometimes three in number.

The Territelariæ approximate on the one side to the *Tubitelariæ (Filistatoidæ* and *Dysderoidæ)* and on the other to the *Citigradæ.* The genus *Catadysas* HENTZ forms an evident transition to this latter sub-order, with which they also in their habits show many analogies. That some of the female Theraphosoidæ carry their young upon their backs, just like species of the genus *Lycosa*, has been long known: LATREILLE states it to be the case with *Nemesia Sauvagesii* (ROSSI) or *Mygale fodiens* WALCK. [1]), and ABBOT has, according to WALCKENAER [2]), observed the same phænomenon in *Actinopus Abbotii* (WALCK.). LINCECUM relates [3]) concerning certain species found in Texas: "Two or three species of *Mygale* carry a sack well filled with eggs *attached to the tip of their abdomen*, and when the young ones hatch out, they take them on their backs and carry them like the *Mygale Hentzii.*"

The European Territelariæ all belong to one family, the *Theraphosoidæ*, all the species of which have *four* pulmonary sacs, and *at least four* spinners. Of the families *Liphistioidæ* and *Catadysoidæ* see pag. 43.

Fam. I. THERAPHOSOIDÆ.

Syn.: 1802. Gen. **Mygale** WALCK., Faune Par., II, p. 241.
1805. Gen. **Theraphosa** ID., Tabl. d. Aran., p. 1.
1817. "**Territèles**" LATR. *in* CUV., Règne Anim., III, p. 79.
1823. **Terrestres** SUND., Gen. Aran. Suec., p. 10.

1) See WALCK., Faune Franç., Arachn., p. 5.
2) Hist. Nat. d. Ins. Apt., I, p. 248.
3) The Tarantula, p. 411.

1825. **Tetrapneumones** LATR., Fam. Nat. du Règne Anim., p. 312.
1830. **Theraphosæ** SUND., Sv. Spindl. Beskr., *in* Vet.-Akad. Handl. f. 1829, p. 203.
1833. **Mygalides** ID., Consp. Arachn., p. 28.

It is well known that CUVIER in the year 1800 gave the name of *Mygale* to a genus of *Mammals*, and that WALCKENAER first in 1802 ("dans un Mémoire lu à la Société Philomatique de Paris": see WALCK., Faune Parisienne, II, p. 249) separated the spiders belonging to the family before us from the others or "spiders properly so called" (*Aranea* WALCK.) under the name of *Mygale*. Some naturalists have curiously enough attempted to avoid the confusion thus introduced, by altering CUVIER'S generic name into *Myogale* or *Myogalea* — which however is only another way of spelling *Mygale* — instead of, in accordance with the law of priority, altering the more recent name or replacing it with another, as reasonableness requires. It can moreover hardly be denied that the name *Mygale*, as that of a genus of spiders, is ill chosen: the Greek word μυγαλῆ, μυγαλέη or μυογάλη signifies a *shrew (Sorex)*, and nothing else. Nevertheless, in spite of the requirements of consistency, we should perhaps not have ventured to exchange this generally known and accepted generic name for another, if the following circumstances had not contributed to induce us to such a step. First and principally the genus *Mygale* has by more recent authors been resolved into several smaller generic groups, by C. KOCH [1]) for inst. into *seven*, so that by him the name of *Mygale* is only retained for a group comprising but two species, *M. Blondii* and *M. Javanensis*, whereas all the other forms described by him bear other generic names — and the matter is accordingly reduced merely to the giving of another name to the above mentioned little group; moreover that other name needs not be a new and previously unknown denomination, for we have at hand an appropriate generic name formed by WALCKENAER himself in 1805, namely *Theraphosa*, which in the original definition of that genus is absolutely synonymous with *Mygale*. This word is not, as has been sometimes supposed, a plural, but a true generic name in the singular number [2]), and has already in 1830 been used by EICHWALD [3]) instead of *Mygale*. In the Tableau des Aranéides WALCKENAER divided "*les Aranéides*" into two great "Divisions",

1) Uebers. d. Arachn.-Syst., 5, p. 72—75.

2) It is so taken by e. g. SUNDEVALL, as is evident from the following words: "WALCKENAER considered that he had sufficient reason to separate the Bird-spiders and the species most nearly allied to them, as a separate genus, *Theraphosa*, from LINNÉ'S *Aranea*." Sv. Spindl. Beskr., *in* Vet.-Akad. Handl. f. 1829, p. 190.

3) Zool. spec., II, p. 73.

Theraphosa and *Aranea* (just as he had before divided them into *Mygale* and *Aranea*), after which each of these great generic groups was subdivided into a number of smaller groups, "genres": *Theraphosa* into *Mygale*, *Oletera* and *Missulena; Aranea* into *Lycosa*, *Dolomedes*, *Ctenus* etc. The name *Aranea* has been entirely abandoned as a generic name, simply because the whole Order of Spiders ought to be called *Araneæ;* but any sound reason for not preserving the name *Theraphosa* for some portion of the forms to which it has once belonged, it would assuredly be hard to assign. We propose therefore with EICHWALD to replace the name *Mygale*, which had already been appropriated by CUVIER, with *Theraphosa*, giving *Th. Blondii* as type of the genus. *Theraphosa* (WALCK.) NOB. is therefore = *Mygale* (WALCK.) C. KOCH 1850.

We also desire to call attention to the following circumstance. When, in 1811, OLIVIER [1]) adopted WALCKENAER'S genus *Mygale* as separate from *Aranea*, he restricted it to "*les Araignées mineuses*", excluding all the other *Mygale*-species or "*les Araignées aviculaires*", which he referred to *Aranea.* He was followed by LAMARCK [2]), who also (in the year 1818) received into the genus *Mygale* only "les Araignées mineuses"; but for "les Araignées aviculaires" this author formed a separate genus, *Avicularia* LAM. [3]). It was not till several years later (1825), that LATREILLE gave to "*les Araignées mineuses*" the name "*Ctenize*", and in opposition to OLIVIER'S and LAMARCK'S limitation of the genus *Mygale*, applied that name to "*les Araignées aviculaires*". It is accordingly evident that *if* the name *Mygale* were to be preserved to any genus of spiders at all, it ought, according to the law of priority, to belong to that genus which is usually called *Cteniza* LATR. (*Nemesia* SAV. et AUD.). — We have preferred the denomination *Theraphosa* to that of *Avicularia* for the species of "les Araignées aviculaires", which in KOCH and SIMON bear the name of *Mygale*, and thus have been by them considered as types of the genus *Mygale* WALCK., partly because *Theraphosa* is the older appellation of the two, partly because the name *Avicularia* ought in our opinion to be reserved for that group of species among "les Mygales aviculaires", which comprises LINNÉ'S *Aranea avicularia.* (Vid. p. 169 sub gen. *Avicularia* (LAM.)).

We divide provisionally "les Mygales aviculaires" into the 4 following genera, which number will however doubtless hereafter, when these animals have been more accurately studied, be considerably augmented:

---

1) Encycl. Méth., VIII, p. 83.
2) Hist. Nat. d. Anim. sans Vertèbres, V, p. 105.
3) Ibid., p. 107.

1. *Theraphosa* (WALCK.) = *Mygale* (WALCK.) C. KOCH; 2. *Avicularia* (LAM.) = *Eurypelma* (C. KOCH); 3. *Trechona* (C. KOCH), and 4. *Diplura* C. KOCH. The first-named two genera together answer to WALCKENAER'S "*Plantigrades*", the latter two to his "*Digitigrades inermes*". "*Les Mygales (Digitigrades) mineuses*" ought to be called *Nemesia* SAV. et AUD. — The family *Mygalides* we call, in conformity with the method, in which we have formed the other family-names, *Theraphosoidæ*.

If the genus *Atypus* have really, as LATREILLE [1]) and DUGÈS [2]) expressly state, *six* spinners, and not only four, as WALCKENAER [3]) says, that genus ought to be made the type of a separate sub-family, *Atypinæ*, in contradistinction to the ordinary Theraphosoidæ *(Theraphosinæ)*, which are provided with only *four* spinners. Also in *Eriodon formidabile* CAMBR. the spinners, according to CAMBRIDGE [4]), are 6 in number. According to LUCAS [5]) however the oldest known species of that genus, *E. occatorium* (WALCK.), has only two pair of spinners (?).

The European genera included in the family Theraphosoidæ are the following:

A. Maxillæ versus basin dilatatæ: palpi dilatationi lateris affixi. Cephalothorax anteriora versus dilatatus. . . . . 1. *Atypus.*

B. Maxillæ angustæ, sub-cylindratæ; palpi apici earum inserti.

a. Area oculorum $2\frac{1}{2}$—3-plo latior quam longior. Cephalothorax antice alte elevatus. Pedes breves, robusti, 3tii paris reliquis breviores. 2. *Cyrtauchenius.*

b. Oculi conferti, eminentiæ communi parvæ impositi; area, quam occupant, c:a dimidio — duplo tantum latior quam longior.

I. Mandibulæ ad apicem dentibus vel lamellis corneis liberis, rastellum vel pecten formantibus, armatæ. Pedes apicem versus plus minus attenuati; ungues ipsi apici tarsorum inserti. . . . . 3. *Nemesia.*

II. Mandibulæ rastello carentes.

* Pedes versus apicem attenuati, unguibus ipsi apici tarsorum insertis.

1. Mamillæ superiores (posteriores) articulis quaternis. . 4. *Diplura.*

2. Mamillæ superiores articulis trinis. . . . . 5. *Trechona.*

[** Pedes robusti, versus apicem vix vel parum attenuati, unguibus supra apicem tarsi insertis, retrahendis. . . . . 6. *Avicularia.*]

---

1) Cuv., Règne Anim., 2e Éd., IV, p. 228.

2) Observ. sur les Aran., p. 197; Cuv., Règne Anim., 3e Éd., Arachn., p. 31, Pl. 5, fig. 2*b*.

3) Hist. Nat. d. Ins. Apt., I, p. 245.

4) Descr. of a new gen. and six new spec. of Spid., p. 267.

5) Observ. sur le genre Eriodon, p. 316.

*Gen. 1. ATYPUS LATR. 1804.

Deriv. *a priv.*, and *τυπόω*, form (accordingly, unshapely; "laid de figure": LATREILLE).

Syn.: 1804. **Atypus** LATR., *in* Nouv. Dict. d'Hist. Nat., XXIV, p. 133.
1804. „ ID., Hist. Nat. d. Crust. et d. Ins., VII, p. 168.
1805. **Oletera** WALCK., Tabl. d. Aran., p. 7.
1861. **Atypus** BLACKW., Spid. of Gr. Brit., I, p. 14.
1864. „ [**Atypa**] SIM., H. N. d. Araignées, p. 83.

Type: *Atypus piceus* (SULZER).

The synonyms show that the nama *Atypus* has the right of priority before *Oletera*, and not *vice versa*, as LUCAS [1]) has supposed.

Gen. 2. CYRTAUCHENIUS N.

Deriv.: *κυρτός*, crooked; *αὐχήν*, neck.

Syn.: 1845. **Cyrtocephalus** LUCAS, Note sur une nouv. esp. d'Aran. appart. au genre Actinopus, p. 58.
1845. „ ID., Explor. de l'Algérie, Arachn., p. 92.
1864. „ [**Cyrtocephala**] SIM., H. N. d. Araignées, p. 81.

Type: *Cyrtauchenius Walckenaerii* (LUCAS).

The name *Cyrtocephalus* having been already disposed of, before it was applied by LUCAS to this genus (conf. p. 36, note 2), I have been obliged to give it a new denomination. — I possess a specimen (a ♀) of a species of *Cyrtauchenius*, from Corfu, given to me by Count KEYSERLING, which is perhaps identical, or at least very closely connected, with *C. lapidarius* (LUC.) from Crete. It is distinguished by the palpi as well as the first *two* pairs of legs being towards the extremities (on the last three joints of the legs and the last two of the palpi), on both sides and for some distance downwards, armed with a band of, especially on the last joint, closely arranged, short, blunt, very strong spines, which undoubtedly make these extremities excellent digging organs. On the 3rd and 4th pairs these joints only show a few sparse spines. Of the palpi of *C. lapidarius* LUCAS [2]) states, that between the hairs that cover them, one may remark "des épines *placées çà et là*", and of the legs of the same species, that it has "le métatarse et le tarse des *trois* premières paires armés d'épines d'un

1) De la man. de vivre etc. de l'Oletera picea, p. CLXX.
2) Anim. artic. de l'ile de Crète, p. 16.

brun rougeâtre". In other respects LUCAS' description accurately corresponds with the spider I have mentioned. Should this spider be found not identical with *C. lapidarius*, it may be called *C. Corcyræus.*

SIMON (loc. cit.) enters under the genus *Cyrtocephalus* [*-a*] a species "*C. lapidaria* ROULIN, Ile de Cuba", which is probably a slip of the pen for "*C. lapidaria* LUCAS, Ile de Crète". He has however not inserted this genus in his Catal. syn. d. Aranéides d'Europe.

The tarsal claws of *Cyrtauchenius* are 3 in number on each tarsus, as in *Nemesia.* The tarsi of the posterior legs are somewhat thicker towards the extremity, almost clublike. The superior or posterior spinners show only 3 distinctly separated joints.

Gen. 3. NEMESIA SAV. et AUD. 1825—27.

Deriv.: *Νεμέσιος* or *Νέμεσις*, mythol. proper name.

Syn.: 1805. **Mygale** WALCK., Tabl. d. Aran., p. 5 (*ad part.:* "3e Fam. Digitigrades mineuses, *Cunicularia*").
1811. „ OLIV., Encycl. Méth., VIII, p. 83.
[1825. "**Ctenize**" LATR., Fam. Nat. du Règne Anim., p. 315].
1825—7. **Nemesia** SAV. et AUD., Descr. de l'Égypte, (2e Éd.:) XXII, p. 302.
1827. **Ctenize** BERTH., LATR. Natürl. Fam. d. Thierr., p. 298.
1829. **Ctenisa** LATR., *in* CUV., Règne Anim., 2e Éd., IV, p. 230.
1864. **Mygalodonta** SIM., Hist. Nat. d. Araignées, p. 75.

Type: *Nemesia cellicola* SAV. et AUD.

The most commonly received name of this genus is not *Nemesia*, but *Cteniza*, which name is first found in LATREILLE'S Familles Naturelles du Règne Animal (1825), where "*les Araignées mineuses*" are brought together under the *French* appellation "*Ctenize*". Whether the *scientific* name was intended to be *Ctenizus*, *Cteniza* or any thing else, it is not possible to see there, for the generic names, even those newly formed, appear in that work only in their French form, whence also follows (Conf. p. 4 note 1), that any right of priority cannot be claimed for the generic names there proposed. It is true that BERTHOLD, in his German translation of LATREILLE'S Familles Naturelles (1827), gave a Latin form to these new generic denominations [1]), and in the cases, in which he was the first who did so

1) He however calls LATREILLE'S "Ctenize" not *Cteniza*, but *Ctenize*, as the genus is also called by for inst. SUNDEVALL (Consp. Arachn., p. 28). That LATREILLE'S meaning was, that the name should end i *a*, is visible in his *subsequent* works, as e. g. in the edition of CUVIER'S Règne Animal published in 1829, and it has since generally received that termination.

(as is undoubtedly the case with the name in question), the time of the name's publication must be reckoned from that translation; but SAVIGNY and AUDOUIN had, if I mistake not, a little before its appearance, given the name of *Nemesia* to a species belonging to the "mining" spiders, and this name, as probably somewhat older, I have considered myself bounden to perfer to *Cteniza* BERTH.

SIMON has exchanged (*Nemesia* and) *Cteniza* for an entirely new name, *Mygalodonta*, and says concerning *Cteniza* (loc. cit. p. 76) that "cette dénomination est restée inconnue". It has therefore escaped his observation, that that name is both known and used in a work that he often cites, KOCH's Die Arachniden, and SIMON even himself cites (p. 453) in his account of his *Mygalodonta fodiens:* "*Cteniza Graja* KOCH".

That the name *Mygale*, if it could be used of a genus of spiders, would by right belong to the genus before us, I have already (p. 163) endeavoured to show.

N. cellicola, according to O. G. COSTA [1]), is met with, though rarely, in the south of Italy, at Naples. COSTA states that it has 3 claws upon the tarsi of the 3rd pair only, the first pair being armed with 2, and the 2nd with but one claw respectively (!). According to SAVIGNY and AUDOUIN [2]) this species has however *three* claws on *each* of the tarsi, like other species of the genus.

* Gen. 4. DIPLURA C. KOCH. 1850.

Deriv.: *διπλόος*, double; *οὐρά*, tail.

Syn.: 1805. **Mygale** WALCK., Tabl. d. Aran., p. 5 ("2e Fam. Les Digitigrades inermes" *ad partem*).
1850. **Diplura** C. KOCH, Uebers. d. Arachn.-Syst., 5, p. 75.
1864. **Mygale**: sub-gen. **Pexionyx** **[Pezionyx]** SIM., H. N. d. Araignées, p. 64, 68 (*ad partem*).

Type: *Diplura macrura* C. KOCH.

This genus, corresponding with those of WALCKENAER'S "*Mygales digitigrades inermes*", which have very elongated superior spinners, consisting of 4 distinct joints, belongs to the European Spider-Fauna at least through *Mygale Calpetana* [*Calpeiana*] WALCK., which, according to WALCKENAER'S description [3]), in this feature agrees with the species, *D. macrura* C.

1) Fauna d. Regno di Napoli, Aracn., p. 20.
2) Descr. de l'Égypte, (2e Éd.:) XXII, p. 304.
3) Hist. Nat. d. Aran., Livr. 1, n:o 8 et 9.

KOCH [1]), given by KOCH as typical of *Diplura*. Also *Mygale luctuosa* LUCAS from Spain, which is said to be very closely allied to *D.* (*M.*) *Calpetana*, and to have the superior spinners about as long as the abdomen, appears to belong to this genus; but LUCAS does not state of how many joints these spinners consist [2]).

Gen. 5. TRECHONA (C. KOCH). 1850.

Deriv.: τρέχω, run.

Syn.: 1805. **Mygale** WALCK., Tabl. d. Aran., p. 59 ("2e Fam. Les Digitigrades inermes" *ad partem*).
1850. **Trechona** C. KOCH, Uebers. d. Arachn.-Syst., 5, p. 74 (*saltem ad max. part.*).
1864. **Mygale**: sub-gen. **Pexionyx** [**Pezionyx**] SIM., H. N. d. Araignées, p. 64, 68 (*ad partem*).
? 1864. „ sub-gen. **Eurypelma** ID., ibid., p. 66 (*ad partem*).

Type: *Trechona Valentina* (DUF.).

Some of the species classed by C. KOCH under this genus, are by SIMON referred to the sub-genus *Eurypelma* ("groups" *Eurypelma* and *Lasiodora*) — whether rightly or not, I cannot venture to decide. In the species which we assign to *Trechona*, as e. g. *T.* (*Mygale*) *Valentina* (DUF.) the superior spinners have but 3 distinct joints [3]), which distinguishes them from the preceding genus, *Diplura*.

[Gen. 6. AVICULARIA (LAM.) 1818.

Deriv.: *avicularius* (bird-keeper), in the signification adopted, bird-catcher.

Syn.: 1805. **Mygale** WALCK., Tabl. d. Aran., p. 5 ("1e Fam. Les Plantigrades" *ad max. part.*).
1818. **Avicularia** LAMARCK, H. N. d. Anim. sans Vertèbres, V, p. 107 (*ad partem*).
1830. **Theraphosa** [**Teraphosa**] EICHW., Zool. spec., II, p. 73 (*ad partem*).
1850. **Eurypelma** C. KOCH, Uebers. d. Arachn.-Syst., 5, p. 73
1850. **Lasiodora** ID., ibid., p. 72
1850. **Scurria** ID., ibid., p. 74.
1850. **Typhochlæna** ID., ibid., p. 75.
(the four 1850 entries above: *saltem ad max. part.*)
? 1850. **Trechona** ID., ibid., p. 74 (*ad partem*).
1864. **Mygale**: sub-gen. **Eurypelma** SIM., H. N. d. Araignées, p. 64, 66 (*ad max. part.*).

Type: *Avicularia vestiaria* (DE GEER).

---

1) Die Arachn., IX, p. 38, Taf. CCC, f. 715.
2) Conf. LUCAS, Note sur une nouv. esp. d'Aran. qui habite l'Esp. mérid., p. 17.
3) DUFOUR, Observ. sur quelques Arachn. quadripulm., p. 100, 102.

As we remarked above (p. 163), LAMARCK divides WALCKENAER'S *Mygale* into two genera, *Avicularia* and *Mygale*, of which the former is synonymous with *Mygale* LATR., the latter with *Cteniza* [LATR.] BERTH. or *Nemesia* SAV. et AUD. As type for *Avicularia* LAM., I propose *Aranea avicularia* LINN. (*Ar. vestiaria* DE GEER, *Avicularia canceridea* LAM.), partly for the sake of the name, and partly because it is the first species entered by LAMARCK under the genus *Avicularia*. As it was for this species and forms nearly related to it, that C. KOCH proposed the genus *Eurypelma*, it will be to the species of that genus that the older name *Avicularia* ought in the first place to be applied. The other new genera cited in our *Syn.*, which KOCH formed at the cost of WALCKENAER'S "*Mygales plantigrades*", may probably for the present be united with *Eurypelma* or *Avicularia*.

I am not convinced, that any species belonging to this genus is met with in Europe. As however SIMON in his sub-genus *Eurypelma* — which he states to have "tarses élargis, garnis de brosses adhérentes; griffes très-retractiles", and which thus by these characteristics agrees with *Avicularia* (LAM.) NOB. — includes e. g. *Mygale (Trechona) icterica* C. KOCH from Greece, which species is to me unknown, I consider that I ought, at least provisionally, to insert here the genus *Avicularia*.]

Sub-ordo V. LATERIGRADÆ.

Syn.: Vid. infra sub Fam. *Thomisoidæ*.

In their peculiar manner of moving — with about as much ease sideways or backwards as forwards, and with their femora depressed and stretched out sideways, the following joints of the legs moving towards the femora in a plane more nearly approaching the horizontal than the vertical plane — the spiders belonging to this sub-order have a distinctive mark, by which, as is well known, they may usually without difficulty be distinguished from all other spiders. Of the European genera, *Micrommata* (LATR.) is the only one, which has not the crab-like appearance that is peculiar to the other Laterigradæ. Many of the great exotic forms of this sub-order (especially those of the genus *Heteropoda*), present a striking analogy with certain *Theraphosoidæ;* but it is to the *Drassoidæ* in the sub-order *Tubitelariæ*, that the Laterigradæ are most nearly related, and between which and them it is most difficult to assign the line of demarcation. Like the

Drassoidæ, they have only *two* claws at the extremity of the tarsi [1]): as in them, the eyes generally form two transversal rows; but these rows usually enclose a crescent-shaped or circular-segmental area, and are but rarely nearly parallel or curved towards each other [2]). Most frequently (also in *Micrommata*) the second pair of legs is longer than the others, which on the other hand, as far as I am aware, is never the case with the *Drassoidæ.* The maxillæ are usually narrow and strongly inclined towards the labium, the mandibles small and conical: nevertheless there are numerous exceptions to this, of which *Heteropoda* and the genera nearly connected with it are striking examples.

The species of this sub-order, at least the European ones, may for the present be united in a single family, *Thomisoidæ*, to which we also refer the wonderful and but little known genus *Anetes* Menge, which is stated to be destitute of both spinners and tarsal claws.

Fam. I. THOMISOIDÆ.

Syn.: 1817. "Latérigrades" Latr., *in* Cuv., Règne Anim., III, p. 91.
1823. Retrogradæ Sund., Gen. Aran. Suec., p. 18.
1825. Laterigradæ Latr., Fam. Nat. du Règne Anim., p. 315.
1833. Thomisides Sund., Consp. Arachn., p. 27.

Latreille in 1804 [3]) formed, at the expense of Linné's *Aranea*, for spiders belonging to this family the genera *Heteropoda*, *Misumena* and *Micrommata.* As the characteristic difference between the two first mentioned,

---

1) A remarkable exception is *Sparassus abnormis* Blackw., which has only "a single claw at the extremity of each tarsus" (Blackw., A list of spiders captured in the South-East region of Equat. Africa, p. 457). This species ought probably to form a separate genus.

2) In *Eripus* Walck. the eyes are arranged in 3 or 4 (?) transversal series. In *Platythomisus* Dolesch. the eyes form two rhomb-like groups, situated far apart at the two corners of the forehead; in *Arcys* Walck., *Heterognatha* Nic. and *Anetes* Menge on the contrary the lateral eyes are far removed from the central eyes, much about as in *Epeira.* In *Stephanopis* Cambr. the eyes are arranged in a ring, in *Diphya* Nic. they have again about the same position as in *Ocyale. Thomisus yolophus* Doum. has but 6 eyes, and ought of course to form a separate genus, for which we propose the name *Daradius* (from *Daradus*, the river Senegal); *Sicarius* Walck. or *Thomisoides* Nic., which, I suspect, belongs to this family, has also only 6 eyes.

3) Nouv. Dict. d'Hist. Nat., XXIV, p. 135.

he adduces the different relative length of the two posterior pairs of legs: they are in *Misumena* "brusquement plus menues et plus courtes que les autres", which is *not* the case in *Heteropoda*. *Micrommata*, according to LATREILLE, differs from both these genera in having the maxillæ straight, not inclined to the labium. The next year WALCKENAER (in Tableau des Aranéides) united *Heteropoda* and *Misumena* in one genus, which he called *Thomisus*, instead of retaining for it, as in justice he ought to have done, one of the Latreillian names. The genus *Micrommata* he adopted unaltered, but gave also to it a new name, *Sparassus* [1]). In the Tabl. des Aran., *Thomisus* is divided into three sections: "les *Hétéropodes*", answering to *Misumena*, and "les *Équipèdes brévirostres*" and "les *Équipèdes longirostres*", both together answering to *Heteropoda* LATR. In Faune franç., Arachn., Livr. 11 et 12 (1825 ?), the French forms of WALCKENAER'S *Thomisus* were by that author again divided between two genera, *Philodromus* and *Thomisus*, the first of which corresponds to a part of *Heteropoda* LATR., the last to *Misumena* LATR. In the same work, a few years later (1830), the genus *Delena* was proposed (p. 110): afterwards WALCKENAER, as is known, created or adopted several new genera formed at the expense of his *Thomisus*: *Selenops*, *Clastes*, *Arcys*, *Eripus*, *Olios* (= *Sarotes* SUND.). — WALCKENAER soon perceived the intimate connexion between *Micrommata* LATR. or *Sparassus* and the spiders, which in his Tabl. d. Aran. form the 8th family of *Thomisus* (*Thom. leucosius* or *Ar. venatoria* LINN. and others, for which he afterwards formed the genus *Olios*): in Faune Franç., loc. cit. we even find these latter referred to *Sparassus*, whereas LATREILLE had united them with the species of *Philodromus*, with which they have far less affinity. — The very different development of the posterior, compared with the anterior extremities in *Misumena* or *Thomisus* on the one side, and *Heteropoda (Philodromus)* and *Micrommata* on the other, probably still affords the best basis for the division of the *Thomisoidæ* into larger groups, after the resolution of these old genera into a number of smaller; this basis has gained increased stability since attention has been called (by DUGÈS, OHLERT, and others) to the presence of *hair-tufts* (claw-brushes, claw-tufts, as I call them) under the tarsal claws in the last two Latreillian genera, and the absence of them in the first-named. SIMON also divides, chiefly on that principle, his family "*Thomisiformes*" into two tribes, "*Philodromiens*" and "*Tho-*

1) LATREILLE soon submitted in part to this usurpation, and himself adopted a couple *(Thomisus, Philodromus)* of the names imposed by WALCKENAER. But this of course does not authorize us here any more than elsewhere to neglect the law of priority.

misiens", uniting *Micrommata (Sparassus)* with the former [1]). The same two groups are also adopted by PRACH [2]), who calls them *Philodromi* and *Cancroides*. According to our method they constitute sub-families, and may be called *Philodrominæ* and *Thomisinæ*. The sub-family *Anetinæ* we have added merely *provisionally* for the as yet too imperfectly known genus *Anetes* MENGE.

The exotic genus *Arcys* WALCK. [3]) ought, it seems to me, to be considered as the type of a separate sub-family, *Arcyinæ*, which shows strong analogies with certain *Epeiroidæ*, as *Gasteracantha* (SUND.) and *Peniza* THOR. [4]). With the Arcyinæ, *Anetes* might *perhaps* also be united.

WESTRING and BLACKWALL have divided the *Thomisoidæ* belonging to the European Fauna, with which they were acquainted, into only three genera, *Thomisus*, *Philodromus* and *Sparassus*. C. KOCH detached from *Thomisus* the genus *Xysticus*, and from *Philodromus* the genera *Artamus* and *Thanatus* [5]), which three new genera have been adopted by SIMON, OHLERT and others. SIMON adds one more European genus, *Oxyptila* [6]). SIMON however in a paper lately published [7]) has abandoned his former division of the Thomisoidæ. Not satisfied with taking the genus *Thomisus* in as extensive a meaning as that which it bears in WALCKENAER'S latest works, he also unites with it *Monastes* LUC. (*Monæses* NOB.), and even wishes to suppress *Philodromus* WALCK., because that genus only differs from *Thomisus*, "by a greater equality between the eight legs." But the greater part of the European genera of e. g. the family *Attoidæ* adopted by SIMON [8]), are most assuredly as nearly connected with each other, and exhibit among themselves quite as evident transitions as the above Thomisoid genera, and it cannot be right in estimating the value of generic characteristics to follow one rule with one family and another with another [9]).

---

1) Hist. Nat. d. Araignées, p. 392. 2) Monogr. d. Thomisiden v. Prag, p. 8 (604).

3) I possess one species of the genus *Arcys* from New Holland, kindly presented by Prof. LEUCKART, which appears to be identical with *A. lancearius* WALCK. Species of that genus have else only been found in South America (Brazil, Chili).

4) Vid. THORELL, Eugenies Resa, Arachn., 1, p. 10.

5) Uebers. d. Arachn.-Syst., 1, p. 25—28.

6) Hist. Nat. d. Araignées, p. 440.

7) Sur quelques Araignées d'Espagne, p. 285.

8) SIMON, Monogr. d. espèces Europ. de la fam. d. Attides, p. 6 (16).

9) The very principle on which SIMON'S view of the connexion of the abovementioned Thomisoid genera appears to rest, viz. that all genera, which gradually pass into each other, ought to be united in one, appears to me quite wrong. The case is just the same with genera as with families, orders, classes, etc., nay even

We arrange the European Thomisoidæ under the following genera:

§ Mamillæ ut et ungues in apice tarsorum adsunt.

* Pedes 4 posteriores reliquis non vel parum graciliores, sæpissime iis non vel parum breviores. Tarsi in apice sub unguibus fasciculis duobus pilorum plus minus dilatatorum instructi. . . . . I. PHILODROMINÆ.

A. Utraque oculorum series ex oculis 4 composita.

a. Oculi medii antici vix vel non longius a margine clypei quam a mediis posticis remoti. Maxillæ plerumque rectæ et parallelæ. (Fasciculi unguiculares spississimi, ex pilis longis, tenuibus, in ipso apice tantum paullo dilatatis constantes).

1. Series oculorum antica paullo recurva, postica, desuper visa, paullo procurva. (Oculi intermedii in trapezium antice angustius dispositi). Genua pedum altius elevata. . . . . 1. *Micrommata.*

---

with the two great main divisions of the organic world, the animal and vegetable kingdoms: all these various kinds of systematic unities have been formed on the strength of a certain, greater or less, number of common features, which the natural productions united under them seem to us to possess, and although we see now a greater, now a less saltus between the most nearly related coordinate groups, yet the differences in this respect do not affect the propriety of considering them as independent genera, families, orders, etc., provided only their *typical* forms show the amount of peculiarities, which one assumes to be necessary for a group to be acknowledged as possessing the significancy of a genus, family, etc., and provided some *sure*, even if insignificant, feature can be pointed out as determining in doubtful cases the limit of the group. The groups, which, like e. g. the *genera Dinopis* and *Hyptiotes* among Spiders, or like this and most other *orders* within the class of Arachnoidea, do not exhibit transitions to any other group, are comparatively few; and how vast differences in this respect are visible between e. g. the different orders of the class *Crustacea* on the one and of the *Arachnoidea* on the other hand! And yet surely no one will deny, that for inst. *Copepoda* and *Branchiopoda* are as natural and rational orders as *Araneæ* and *Opiliones*, although the boundary between the former is not so sharply defined, but that the same genus (e. g. *Argulus*) is referred by some authors to the Copepoda and by others to the Branchiopoda. Precisely similar to the relation between these two *orders*, is that between many *genera*, and among them that between *Thomisus*, *Monæses* and *Philodromus:* transitions there are, it is true, but the groups are *on the whole* and *in their typical forms* sufficiently different, to deserve their separate denominations and the rank in the system, which it has hitherto been customary to give them. — The more new forms (especially fossile ones) are discovered, the more the intervals between a number of genera and of higher groups, which had previously been considered as widely separated, are filled up. If we were fully acquainted with the entire animal and vegetable world, both the now living and the extinct, all such gaps would assuredly be filled up, and the truth of the old adage: *natura non facit saltus*, would stand out in all its grandeur.

2. Series oculorum antica paullo recurva, postica, desuper visa, subrecta. (Oculi intermedii plerumque fere in rectangulum dispositi). Femora sub-librata, genubus parum elevatis. . . . 2. *Sparassus*.

[3. Series oculorum antica sub-procurva vel recta, postica paullo recurva vel sub-recta. Oculi laterales antici mediis anticis non manifeste majores. . . . . 3. *Heteropoda*.]

b. Oculi medii antici evidenter longius a margine clypei quam a mediis posticis remoti. Maxillæ in labium inclinatæ. (Pili fasciculorum unguicularium breviores, compressi, in formam fere spathæ dilatatæ).

α. Pedum proportio 2, 1, 4, 3 (vel 2, 1, 3, 4). Cephalothorax breviter ovatus vel sub-orbiculatus.

1. Series oculorum antica modice, postica levius recurva, oculi laterales inter se paullo minus quam medii antici a mediis posticis distantes. Oculi laterales mediis paullo majores. Abdomen depressum, breviter et inverse ovatum vel sub-pentagonum. . . . . 4. *Artanes*.

2. Series oculorum ambæ modice et æqualiter recurvæ; laterales inter se spatio non minori distantes quam quo distant medii antici a mediis posticis. Abdomen plerumque ovatum vel inverse ovatum. . . . . 5. *Philodromus*.

β. Pedum proportio 2, 4, 1, 3 vel 2, 4, 3, 1: series oculorum ambæ fortiter recurvæ. Cephalothorax et abdomen oblonga. 6. *Thanatus*.

B. Series oculorum antica ex oculis 6, postica ex 2 tantum oculis constat. . . . . 7. *Selenops*.

** Pedes 4 posteriores reliquis graciliores et breviores multo. Tarsi fasciculis unguicularibus carent. . . . . II. Thomisinæ.

A. Frons cum mandibulis declivis, sub-porrecta; oculi medii antici a margine clypei longius distantes quam a mediis posticis.

1. Series oculorum antica levius, postica fortius recurva; laterales antici evidenter majores quam medii antici. (Abdomen postice in tuberculum elevatum vel acuminato-productum). . . . . 8. *Monæses*.

2. Series oculorum antica fortius, postica levius recurva, laterales antici non majores quam medii antici. . . . . 9. *Thomisus*.

B. Frons et mandibulæ sub-verticales; oculi medii antici non longius a margine clypei quam a mediis posticis remoti.

a. Series oculorum antica plus minus recurva.

α. Oculi laterales postici vix vel non majores quam medii postici. (Oculi 4 medii plerumque in trapezium antice angustius dispositi). Aculei tibiarum graciles.

1. Series oculorum anticorum fortius, posticorum levius recurva; oculi laterales antici non vel parum majores quam intermedii antici. . . . . 10. *Misumena*.

2. Series oculorum anticorum levius, posticorum fortius recurva; laterales antici manifeste majores quam intermedii antici. 11. *Diæa.*

β. Oculi laterales postici evidenter majores quam medii postici; laterales antici multo majores quam intermedii antici; laterales inter se vix vel non longius remoti quam medii antici a mediis posticis. (Oculi 4 medii sæpius in rectangulum dispositi). Tibiæ et metatarsi anteriores subtus aculeis robustis armati. . . . . 12. *Xysticus.*

b. Series oculorum antica sub-recta; oculi laterales inter se manifeste longius distantes quam medii antici a mediis posticis; oculi 4 medii in rectangulum latiorem quam longiorem dispositi. Corpus valde depressum. 13. *Coriarachne.*

§§ Mamillæ et ungues desunt. . . . . III. ANETINÆ.

1. Oculi laterales a mediis longe remoti. . . . . 14. *Anetes.*

Sub-fam. I. PHILODROMINÆ.

The powerful development of the posterior extremities gives the spiders of this sub-family that quickness and lightness of motion in which they so remarkably excell the *Thomisinæ.* — The claws are long and slender, generally straight or somewhat sinuated (i. e. slightly curved in the form of an ∽) the greater part of their length, with only the extremity bent down to a hook. The claw-tufts vary in length and density, but are always present. — We assign the genus *Selenops* to this sub-family; by SIMON it is referred to the *Thomisinæ*, because the eyes in that genus are of *different* sizes, which he considers as one of the features by which the Thomisinæ are distinguished from the Philodrominæ. This is however no reliable characteristic, and indeed SIMON himself, in his description of the genus *Thomisus*, says: "yeux égaux" 1).

Gen. 1. MICROMMATA (LATR.) 1804.

Deriv.: *μικρόμματος*, small-eyed (*μικρός*, small; *ὄμμα*, eye).

Syn.: 1804. Micrommata [Micromata] LATR., *in* Nouv. Dict. d'Hist. Nat., XXIV, p. 135 (*ad partem*).

1) The exotic genus *Delena* WALCK. also we refer to the Philodrominæ, and not, as is done by SIMON, to the Thomisinæ. Its hinder pairs of legs are indeed not inconsiderably shorter than the fore legs, but they are about equal to them *in strength;* and by the presence of strong claw-brushes, by the form of the claws themselves, and the powerfully developed scopulæ under the metatarsi and tarsi, as well as by its general appearance, *Delena* betrays a close affinity to *Heteropoda.*

1805. **Sparassus** WALCK., Tabl. d. Aran., p. 39 (*ad part.:* "1e Fam. Les Mycromates, *Mycromatæ*").
1806. **Micrommata** LATR., Gen. Crust. et Ins., I, p. 115.
1861. **Sparassus** WESTR., Aran. Suec., p. 405.
1861. „ BLACKW., Spid. of Gr. Brit., I, p. 101.
1864. „ SIM., H. N. d. Araignées, p. 396 (*ad partem*).

Type: *Micrommata virescens* (CLERCK).

LATREILLE, in Nouv. Dict. d'Hist. Nat., l. c., states that his *Micrommata* comprises the spiders that WALCKENAER calls "les Grottiformes" (Faune Par., II, p. 225), i. e. *Aranea smaragdula*, *ornata*, *rosea* (and *A. accentuata*, which is placed there by mistake). In Gen. Crust. et Ins., *Micrommata smaragdula* (*Ar. virescens* CLERCK.) is expressly adduced as the type of the genus. — The more recent synonym *Sparassus* we reserve for those species of *Micrommata* LATR. or *Sparassus* WALCK. for which WALCKENAER formed the family "*les Opticiennes*", and which in the whole of their appearance approach far nearer to WALCKENAER'S *Olios* (*Heteropoda* (LATR.) NOB.) than to the 1st family of his *Sparassus*.

By some authors, e. g. WESTRING, *Micrommata* is referred to the *Drassoidæ*. It certainly differs considerably in general appearance from the more typical Thomisoidæ, the knees being so little depressed, that the animal can hardly be called laterigrade; but the intimate relationship of *Micrommata* with the evidently laterigrade species of the next genus, *Sparassus* (WALCK.) NOB., is too palpable to allow of its being separated from the family before us and transferred to the Drassoidæ, although it may be considered as forming the transition to these. — The form of the claws and claw-brushes is precisely that of the next following genus.

The spiders united by HENTZ [1]) under the name of *Micrommata*, cannot belong to this genus, for they all have the posterior row of eyes strongly curved backwards, and the anterior row straight or curved forwards. They seem to approach much nearer to *Dolomedes* or to *Dendrolycosa* DOLESCH., than to *Micrommata*, as far at least as we can judge from the position of the eyes as described and figured by HENTZ.

Gen. 2. SPARASSUS (WALCK.) 1805.

Deriv.: σπαράσσω, tear sunder.

Syn.: 1805. **Sparassus** WALCK., Tabl. d. Aran., p. 39 ("2e Fam. Les Opticiennes, *Optices*", *saltem ad part.*).

1) Aran. of the United States, *in* Bost. Journ. of Nat. Hist., V, p. 192.

1818. **Micrommata** LATR., *in* Nouv. Dict. d'Hist. Nat., 2e Éd., XX (*ad part.; sec.* WALCK.).
† 1838. **Ocypete** C. KOCH, Die Arachn., IV, (*ad part.:*) p. 83.
1864. **Sparassus** [Sparassa] SIM., H. N. d. Araignées, p. 396 (*ad partem*).

Type: *Sparassus Argelasii* WALCK.

The species we have proposed as type for this new genus has, it appears to us, been referred by C. KOCH to his *Ocypete* (*Olios* WALCK., *Heteropoda* (LATR.) NOB.), and described under the appellation of *O. tersa* (loc. cit.), although it has by all other writers, who have treated on it, been considered as a *Micrommata* or *Sparassus*. From *Micrommata*, as that genus has been limited by us, it differs in the strongly marked laterigrade position of the legs, in its more dense scopulæ, in the closer vicinity to each other of the two rows of eyes, etc. The eyes are moreover larger, and the anterior central eyes at least as large as the anterior lateral ones. From the next following genus, *Heteropoda*, it differs in that the anterior row of eyes is curved backward instead of being straight or curved somewhat forward. For this genus we have assumed the name *Sparassus*, which has previously been synonymous with *Micrommata*, and under which the typical species was first described.

In *Sparassus Argelasii* the tarsal claws are very long and slender (somewhat longer still than in *Micrommata*), straight, only a little sinuated towards the middle, and with the extremity turned down into a hook. The teeth are short, blunt and pretty close together, gradually longer towards the extremity of the claw, their points forming an almost straight line; they are about 16 in number on the inner, and a couple less on the outer claw. The female's palpal claw has about 8 tolerably strong, close-set comb-teeth, gradually increasing in length. The hairs in the thick claw-brushes are long and fine, with the extremity compressed, somewhat dilated, and bifid.

[Gen. 3. HETEROPODA (LATR.) 1804.

Deriv.: ἑτερόπους, with dissimilar feet (ἕτερος, other; πούς, foot).

Syn.: 1804. **Heteropoda** LATR., *in* Nouv. Dict. d'Hist. Nat., XXIV, p. 135 (*ad partem*).
1805. **Thomisus** WALCK., Tabl. d. Aran., p. 28 (*ad part.:* "8e Fam. Les Robustes, *Robustæ*").
1830. **Sparassus** ID., Faune Franç., Arachn., p. 102 (*ad partem*).
1833. **Sarotes** SUND., Consp. Arachn., p. 28 (*ad partem*).
† 1837. **Ocypete** C. KOCH, Uebers. d. Arachn.-Syst., 1, p. 27 (*ad max part.*).

1837. Olios WALCK., H. N. d. Ins. Apt., I, p. 563 (*ad part.: saltem* "1e Fam. Les Robustes, *Robustæ*").

1864. „ SIM., H. N. d. Araignées, p. 409 (*ad partem*).

Type: *Heteropoda venatoria* (LINN.).

The only species provided with specific name, that LATREILLE takes up loc. cit. as an example under his genus *Heteropoda*, is *Aranea venatoria* LINN. (*Olios leucosios* WALCK.) [1]), which must accordingly be considered as the type of the genus. According to the *characteristics* given by LATREILLE to *Heteropoda*, it answers to the whole sub-family of Philodrominæ, quite as *Misumena* LATR. answers to the sub-family Thomisinæ. — That a whole *Class* (of *Mollusca*) several years afterwards (1812) should have received the name of *Heteropods (Heteropoda)*, is certainly unfortunate, but this circumstance, it seems to us, cannot hinder the use of the *singular* form *Heteropoda* as a generic name, any more than the circumstance, that this name would have been much more suitable to a Thomisine than to a Philodromine genus, since at any rate it is not *false* as applied to this last. (Conf. p. 10, note 3).

The tarsal claws of *H. venatoria* are very long and slender, only at the end bent downwards, with (on the 2nd pair) about 12 comb-teeth on the inner claw; those nearest the base (the interior) are very close to each other, parallel, the exterior coarse and divergent; all are rather short, gradually increasing a little in length towards the extremity of the claw; on the outer claw they are less numerous and more sparse. The claw-brushes are long and thick, every separate hair very fine and somewhat incrassated just at the apex: seen in profile it there appears to be serrated on the underside.

The genus *Olios* WALCK. seems to us to contain forms too heterogeneous to allow of its remaining long undivided. Its "1re Famille", and perhaps a couple more, belong to *Heteropoda*, as we have in p. 174 determined the limits of that genus. The same generic group, which WALCKENAER calls *Olios*, had been previously characterized by SUNDEVALL under the name of *Sarotes*. That name, the oldest synonym of *Heteropoda*, ought to be made use of, if ever the genus comes to be divided into smaller generic groups. The Walckenaerian name is so incorrectly formed — it is said to be derived from *ὀλοός*, *ὀλοιός*, *destructive*, and accordingly should be writ-

1) *Ar. venatoria* FABR., Ent. Syst., II, p. 407 = *Ar. nidulans* ID., Mant. Insect., p. 343 (1787), is a Theraphosoid *(Nemesia)*, and therefore altogether different from *Ar. venatoria* LINN.

ten *Olous* or *Olæus* — that on that account alone it ought to be discarded. *Ocypete*, as the genus *Heteropoda* has been denominated by C. KOCH, is a name already in 1815 assigned by LEACH to a genus of *Acari*.

It is with doubt that I include this genus among those of Europe. Of the four species of *Olios* or *Ocypete* stated to belong to the European Fauna, one, the *Ocypete tersa* C. KOCH [1]), is undoubtedly identical with *Sparassus Argelasii*, of which species I have found a specimen at Nizza, and have received another from Spain from Mr. SIMON. The second, *Ocypete vulpina* (HAHN) C. KOCH, described by HAHN as an *Epeira*, has according to KOCH [2]) its front row of eyes evidently curved backwards, and is therefore surely a *Sparassus* (WALCK.) NOB. The third species, which, as well as the preceding, is unknown to me, *Olios spongitarsis* (DUF.) WALCK. [3]), is referred by DUFOUR [4]) to *Micrommata* (*Sparassus* WALCK.), and probably also belongs to *Sparassus* NOB. A fourth species, from Naples, described by CANESTRINI and PAVESI [5]), is called *Ocypete nigritarsis:* it is perhaps also a *Sparassus*.]

Gen. 4. SELENOPS DUF. 1820.

Deriv.: σελήνη, moon; ὤψ, eye.

Syn.: 1820. **Selenops** DUF., Descr. de six Arachn. nouv., p. 361.
1839. **Hypoplatea** (sub-gen. of **Selenops**) MAC LEAY, On some new forms of Arachn., p. 6.
1864. **Selenops** SIM., H. N. d. Araignées, p. 420.

Type: *Selenops homalosoma* DUF.

The typical European species is to me unknown. — In a species from Asia Minor (Caramania), belonging to the "3me Fam. Les *Aphartères*" of the genus in WALCKENAER (Ins. Apt., I, p. 548), and which I have received from Count KEYSERLING, the claws differ in appearance from those of all other Thomisoidæ known to me. They are indeed very long and slender, like those of the Philodrominæ in general, but they are pretty *uniformly curved*, not straight the greatest part of their length, and entirely *destitute of teeth*. Under the claws are two strong, very thick claw-brushes, the hairs of which are long and fine, slightly dilated at the end, as in

1) Die Arachn., IV, fig. 305; *ibid.*, XII, p. 39, figg. 980, 981.
2) Ibid., XII, p. 30, fig. 974.
3) Hist. Nat. d. Ins. Apt., 1, p. 574.
4) Descr. de six Arachn. nouv., p. 12 (366); Sur la Micr. spongitarsis, p. LIV.
5) Aran. ital., p. 133.

Heteropoda, *Micrommata*, etc. The thick *scopula* under the tarsus and metatarsus of these genera is *absent* in *Selenops*, which genus thus is distinguished not by its peculiar position of the eyes alone.

It is possible that this genus may have been created already by LATREILLE, in the 2nd Edit. of Nouv. Dict. d'Hist. Nat., which I have not had the opportunity of consulting (Conf. DUFOUR, loc. cit.). In his later works however LATREILLE calls it: "*Selenops* DUFOUR."

Gen. 5. ARTANES N.

Deriv.: *Ἀρτάνης*, proper name.

Syn.: †1837. **Artamus** C. KOCH, Uebers. d. Arachn.-Syst., 1, p. 27.
1861. **Philodromus** WESTR., Aran. Suec., p. 445 (*ad partem*).
1861. „ BLACKW., Spid. of Gr. Brit., I, p. 91 (*ad partem*).
1864. **Artamus** [**Artama**] SIM., H. N. d. Araignées, p. 415.

Type: *Artanes margaritatus* (CLERCK).

Artamus being the well known and accredited name of a genus of birds, so named by VIEILLOT as early as 1816, I have been obliged to give the spider-genus *Artamus* a new appellation.

In this and the two following genera, the hairs of the claw-tufts have an appearance quite different from that presented in any of the genera of the family, we have as yet described: these hairs are in fact beyond all comparison shorter and broader, flattened, spade-like or feather-like, and far less numerous (especially in *Thanatus*). The claws are often shorter, especially in *Thanatus*, but of the same form; the teeth usually far more numerous on the inner than on the outer claw, in *Philodr. aureolus*, for inst., about 5 on the outer and about 14 on the inner claw; in *Thanatus oblongus* about 3 on the outer and about 10 on the inner; but in *Th. formicinus* about 5 on the outer and 8 on the inner. The number of teeth on the claws is here, as usual, frequently very different not only on the different pairs of legs of the same individual, but on the same pair in different individuals of the same species, and accordingly the number observed by me in the various specimens that I have examined, frequently differs considerably from that given by OHLERT.

Gen. 6. PHILODROMUS (WALCK.) 1820—26.

Deriv.: *φιλέω*, love, like; *δρόμος*, course, run.

Syn.: 1825 (?) **Philodromus** WALCK., Fauna Franç., Arachn., p. 86 (*ad partem*).
1837. „ C. KOCH, Uebers. d. Arachn.-Syst., 1, p. 28.

1861. **Philodromus** Westr., Aran. Suec., p. 445 (*ad partem*).
1861. „ Blackw., Spid. of Gr. Brit., I, p. 91 (*ad partem*).
1864. „ [**Philodroma**] Sim., H. N. d. Araignées, p. 406.

Type: *Philodromus aureolus* (Clerck).

Walckenaer refers *Thaumasia senilis* Perty [1]) to the genus *Philodromus*, without doubt erroneously: it is not certain that *Thaumasia* is even a *Thomisoid:* Perty himself refers it, though doubtfully, to the *Tubitelariæ*.

Gen. 7. THANATUS C. Koch. 1837.

Deriv.: θανατός, death.

Syn.: 1837. **Thanatus** C. Koch, Uebers. d. Arachn.-Syst., 1, p. 28.
1861. **Philodromus** Westr., Aran. Suec., p. 445 (*ad partem*).
1861. „ Blackw., Spid. of Gr. Brit., I, p. 91 (*ad partem*).
1864. **Thanatus** [**Thanata**] Sim., H. N. d. Araignées, p. 401.

Type: *Thanatus formicinus* (Clerck).

Sub-fam. II. THOMISINÆ.

This sub-family includes the forms which are strictly speaking typical of the whole family — the κατ' ἐξοχήν "crab-spiders". Their two pairs of back legs are always weaker and much shorter than the fore legs; single feather- or spade-like hairs are sometimes found under the claws, but they do not form claw-tufts or claw-brushes as in the Philodrominæ. The tarsal claws are (at least in the females) broad at the base, short and strong, and curved almost from the base, with rather long, closely set comb-teeth. In *Xysticus* and *Coriarachne* they are particularly coarse and blunt, in *Misumena* and other genera slenderer and more pointed.

From the genus *Thomisus* Walck., in the compass given to it in the "Faune Française", and which is still received by for example Westring and Blackwall, i. e. as identical with *Misumena* Latr., C. Koch already in 1835 detached his genus *Xysticus*, which is very natural, and has been adopted by many arachnologists. The remaining species of *Thomisus* Walck., for which C. Koch preserved that latter generic name, are on the contrary too heterogeneous to be allowed to remain united under a common name. One is obliged either to preserve *Misumena* Latr. (*Thomisus* Walck.) un-

1) Delect. Anim. Art. Bras., p. 192, Tab. XXXVIII, fig. 5.

divided, or else to break up *Thomisus* C. KOCH into some few minor genera. I, for my part, have preferred the latter alternative. SIMON [1]) has already divided *Thomisus* C. KOCH, SIM., into 4 "sub-genera", *Phlœoides*, *Thomisus*, *Pachyptila* and *Synœma*, and the sub-genus *Thomisus* again into three "groups", *Thomisus*, *Cirrofera* (= *Platythomisus* DOLESCHALL [2]) saltem ad partem) and *Diana*, and has furthermore proposed the new genera *Oxyptila* and *Phrynoides* (*Phrynarachne* NOB.: vid. sup. p. 37). The last-named (exotic) genus appears to me to merit preservation, but the other, European, groups, to which he has assigned generic names, I cannot, in the very vague limits of SIMON'S definitions, accept as genera.

Gen. 8. MONÆSES N.

Deriv.: *Μοναίσης*, proper name.

Syn.: † 1845—47. **Monastes** LUC., Explor. de l'Algérie, Arachn., p. 192.
1847. „ WALCK., H. N. d. Ins. Apt., IV, p. 432.
1838. **Xysticus** C. KOCH, Die Arachn., IV, (*ad part.*:) p. 79.
1864. **Monastes** SIM., H. N. d. Araignées, p. 418.
1864. **Xysticus** ID., ibid., p. 524 (*ad partem*).
1868. **Thomisus** ID., Sur quelques Araignées d'Espagne, p. 284 (*ad partem*).

Type: *Monœses paradoxus* (LUC.).

The genus *Monastes* — or *Monœses*, as I have called it, the name *Monastes* being already appropriated (vid. p. 37) — was formed by LUCAS for two remarkable Thomisinæ from Algeria, and has lately been increased by SIMON with a third and European species, *Thomisus Piochardi* SIM. from Spain. As a fourth species I refer to this genus *Xysticus cuneolus* C. KOCH, which also belongs to the Fauna of Europe. SIMON in the latter of his works above cited has suppressed this genus and united it with Thomisus, which appears to us by no means a happy step. On this subject se farther p. 172.

The genus *Sylvia* NIC. [3]) seems to me, judging from the figures, to be nearly related to, perhaps identical with *Monœses*, although the species of that genus are said not to be laterigrade, and to have vertical mandibles. — *Sylvia* is the old Linnæan name of a genus of birds.

In *M. cuneolus* the tarsal claws present very nearly the same construction as in *Misumena* and *Diœa;* the inner claw has about 12 long,

---

1) Hist. Nat. d. Araignées, p. 432.
2) Tweede Bijdr. t. de Kenn. d. Arachn. v. d. Ind. Arch., p. 59.
3) GRAY, Hist. fis. e pol. de Chile, Zool., III, p. 465.

parallel comb-teeth, of which those nearest the base are considerably finer and very close-set; the outer has about 8 somewhat equal, coarse teeth. The palpal claw is small, with about 4 pretty long comb-teeth.

Gen. 9. THOMISUS (WALCK.) 1805.

Deriv.: perhaps θωμίσσω, bind, whip.

Syn.: 1805. **Thomisus** WALCK., Tabl. d. Aran., p. 28 (*ad partem*).
1825(?). „ ID., Faune Franç., Arachn., (*ad part.:*) p. 70.
1837. „ C. KOCH, Uebers. d. Arachn.-Syst., 1, p. 24 (*ad partem*).
1861. „ BLACKW., Spid. of Gr. Brit., I, p. 66 (*ad partem*).
1864. „ [**Thomisa**]: sub-gen. **Phlœoides** SIM., H. N. d. Araignées, p. 431 (*ad partem*).

Type: *Thomisus abbreviatus* (WALCK.).

We preserve WALCKENAER'S generic name *Thomisus* for the large and remarkable species, which WALCKENAER called *Th. abbreviatus* and HAHN *Th. diadema*, and which also in the works of C. KOCH retains the generic name *Thomisus*. — By its high and sloping clypeus and its somewhat protruded mandibles, this spider, like the species of *Monæses*, in some degree resembles the Philodrominæ, but the entire general appearance of the animal, as well as the presence of the characters that distinguish the sub-family Thomisinæ, gives it an undoubted place in the last named group.

The tarsal claws of *Th. abbreviatus* ♀ are small, but coarse, not so blunt however as those of *Xysticus*, with about 8 tolerably long, somewhat curved comb-teeth on the inner and 4 on the outer claw. The female's palpal claw is almost straight throughout half its length, then sharply curved, with a long point and about 5 long comb-teeth, of which that nearest the base is considerably smaller than the rest.

Gen. 10. MISUMENA (LATR.). 1804.

Deriv.: μισούμενος, hated (μισέω, hate).

Syn.: 1804. **Misumena** LATR., *in* Nouv. Dict. d'Hist. Nat., XXIV, p. 135 (*ad partem*).
1805. **Thomisus** WALCK., Tabl. d. Aran., p. 28 (*ad partem*).
1837. „ C. KOCH, Uebers. d. Arachn.-Syst., p. 24 (*ad partem*).
1861. „ WESTR., Aran. Suec., p. 410 (*ad partem*).
1861. „ BLACKW., Spid. of Gr. Brit., I, p. 66 (*ad partem*).
1864. „ [**Thomisa**]: sub-gen. **Phlœoides**, **Thomisus** et **Pachyptila** SIM., H. N. d. Araignées, p. 431 (*ad partem*).

Type: *Misumena vatia* (CLERCK).

Misumena LATR. 1804 is, as we already know, synonymous with *Thomisus* WALCK. 1805 *ad part.*, and accordingly has the right of priority in preference to that later name. As LATREILLE loc. cit. names *Aranea citrea* DE GEER (*Aran. vatius* CLERCK) as the type of *Misumena*, this *oldest* generic name must be reserved for that one of the smaller genera, into which *Misumena* or *Thomisus* has by later authors been resolved, that includes *Ar. vatius* CLERCK. To *Misumena*, besides *Ar. vatius*, I reckon among others *Ar. truncata* PALL. (*horrida* FABR.), *Thom. lateralis* C. KOCH, as also *Thom. villosus* LATR., for which SIMON has formed the sub-genus *Pachyptila*. In order that the generally known name *Thomisus* may not be altogether lost, I have preserved it for a genus formed by myself, of which the type is *Thomisus abbreviatus* WALCK. See preceding genus.

Gen. 11. DIÆA N.

Deriv.: *Διαῖος*, proper name.

Syn.: 1805. **Thomisus** WALCK., Tabl. d. Aran., p. 28 (*ad partem*).
1837. „ C. KOCH, Uebers. d. Arachn.-Syst., 1, p. 24 (*ad partem*).
1861. „ WESTR., Aran. Suec., p. 410 (*ad partem*).
1861. „ BLACKW., Spid. of Gr. Brit., I, p. 66 (*ad partem*).
1864. „ [**Thomisa**]: sub-gen. id.: "*groupe*" **Diana**, et sub-gen. **Synæma** [**Synema**] SIM., H. N. d. Araignées, p. 431 (*saltem ad partem*).

Type: *Diæa dorsata* (FABR.).

The spiders belonging to this genus, which are usually referred to the same genus (*Thomisus* C. KOCH) as *Misumena vatia* (CLERCK), differ from that and from other species of *Misumena* by having the anterior row of eyes less curved than the posterior, as also the anterior lateral eyes evidently larger than the anterior central ones. In that respect they more nearly approach *Xysticus* than *Misumena*, which latter genus however they most closely resemble in their weak extremities, armed with fine spines, and their usually lively colours. This genus appears very nearly to coincide with the "group" *Diana* of SIMON'S *Thomisus;* but the name given by SIMON being previously engaged (vid. p. 36), I have replaced it with *Diæa.* — *Ar. globosa* FABR., which appears to be the type of the sub-genus *Synæma* SIM., may, although in its appearance tolerably different from *Diæa dorsata*, *D. tricuspidata* (*Thom. Diana* WALCK.) etc., perhaps for the present be united with *Diæa.*

Gen. 12. XYSTICUS (C. KOCH). 1835.

Deriv.: probably ξυστικός, scraping (ξύω, scrape, polish).

Syn.: 1835. **Xysticus** C. KOCH, *in* HERR.-SCHÆFF., Deutschl. Ins., 129, 16, 17.
1837. „ ID., Uebers. d. Arachn.-Syst., 1, p. 25 (*ad partem*).
1861. **Thomisus** WESTR., Aran. Suec., p. 410 (*ad partem*).
1861. „ BLACKW., Spid. of Gr. Brit., I, p. 66 (*ad partem*).
1864. **Xysticus** [**Xystica**] SIM., H. N. d. Araignées, p. 427 (*ad max. part.*).
1864. **Oxyptila** [**Ozyptila**] ID., ibid., p. 440.
1867. **Xysticus** OHL., Aran. d. Prov. Preuss., p. 108.

Type: *Xysticus Kochii* N. = *X. viaticus* C. KOCH [1]).

The genus *Oxyptila* SIM., formed for *Thomisus claveatus* WALCK., appears to me to differ from *Xysticus* only by the bristles on the body being incrassated at the extremity; this is also the case in *Thom. scabriculus* WESTR., which species I cannot generically distinguish from e. g. *Xyst. brevipes*, in which the bristles display, though in a less degree than in *Th. claveatus* and *scabriculus*, a tendency to become thicker towards the end. The name *Oxyptila* can moreover hardly be retained, on account of its signification (from ὀξύς, sharp and πτίλον, bristle), which is absolutely the reverse of the characteristic feature (the club-like thickening of the bristles towards the apex) which seems to constitute the principal claim of this group to be considered as a separate genus.

In the genus *Xysticus* the tarsal claws are very different in the two sexes. In *X. cristatus* for ex. they are in the female short and strong, pretty regularly curved, with 4 or 5 strong comb-teeth and frequently also a finer tooth near the base. In the *male* the claws are weaker, rather long and slender: they are but slightly curved for the greatest part of their length, almost straight, with the point turned downwards; the outer claw has about 5 sparse and coarse teeth; on the inner claw the teeth are more numerous, for where in the outer claw the inmost tooth is posited, we find in the inner a group of about 5 closely set, fine teeth.

---

1) *Aranea viatica* LINN. or *A. cristatus* CLERCK, which C. KOCH considers to be the same as his *Xysticus viaticus*, is an entirely different species, and = *X. audax* C. KOCH. — In both species the genital bulb is on the underside, nearer the base, provided with two processes: in *X. cristatus* that nearest the base is *broad*, *compressed*, claw-like, the other is slender and has almost the form of a ⊥ or an *anchor;* in *X. Kochii*, *both* processes are slender and of about the same substance: that nearest the base is bent almost in the form of a *boot*, the other process has its short, *blunt* extremity curved against the foot of the boot. — *X. Kochii* has not as yet been found in Sweden.

Gen. 13. CORIARACHNE N.

Deriv.: *κόρις*, bug; *ἀράχνη*, spider.

Syn.: 1837. **Thomisus** C. KOCH, Uebers. d. Arachn.-Syst., 1, p. 25 (*ad partem*).
1838. **Xysticus** ID., Die Arachn., IV (*ad part.:*) p. 67.
1850. **Thomisus** ID., Uebers. d. Arachn.-Syst., 5, p. 37 (*ad partem*).
1861. „ WESTR., Aran. Suec., p. 410 (*ad partem*).
1864. **Xysticus** SIM., H. N. d. Aran., p. 427 (*ad partem*).

Type: *Coriarachne depressa* (C. KOCH).

That the spider C. KOCH has in the above cited passage of "Die Arachniden" described under the name of *Xysticus depressus*, cannot permanently be considered as belonging to the genus *Xysticus*, he has himself seen, and has accordingly in Uebers. d. Arachn.-Syst., 5, loc. cit. moved it to his *Thomisus*. But that he is still dissatisfied with the position he has thus assigned to this remarkable species, appears from his appending the remark: "*Allen Formen nach eine eigene Gattung.*" In fact this spider, which in the particularly depressed form of its body resembles certain species of *Delena* and *Heteropoda*, must be considered as the type of a special genus, the nearest neighbour to *Xysticus*, but distinguished from that genus, not only by its flattened body, but by having the anterior row of eyes straight, while the posterior row is sensibly curved backward. — The claws have much the same appearance as those of *Xysticus*.

Sub-fam. III. ANETINÆ.

* Gen 14. ANETES MENGE. 1850.

Deriv.: *α priv.*; *νέω*, spin.

Syn.: 1850. **Anetes** MENGE, Verzeichn. Danz. Spinn., p. 71.

Type: *Anetes cœletrum* MENGE.

All that is known about this remarkable genus is contained in the following lines. "Lastly I mention here a spider, which I look upon as new both as to genus and species, and which I shall call Anetes cœletron. Eyes posited as in Epeira. Abdomen oblong heart-formed, flat, terminating in a hard point posteriorly; on the underside of the belly a triangular, bordered *(umsäumte)* depression, in which I have not been able to discover any spinners. Tarsi destitute of claws. Length about 2 lines. Cephalothorax brownish, abdomen white, densely sprinkled with dark-brown points. Legs yellowish-white, mottled with brown. Lives in decaying matter. Unfortunately I possess no more than one female specimen. Appears to be nearly related to *Arkys lancearius* WALCK. Apt., I, 497, pl. 13, Fig. 3." (MENGE, loc. cit.).

Sub-ordo VI. CITIGRADÆ.

Syn.: 1817. "**Citigrades**" LATR., *in* CUV., Règne Anim., III, p. 95.
1823. **Cursores** SUND., Gen. Aran. Suec., p. 20.
1825. **Citigradæ** LATR., Fam. Nat. du Règne Anim., p. 316.
1833. **Lycosides** SUND., Consp. Arachn., p. 23.
1852. **Venatores** DOLESCH., Syst. Verzeichn. etc., p. 8.

This perfectly natural and universally acknowledged group, almost identical with WALCKENAER'S "*Coureuses*" and SIMON'S "*Lycosiformes*", and characterised by its high, almost prismatic cephalothorax, with narrow back, its eyes, which are arranged in 3 or 4 transversal rows (rarely in 2, and, when so, the posterior row strongly curved backwards), its 3 tarsal claws, its wandering habits, etc., has but few points of connexion with other suborders. The *Lycosoidæ* however show (through *Dolomedes*) a relationship with the *Agalenoidæ (Textrix)* and *Drassoidæ (Zora)*, but may, as far at least as regards the European forms, be easily distinguished from them by differences in the form of the cephalothorax, as also by the the position of the spinning tubes on the spinners, or by the number of the claws. They also show a certain affinity to the *Hersilioidæ*, but these are without difficulty distinguished by their long superior spinners, garnished with spinning tubes all along the under side, by the form of the parts of the mouth, etc. (Conf. p. 114). The genus *Catadysas* HENTZ (vid. p. 43, 161) is a connecting link between the *Lycosoidæ* and the *Theraphosoidæ* [1]). The *Oxyopoidæ* show evident analogies with the *Attoidæ;* both the *Oxyopoidæ* and *Ocyale* resemble in their general appearance certain *Philodrominæ (Thanatus);* but the form of the cephalothorax and the number of the claws is different, and the dif-

1) Like several of these latter, many Lycosoidæ dig with their mandibles deep holes or galleries in the ground: HENTZ (Aran. of the United States, *in* Bost Journ. of Nat. Hist., IV, p. 229) even states that he once found such a hole, in the winter, which was supplied with a *lid*. Also of the European *Tarentula Apuliæ* it has been said that it closes the orifice of its gallery for hibernation; but this is an error: Conf. BERGSØE, Iagttagelser om den Italienske Tarantel etc., p. 255.

ference in the *form* of the claws between any of these Citigradæ on the one side and the *Philodrominæ* and *Attoidæ* on the other is still greater.

SIMON divides his "*Lycosiformes*" into 3 tribus, *Herséliens*, *Lycosiens*, and *Dolomédiens*. The first of these answers to our *Hersilioidæ*, which appear to us to belong to the sub-order *Tubitelariæ*, and not to the Citigradæ. The other two, which are distinguished, the "*Lycosiens*" by having "yeux inégaux, corps court et ramassé, membres robustes et courts", whereas the "*Dolomédiens*" have "yeux peu inégaux, corps étroit et allongé, membres fins, longs et allongés", I cannot consider even as sub-families, for these characteristics do not appear to me to hold good: *Dolomedes* for example cannot surely be said to have a slenderer body and finer extremities than e. g. *Lycosa*. *Oxyopes* LATR. on the other hand is already by the position of the eyes so distinctly separated from other *Citigradæ*, that that genus may reasonably be considered as the type of a separate family.

We accordingly divide the European *Citigradæ* into two families, *Lycosoidæ* and *Oxyopoidæ*, in the following manner:

1. Oculi in series transversas tres vel duas dispositi: oculi 4 posteriores in trapezium postice latius, vel in lineam fortiter recurvam dispositi. I. *Lycosoidæ*.
2. Oculi in series transversas quattuor vel tres dispositi; oculi 4 posteriores in trapezium postice angustius vel in seriem procurvam dispositi. II. *Oxyopoidæ*.

Fam. I. LYCOSOIDÆ.

Syn.: 1833. Lycosides SUND., Consp. Arachn., p. 23 (*ad max. part.*).

In this family we include all genera belonging to the *Citigradæ*, with the exception of *Oxyopes* LATR. or *Sphasus* WALCK. and *Pasithea* BLACKW. or *Peucetia* NOB. — The claws in this family are very nearly similar to those of the *Agalenoidæ:* the superior tarsal claws are strong, broad at the base, pectinated; the inferior claw is bent suddenly downwards, but, unlike what is usually the case with the Agalenoidæ, is generally toothless; occasionally it is furnished with one or two pointed teeth. The palpal claw of the female is also pectinated, but has usually only a few teeth. In ♂ of many species, especially within the genera *Lycosa* and *Trochosa*, the palpus is, as OHLERT has shown [1]), provided at the extremity with an appendage more or less resembling a claw, which however can only be considered as a coarse spine, in as much as that it is not, like a real claw, broader at

1) Klauenbild. d. Preuss. Spinn., p. 12.

the base, moveable and articulated to the tarsus; sometimes two or even three such spines are found situated close to each other at the extremity of the palpus. In *Dolomedes* (at least *D. fimbriatus*), the palpus of the male (as has been discovered by OHLERT, loc. cit.) is provided with *a genuine pectinated claw* at its extremity, which is not the case in any other genus that I know of among the Citigradæ (not even in *Ocyale*), and has only been observed in one spider beside *Dolomedes*, namely in *Hersiliola oraniensis* (Conf. p. 116).

The European genera accepted by us as belonging to this family are as follows:

§ Series oculorum antica ex oculis 4 formata.

* Oculi medii seriei anticæ a margine clypei spatio remoti quod diametrum oculorum non vel paullo tantum superat.

A. Mamillæ superiores reliquis saltem dimidio longiores. Facies alta, subquadrata, fronte prominenti; series oculorum antica procurva. 1. *Aulonia.*

B. Mamillæ superiores reliquis vix vel non longiores.

a. Series oculorum antica paullo brevior quam media. Area oculorum æque saltem longa atque lata. Facies alta.

1. Facies sub-quadrata, versus mandibulas non vel parum latior, lateribus rectis. Pedes extus tenues. . . . . 2. *Lycosa.*

2. Facies versus mandibulas multo latior, lateribus fortiter convexis. Pedes plerumque robusti et extus parum attenuati. . 3. *Tarentula.*

b. Series oculorum antica plerumque paullo longior, saltem non brevior quam media. Area oculorum plerumque latior quam longior. Facies humilis, lateribus convexis; oculi medii seriei anticæ vix longius quam diametro suo a margine clypei remoti.

1. Oculi medii seriei anticæ majores, vix vel non minores quam oculi seriei posticæ: series oculorum anticorum evidenter longior quam series media. Cephalothorax plerumque densius appresso-pubescens. . . . . 4. *Trochosa.*

2. Oculi medii seriei anticæ parvi, evidenter minores quam oculi seriei posticæ: cephalothorax parce pubescens. . . . . . 5. *Pirata.*

** Oculi medii seriei anticæ a margine clypei spatio remoti, quod diametro oculorum maximorum duplo saltem majus est.

1. Oculi 4 seriei anticæ sub-æquales. Pedes robustiores. . . 6. *Dolomedes.*

2. Oculi 2 laterales seriei anticæ evidenter majores quam medii ejusdem seriei. Pedes graciles. . . . . , ... 7. *Ocyale.*

[§§ Series oculorum antica ex duobus tantum oculis constans. Oculi laterales seriei mediæ ab oculis duobus seriei posticæ longe disjuncti. . . . . 8. *Ctenus.*]

Gen. 1. AULONIA C. KOCH. 1848.

Deriv.: *ἀυλών*, defile, valley.

Syn.: 1805. **Lycosa** WALCK., Tabl. d. Aran., p. 10 (*ad part.:* "3e Fam. Les Porte-Queues, *Caudatæ*").
1848. „ sub-gen. **Aulonia** C. KOCH, Die Arachn., XIV, p. 97.
1864. **Lycosina** SIM., H. N. d. Araignées, p. 369.

Type: *Aulonia albimana* (WALCK.).

In this genus, which is especially distinguished by its long superior spinners, the claws are of the form usual within the family. The typical species, of which I found several examples at Kissingen, has about 7 or 8 gradually increasing comb-teeth on the superior tarsal claws, and *two* fine, rather long teeth on the inferior claw. The palpal claw has 3 or 4 teeth gradually increasing in length.

Gen. 2. LYCOSA (LATR.). 1804.

Deriv.: *λυκόω*, tear like a wolf (*λύκος*, wolf).

Syn.: 1804. **Lycosa** LATR., *in* Nouv. Dict. d'Hist. Nat., XXIV, p. 135 (*ad partem*).
1805. „ WALCK., Tabl. d. Aran., p. 10 ("1e Fam. Les Terricoles, *Terricolæ*" *ad part.*).
1833. „ : sub-gen. **Lycosa** SUND., Consp. Arachn., p. 24.
1848. „ : sub-gen. **Pardosa** C. KOCH, Die Arachn., XIV, p. 96.
1848. „ : sub-gen. †**Limonia** [**Leimonia**] ID., ibid., p. 99.
1861. „ WESTR., Aran. Suec., p. 467 (*ad partem*).
1861. „ BLACKW., Spid. of Gr. Brit., I, p. 16 (*ad partem*).
1864. „ : sub-gen. **Limonia** [**Leimonia**] et **Lycosa** SIM., H. N. d. Araignées, p. 349, 351, 352.
1867. **Pardosa** OHL., Aran. d. Prov. Preuss., p. 127, 136.
1867. **Limonia** [**Leimonia**] ID., ibid., p. 127, 133.

Type: *Lycosa lugubris* WALCK.

When in 1848 C. KOCH divided the genus *Lycosa* of LATREILLE into several sub-genera (as SUNDEVALL had already done in 1833), he gave new names to them all, without preserving to any the old name *Lycosa.* His sub-genus *Pardosa* appears to us to embrace the forms, in which the type of the Lycosoidæ is best and most fully developed, and SIMON has therefore done rightly in preserving to that sub-genus the old generic name *Lycosa.* As type of the genus we select the well-known *L. lugubris* WALCK. (= *L. silvicola* SUND., *L. alacris* C. KOCH).

As regards *Limonia [Leimonia]* C. KOCH, the pecularities in the form of the head and position of the eyes of this sub-genus do not appear to us of sufficient importance to require a generic separation from *Pardosa* or *Lycosa*, and we therefore give to the last named genus the compass assigned by SUNDEVALL in the Consp. Arachn. to his sub-genus *Lycosa.* The difference in the habits of *Pardosa* and *Limonia*, mentioned by C. KOCH, is not universal, for e. g. *L. lignaria* (CLERCK), which is evidenly a *Limonia*, lives in dry, sunny places (especially in pine-woods), not in wet localities. In the structure of the claws there is no difference: also the form of the cocoon is the same in *Pardosa* and *Limonia.* — The name *Leimonia* had already in 1816 been given by HÜBNER to a genus of Lepidoptera.

BLACKWALL and WESTRING preserve WALCKENAER'S *Lycosa* undivided, and it must be admitted, that the characteristic distinctions, on the strength of which it has by some modern arachnologists been divided into several genera, are by no means so sharp as could be desired. They show themselves more in the animals' habits, in the form given to their cocoons, and in the disposition of the colours, than in distinctly marked differences in the form of the various parts of the body.

The superior tarsal claws in *Lycosa* have ordinarily from 5 to 7 coarse, thinly set, somewhat divergent teeth; the inferior claw is usually unarmed, but, according to OHLERT, is now and then provided with a very small tooth. In the species examined by me the palpal claw is furnished with two or three coarse teeth.

Gen. 3. TARENTULA (SUND.). 1833.

Deriv.: *Tarentum*, proper name of the city now called *Taranto.*

Syn.: 1805. **Lycosa** WALCK., Tabl. d. Aran., p. 10 ("1e Fam. Les Terricoles, *Terricolæ*" *ad partem*).
1833. „ : sub-gen. **Tarentula** SUND., Consp. Arachn., p. 24 (*ad partem*).
1848. „ : sub-gen. **Tarantula** C. KOCH, Die Arachn., XIV, p. 96.
1861. „ WESTR., Aran. Suec., p. 467 (*ad partem*).
1861. „ BLACKW., Spid. of Gr. Brit., I, p. 16 (*ad partem*).
1864. „ : sub-gen. **Tarantula** SIM., H. N. d. Araignées, p. 349, 350.
1867. **Tarantula** OHL., Aran. d. Prov. Preuss., p. 127, 138.

Type: *Tarentula Apuliæ* (WALCK.).

The tarsal claws are similar in form and armature to those of *Lycosa*: the superior have most generally from 5 to 7 teeth (usually 6—8 on the 4th pair), the inferior is destitute of teeth. The palpal claw has about

4 teeth. In the large burrowing species, e. g. *T. melanogaster* (LATR.) or *Narbonensis* (WALCK.), the free point of the claw is longer and bent more deeply downward than in the smaller species found in north and central Europe. In *T. melanogaster*, the palpal claw has 4, the superior tarsal claws 5 or 6 teeth in the first half of their length. Also in *T. Apuliæ* these latter claws are armed with 5 strong comb-teeth, according to BERGSØE [1]).

Gen. 4. TROCHOSA (C. KOCH). 1848.

Deriv.: τροχάω = τρέχω, run.

Syn.: 1805. **Lycosa** WALCK., Tabl. d Aran., p. 10 ("1e Fam. Les Terricoles, *Terricolæ*" *ad partem*).
1833. „ : sub-gen. **Tarentula** SUND., Consp. Arachn., p. 24 (*ad partem*).
1848. **Trochosa** C. KOCH, Die Arachn., XIV, p. 95.
1848. **Arctosa** ID., ibid., p. 94.
1861. **Lycosa** WESTR., Aran. Suec., p. 467 (*ad partem*).
1861. „ BLACKW., Spid. of Gr. Brit., I, p. 16 (*ad partem*).
1864. **Trochosa** SIM., H. N. d. Araignées, p. 345.

Type: *T. ruricola* (DE GEER).

I have not been able to discover any feature depending on difference of *form*, whereby *Arctosa* C. KOCH may with certainty be distinguished from *Trochosa* ID., and I therefore follow SIMON in uniting these two genera in one, under the latter name. A difference might perhaps be shown to exist in the form of the claws, but it does not appear to me advisable to found a genus on a characteristic, that can only be discerned by the aid of the microscope. In the species of *Arctosa* that I have had the opportunity of examining (*A. cinerea* C. KOCH, *A. picta* ID., *Lyc. leopardus* SUND.), the superior tarsal and the palpal claws have their toothless extremity considerably longer, and curved more deeply downwards, than in most other Lycosoidæ, at least on the fore legs: the superior tarsal claws are provided with teeth throughout their first half only, and on the palpal claw the teeth are seated still nearer the base. This form of the claws is, I suppose, connected with these spiders' more fully developed ability of digging themselves cylindrical holes or galleries in the earth. (Conf. preceding genus, *Tarentula*). In *T. (A.) cinerea* I have met with about 10 teeth on the superior tarsal claws of the 1st, and 12 on the 4th pair of legs, those most external being bent somewhat forward, all of about equal length; the inferior claw is small and

1) Iagttag. om den Ital. Tarantel etc., p. 245.

destitute of teeth; the palpal claw has 3 or 4 small comb-teeth close to the base. In *T. (A.) picta* the claws are somewhat shorter, with about 8 teeth on the superior tarsal claws, in *T. (A.) leopardus* with 7: in this last species I have seen *one* tooth on the inferior claw, and a very small point just behind it, at least on the 4th pair of legs. — Also in *T. intricaria* C. KOCH the free extremity of the claws is very long; the superior tarsal claws have but 4 parallel teeth, of which the three outer are very coarse; the palpal claw has also 4 teeth, the innermost much smaller than the others. This species is also distinguished by the trapezoid formed by the 4 posterior eyes being twice as broad behind as in front, whereas in the typical species of the genus it is only 1½ time as broad behind: moreover the anterior row of eyes is longer in comparison with the middle row than in the other species of the genus. But it does not appear to me necessary on account of these deviations to form a new genus for *T. intricaria*.

T. ruricola has 5—6 comb-teeth on the superior tarsal claws; the inferior claw is without teeth; the palpal claw has four gradually increasing teeth. In this species the spine, which is so frequently met with among the Lycosoidæ at the end of the male's palpus, is pointed and somewhat curved at the extremity, and thus very like a toothless claw; it is absent in *T. terricola* THOR., in which species the female's palpal claw is generally furnished with 2 coarse teeth, and a 3rd small tooth behind them.

Gen. 5. PIRATA SUND. 1833.

Deriv.: πειρατής, pirate.

Syn.: 1805. **Lycosa** WALCK., Tabl. d. Aran., p. 10 (*ad part.:* "2e Fam. Les Corsaires, *Piraticæ*").
1833. „ : sub-gen. **Pirata** SUND., Consp. Arachn., p. 24.
1848. „ : sub-gen. †**Potamia** C. KOCH, Die Arachn., XIV, p. 98.
1861. „ WESTR., Aran. Suec., p. 467 (*ad partem*).
1861. „ BLACKW., Spid. of Gr. Brit., I, p. 16 (*ad partem*).
1864. „ : sub-gen. **Potamia** SIM., H. N. d. Araignées, p. 349, 352.
1867. **Potamia** OHL., Aran. d. Prov. Preuss., p. 126, 132.

Type: *Pirata piraticus* (CLERCK).

The name *Pirata* SUND. has right of priority in preference to *Potamia* C. KOCH, which latter moreover had been already several times appropriated, before KOCH in 1848 applied it to the genus before us.—Vid. p. 37.

P. piraticus has about 7 long teeth on the superior tarsal claws, and one fine tooth with the rudiment of a second on the inferior claw. The palpal claw has 3 teeth. In *P. uliginosus* THOR. there are on the superior tarsal claws about 8, on the inferior 1, and on the palpal claw 4 or 5 teeth.

Gen. 6. DOLOMEDES (LATR.). 1804.

Deriv.: *δολομήδης*, wily (*δόλος*, cunning, *μήδομαι*, devise).

Syn.: 1804. Dolomedes LATR., *in* Nouv. Dict. d'Hist. Nat., XXIV, p. 135.
1805. „ WALCK., Tabl. d. Aran., p. 15 (*ad part.:* "1e Fam. Les Riverines, *Ripuariæ*").
1833. Lycosa: sub-gen. Dolomedes SUND., Consp. Arachn., p. 24.
1861. Dolomedes WESTR., Aran. Suec., p. 534.
1861. „ BLACKW., Spid. of Gr. Brit., I, p. 37 (*ad partem*).
1864. „ SIM., H. N. d. Araignées, p. 374.

Type: *Dolomedes fimbriatus* (CLERCK).

On the superior tarsal claws of the typical species I have found 8—10 teeth; the inferior claw has a long curved tooth and a fine short point behind it; the female's palpal claw is more powerful and more sharply curved than in the preceding genera, and armed with 5 or 6 teeth. The male's palpal claw has, according to OHLERT, 5 teeth.

Under the generic name of *Dolomedes* several species are by some writers included, which by no means belong to that genus as defined by the limits which we, together with C. KOCH, WESTRING and others, have assigned it. Of the species of *Zora* (C. KOCH), which WALCKENAER refers to *Dolomedes*, we have elsewhere spoken (p. 140), as also of *Dolomedes agalenoides* LUC. (p. 121). — The East Indian genus *Dendrolycosa* DOLESCH.[1]) appears to differ from *Dolomedes* chiefly in having all the eyes small and of equal dimensions.

Gen. 7. OCYALE SAV. et AUD. 1825—27.

Deriv.: *ὠκύαλος*, moving rapidly on the sea (*ὠκύς*, swift, *ἅλς*, sea).

Syn.: 1805. Dolomedes WALCK., Tabl. d. Aran., p. 15 (*ad part.:* "2e Fam. Les Sylvines, *Sylvariæ*").
1825—27. Ocyale SAV. et AUD., Descr. de l'Égypte, (Éd. 2:) XXII, p. 372.
1861. Ocyale WESTR., Aran. Suec., p. 536.
1861. Dolomedes BLACKW., Spid. of Gr. Brit., I, p. 37 (*ad partem*).
1864. Ocyale [Ocyala] SIM., H. N. d. Araignées, p. 381.

Type: *Ocyale mirabilis* (CLERCK).

1) Tweede Bijdr. t. de Kenn. d. Arachn. v. d. Ind. Arch., p. 51.

This genus, which BLACKWALL, following WALCKENAER, has united with *Dolomedes*, differs even in the whole of its general appearance from the preceding genera, which are more typical of the family. — On the first pair of legs the superior tarsal claws are armed with about 12 teeth, the inferior with *one* tooth; on the 4th pair there are about 9 teeth on the superior and *two* on the inferior claw; and of these last the foremost is rather long and curved, the back tooth small. The palpal claw is strong, with about 7 teeth gradually increasing in length.

[* Gen. 8. CTENUS (WALCK.). 1805.

Deriv.: probably *κτῆνος*, live stock, cattle, a head of cattle.

Syn.: 1805. Ctenus WALCK., Tabl. d. Aran., p. 18.
1837. „ ID., Hist. Nat. d. Ins. Apt., I, p. 363 (*excl.* "3e Fam. Les Phoneutres, *Phoneutriæ*").
1864. Ctenus [Ctena]: sub-gen. *id.* SIM., H. N. d. Araignées, p. 377.

Type: *Ctenus dubius* WALCK.

This genus was originally formed by WALCKENAER for the species we have adduced as its type. To it he afterwards referred — according to a figure and short notice, left by the painter OUDINOT, and representing a spider found by him near Paris — the species *C. Oudinotii* WALCK. WALCKENAER had however not himself seen this spider, and no *Ctenus* has since been met with in France, so that one may reasonably doubt whether *C. Oudinotii* be really a *Ctenus*. WALCKENAER also considered a spider described and figured by ALBIN (Nat. Hist. of Spid., p. 51, Pl. XXXIV, Fig. 167 [1])) as belonging to this genus, probably on the strength of a certain similitude in the position of the eyes (which in ALBIN'S figure are arranged in 2 lines, the first consisting of 2, the other, which is much curved backwards, of 6 eyes); but ALBIN'S figures, perhaps more especially those which represent the positions of the eyes, are in general so faulty, that it is impossible to place any confidence in them; and I am the less inclined to believe that the figure in question really represents a *Ctenus*, since subsequent English arachnologists have never found any species of that genus in their country. It appears therefore to me more than doubtful that the genus *Ctenus* is anywhere represented in the Fauna of Europe.]

---

1) This figure probably represents a *Thanatus oblongus* (WALCK.).

Fam. II. OXYOPOIDÆ.

The spiders of this family, as is known, exhibit certain analogies with both *Attoidæ* and *Philodrominæ*, and seem to form a connecting link between the *Lycosoidæ* and these groups. They resemble the *Attoidæ* in their, comparatively with the Lycosoidæ, broader back of the cephalothorax, and frequently display a remarkable similitude with the *Philodrominæ* in their whole general appearance, and even in the position of the eyes (compare e. g. *Peucetia* and *Eripus*). But the *Lycosoidæ* are, as is generally admitted, their nearest relations, and it is also with them that they most closely agree in the structure of the *claws*. The tarsal claws are however usually longer than in the Lycosoidæ, with a shorter extremity and more teeth; the inferior claw has, in the species that I have examined, *two* or *three* teeth. The males have no claw at the end of the palpus. — To this family I refer two European genera, *Peucetia* and *Oxyopes*.

1. Oculi in series tres, sectorem circuli fere formantes, ordinati: 4 posteriores seriem paullo procurvam designant; medii eorum cum oculis duobus seriei 2^{dæ} in trapezium postice multo angustius, vix longius quam latius, dispositi. . . . . 1. *Peucetia.*
2. Oculi in series quatuor ordinati: 4 posteriores trapezium breve formant: oculi seriei 2^{dæ} et 4^{tæ} fere in rectangulum, evidenter longiorem quam latiorem, dispositi. . . . . 2. *Oxyopes.*

Gen. 1. PEUCETIA n.

Deriv.: *Πευκετίος*, proper name.

Syn.: † 1858. Pasithea Blackw., Descr. of six newly disc. Spid. and a new gen. of Aran., p. 427.

1866. Oxyopes Sim., Sur quelques Araignées d'Espagne, p. 287 (*ad partem*).

Type: *Peucetia viridis* (Blackw.).

The type of this genus is *Oxyopes littoralis* Sim. (loc. cit.), but this species appears to me to be identical with *Pasithea viridis* Blackw. (loc. cit.), which was first by Blackwall aggregated to the *Laterigradæ*, but afterwards [1]) rightly to the *Citigradæ*. — *P. viridis* differs from *Oxyopes*, to which genus it is referred by Simon, not only in the position of the eyes,

1) Blackwall, Descr. of recently disc. spec. etc. from the East of Central Africa, p. 6.

but also by its long, slender maxillæ dilated at the base, etc. The superior spinners are distinctly longer than the inferior. The claws are more powerful than in the genus *Oxyopes:* the superior tarsal claws have only about 7 long, strong, pointed comb-teeth, and the inferior has *three*, of which the outermost two are long and curved. — Of this handsome spider, which has been found in Algeria and Spain, I am acquainted only with the male, of which Mr. SIMON kindly sent me a specimen. The name *Pasithea* being already appropriated, I have substituted a new (vid. p. 36, 37).

Gen. 2. OXYOPES LATR. 1804.

Deriv.: ὀξυωπής, sharp-eyed (ὀξύς, sharp; ὤψ, eye).

Syn.: 1804. **Oxyopes** LATR., *in* Nouv. Dict. d'Hist. Nat., XXIV, p. 135.
1805. **Sphasus** WALCK., Tabl. d. Aran., p. 19.
1861. „ WESTR., Aran. Suec., p. 538.
1861. „ BLACKW., Spid. of Gr. Brit., I, p. 43.
1864. **Oxyopes** [**Oxyopa**] SIM., H. N. d. Araignées, p. 386.

Type: *Oxyopes variegatus* LATR.

On the upper tarsal claws of *O. variegatus* I have counted, on the outer about 17, and on the inner about 14, long, fine, close, parallel comb-teeth; the inferior claw terminates in a long, fine, straight point, and has *two* fine, long, curved teeth at the base. The palpal claw is small, with 10 close-set, fine comb-teeth. *O. italicus* has but about 10 teeth on the superior tarsal claws; on the inferior claw it has two powerful, curved teeth, and on the palpal claw about 8 long teeth.

The Brazilian genus *Idiops* PERTY [1]) is by WALCKENAER [2]) taken up as synonymous with *Sphasus* or *Oxyopes:* it has, it is true, a certain resemblance to that genus in the position of the eyes; but the direction of the mandibular claw, which is articulated *longitudinally*, as in the *Territelariæ*, appears to us to show, that *Idiops* belongs to that sub-order, to which it is also referred by PERTY. The species described by him, *I. fusca* [3]), shows in the form of the male's palpi an evident analogy with the genus *Actinopus* PERTY among the *Theraphosoidæ*, from which genus *Idiops* in other respects would seem to be widely separated. The form of its cephalothorax displays some resemblance to that of *Filistata* LATR.

1) Delect. Anim. Art. Bras., p. 197.
2) Hist. Nat. d. Ins. Apt., I, p. 379.
3) Delect. Anim. Art. Bras., p. 198, Pl. XXXIX, fig. 5.

Sub-ordo VII. SALTIGRADÆ.

Syn.: 1804. Gen. **Salticus** Latr., *in* Nouv. Dict. d'Hist. Nat., XXIV, p. 135.
1817. "**Saltigrades**" id., *in* Cuv., Règne Anim., III, p. 98.
1823. **Saltatores** Sund., Gen. Aran. Suec., p. 20.
1825. **Saltigradæ** Latr., Fam. Nat., du Règne Anim., p. 317.
1833. **Attides** Sund., Consp. Arachn., p. 25.
1843. **Saltioidæ** Blackw., The differ. in the numb. of eyes, etc., p. 616.

The spiders belonging to this group are, as is known, distinguished by their high cephalothorax, which has almost vertical sides and a very broad back, by their usually short and thick extremities, and by the peculiar position of their eyes, which most nearly approaches that of the Lycosoidæ: 4 eyes in fact form a first row, and the remaining 4 a second and third. An exception in the disposition of the eyes is presented by the exotic family *Otiothopoidæ*, in which the eyes form only *two* transversal rows, converging at the ends, and by *Lyssomanes* among the *Attoidæ*, which genus has its eyes arranged in *four* transversal rows. In the *Myrmecioidæ* the eyes may be as truly said to form *two* rows divergent at the ends, as *three;* in *Palpimanus* also they are arranged in two rows, *both* greatly curved in opposite directions, so that one might even say that the eyes of that genus form *four* rows. The family *Dinopoidæ*, which we, though with doubt, refer to this sub-order, differs especially in its very long and fine extremities from other Saltigradæ. Also in certain other genera, as for example *Myrmecium*, *Salticus* and *Leptorchestes*, the extremities are fine, though somewhat short. — The spinners, as far as is known, are six in number, usually not very long. There are generally but two claws on each tarsus, and in this case there is also, except in *Palpimanus* (and *Otiothops?*) a tuft of hairs dilated at the end immediately under the claws; *Eresus* (as well as *Dinopis?*) has 3 claws on each tarsus, as also a claw at the termination of the female's palpus, which is absent in at least *Attoidæ* and *Palpimaninæ*. Most Saltigradæ leap actively, whence the name.

We resolve the European Saltigradæ into two families, *Eresoidæ* and *Attoidæ*, according to the following distinctive features:

1. Cephalothorax antice valde elevato-convexus. Oculi 2 postici inter se multo longius distantes quam sunt duo proxime antecedentes. Tarsi unguibus trinis aut binis instructi, fasciculo unguiculari carentes. . . . . . . I. *Eresoidæ.*
2. Cephalothorax deplanatus, parte cephalica non vel paullo tantum altiore quam parte thoracica. Oculi 2 postici inter se non multo longius quam 2 antece-

dentes remoti. Oculi 4 anteriores inter se proximi: medii (antici) eorum reliquis omnibus multo majores. Tarsi unguibus tantum binis et fasciculo unguiculari instructi. (Palpus feminæ ungui caret). . . . . II. *Attoidæ.*

Fam. I. ERESOIDÆ.

Syn.: 1850. **Eresides** C. KOCH, Uebers. d. Arachn.-Syst., 5, p. 70.

The two sub-families, into which we divide this family, certainly agree in the structure of the cephalothorax, the position of the eyes, and in their whole general appearance very closely with each other, but present the remarkable difference, that whereas the *Eresinæ* are provided with infra-mammillary organ and calamistrum, the *Palpimaninæ* are without these organs. The two genera *Eresus* and *Palpimanus (Chersis)* had already by WALCKENAER[1]) and DUFOUR[2]) been placed in the closest connexion with each other and with *Attus;* SUNDEVALL[3]) and C. KOCH[4]) received them into the family *Attides*, and when the latter afterwards detached them from that family, he united them with the new-formed family *Eresides*, which received a place immediately after the *Attides*[5]). — CANESTRINI and PAVESI[6]) who unite *Eresus* with the Attoidæ, have formed a separate family, *Chersidæ*, for *Palpimanus*, a view which I cannot approve. How SIMON[7]) could refer *Eresus* to the *Epeiroidæ* and *Palpimanus* to the *Myrmecioidæ* is to me inexplicable. — We characterize the two sub-families and thereto belonging European genera as follows:

I. Organum infra-mamillare et calamistrum adsunt. . . . . . . I. ERESINÆ.
   1. Oculi seriei tertiæ longe pone reliquos siti; laterales seriei 1mæ ab intermediis ejusdem seriei longissime remoti. Tarsi omnes unguibus trinis instructi. (Palpus feminæ ungui armatus). Mamillæ breves. . 1. *Eresus.*

II. Organum infra-mamillare et calamistrum desunt. . . . . II. PALPIMANINÆ.
   1. Oculi seriei 3tiæ paullo tantum pone oculos 2dæ seriei siti, cum iis seriem recurvam formantes. Tarsi pedum 6 posteriorum unguibus tantum binis armati. (Palpus feminæ ungui caret). . . . . 2. *Palpimanus.*

---

1) Tabl. d. Aran., p. 21; Mém. sur une nouv. Classif. d. Aran., p. 438; Hist. Nat. d. Ins. Apt., IV, p. 525.
2) Descr. de six Arachn. nouv., p. 364.
3) Consp. Arachn., p. 27.
4) Uebers. d. Arachn.-Syst., 1, p. 34.
5) Ibid., 5, p. 70.
6) Aran. ital., p. 75—76.
7) Hist. Nat. d. Araignées, p. 299, 448.

Sub-fam. I. ERESINÆ.

This sub-family includes for the present 2 genera, *Eresus* WALCK. and *Dorceus* C. KOCH (exotic and distinguished by long, three-jointed mamillæ). C. KOCH has indeed divided *Eresus* into two genera, *Erythrophora* and *Eresus* [1]), but as the genus *Erythrophora* can hardly be distinguished from *Eresus* by anything else than a difference of colour, it seems to me not deserving of preservation.

Gen. 1. ERESUS WALCK. 1805.

Deriv.: probably ἐρείδω, press against, inflict, attack.

Syn.: 1805. **Eresus** WALCK., Tabl. d. Aran., p. 22.
1837. **Chersis** ID., H. N. d. Ins. Apt., I, p. 390 (*ad partem*).
1850. **Eresus** C. KOCH, Uebers. d. Arachn.-Syst., 5, p. 50.
1850. **Erythrophora** ID., ibid.
1861. **Eresus** BLACKW., Spid. of Gr. Brit., I, p. 45.
1864. „ [**Eresa**] SIM., H. N. d. Araignées, p. 299 (*ad max. part.*).

Type: *Eresus cinnaberinus* (OLIV.).

In the few species of this genus known to me, the calamistrum is but slightly developed. In a ♀ of *E. lineatus* LATR. or *E. acanthophilus* DUF. [2]), which has the upperside of the two posterior metatarsi somewhat flattened, the calamistrum is plainly visible on the external edge; but in the male of *E. cinnaberinus*, in which these metatarsi are cylindrical as in the other legs, I cannot perceive any calamistrum distinguishable from the adjacent fine hair. The infra-mammillary organ is on the contrary easily seen in both species: in *E. lineatus* it forms a very narrow, uniformly broad, transversal area, which appears to be divided into two by a middle suture, and exhibits *two rounded foveæ* [3]), one on each side, and a small depression behind these, near the spinners.

The tarsal claws of *Eresus* are short, but extremely broad and strong,

1) Uebers. d. Arachn.-Syst., 5, p. 70.

2) This species was first described by LATREILLE in the 2nd Edition of Nouv. Dict. d'Hist. Nat., X, p. 393 — which I have not been able to consult — under the name of "*Erèse rayé*" (see for inst. WALCK., Ins. Apt., I, p. 399), probably also with the Latin name *Eresus lineatus:* at least it is by AUDOUIN, in Dict. class. d'Hist. Nat., VI, p. 253, called "*Eresus lineatus* LATREILLE".

3) Conf. note, p. 30.

uniformly and much curved, pectinated. In *E. lineatus* ♀ the superior claws of the first pair of legs are from the base to near the apex armed with about 12 long, strong comb-teeth, and the inferior claw with 3 long teeth. On the 4th pair the teeth are less numerous, 7 or 8 on the superior claws, while the inferior claw seems to be without teeth. The female's palpal claw is also short, very strong, and provided with about 9 coarse teeth. *E. cinnaberinus* ♂ has about 16 (and 14) teeth on the superior claws, and 2 on the inferior.

The *Aranea nigra* of PETAGNA [1]), to which WALCKENAER has given the name *Chersis dubius* [2]) is most certainly an *Eresus* (perhaps but a variety of *E. cinnaberinus*) and not a *Chersis (Palpimanus)*.

Sub-fam. II. PALPIMANINÆ.

Syn.: 1869. **Chersidæ** CANESTR. et PAV., Aran. Ital., p. 75.

Of this family only one genus is known, that namely formed by L. DUFOUR under the name of *Palpimanus*.

Gen. 2. PALPIMANUS DUF. 1820.

Deriv.: *palpare*, caress, touch; *manus*, hand.

Syn.: 1820. **Palpimanus** DUF., Descr. de six Arachn. nouv., p. 12.
1825—27. **Platyscelum** SAV. et AUD., Descr. de l'Égypte, (2 Éd.:) XXII, p. 401.
1837. **Chersis** WALCK., H. N. d. Ins. Apt., I, p. 390 (*ad max. part.*).
1864. „ SIM., H. N. d. Araignées, p. 448 (*ad max. part.*).

Type: *Palpimanus gibbulus* DUF.

Palpimanus is, as may be seen from the synonyms, the oldest name of the genus, and there is no plausible reason for abandoning it. That certain *Attoidæ* also have thicker fore-legs, which appear to serve as organs of touch (whence the name *Palpimanus*), and that SAVIGNY *intended* to call it *Chersis* [3]), can of course be no reason for cashiering the name *Palpima*-

1) Spec. Ins. Ulter. Calabriæ, p. 34 (of the Ed. printed "Francofurti et Moguntiæ, 1787").

2) Hist. Nat. d. Ins. Apt., I, p. 392.

3) Conf. WALCK., Ins. Apt., I, p. 393. — SIMON considers that the name *Palpimanus* must mean that the *palpi* resemble *hands;* but this is not the case: *manus* here signifies the *fore-legs*, not the *palpi*.

nus, that name not being unfit for the animal to which it has been applied, and having been published long before the name *Chersis*.

The genus *Palpimanus* is extremely interesting, not only on account of the well known singular form of its first pair of legs, but also for certain characteristics, which mark it as a connecting-link between *Eresinæ* and *Attoidæ*. The agreement with these last in the absence of an infra-mammillary organ and calamistrum, we have already mentioned: also the close position of the 4 anterior eyes, of which the 2 central ones are larger than the other 6, shows a tendency to similitude to the *Attoidæ*. The female's palpi are incrassated outwards, flattened on the underside, and, like those of the last-mentioned spiders, destitute of a claw at the extremity. In the *Eresinæ*, as we have already observed, the fasciculus unguicularis or claw-tuft usually found in the *Attoidæ*, is wanting: it is also absent in *Palpimanus;* but the peculiarly formed hairs of which it is composed are found in that genus, though they have been transferred to another place. The broad compressed metatarsus has in fact (in *P. gibbulus*) both its superior and inferior edge covered with hairs which rapidly dilate to oval or spade-like blades, and a band of such hairs, enclosed by longer, pointed bristles, is continued also along the upper edge of the tibia and patella. These hairs are longer on the upper edge of the metatarsus, where they are mixed with numerous longer, pointed bristles, than on its underside, where they are closer, shorter and of uniform length, and where only a few longer, pointed bristles occur; they accordingly here form a *scopula*, which is continued under a part (the base) of the tarsus itself. This joint is else only covered with pointed hairs and bristles.

The claws, as is known, are but 2 in number on the tarsi of the six posterior legs. They are weaker than those of the *Eresinæ*, and stouter than those of the *Attoidæ*. They are rather large, of uniform breadth, and curved in the form of a semicircle; the outer claw has (in *P. gibbulus*) on the 4th pair of legs about 7 teeth, the inner 6; on the 2nd and 3rd pairs the teeth are less numerous (5 and 4 on the 2nd pair). These teeth are conical, rather short and far apart.

According to Dufour's frequently repeated statement, *P. gibbulus* differs from all other spiders by having *no claws on the first pair of legs.* This is nevertheless so far from being the case, that this spider has really no less than *three claws on the first pair of legs, but only two on the succeeding pairs!* In this respect *Palpimanus* probably stands quite alone in the order of spiders. The claws on the 1st pair are however so small that they are quite concealed by the hairs at the extremity of the tarsus, and can only be clearly seen with a good microscope. The superior ones are similar in form to those of the following legs, except that they are less curved and have only about 3 conical teeth; the inferior claw has the form of a very small hook, sharply bent downwards, with a long fine extremity, and seems to be armed on the underside with *one* long fine tooth. Thus the number of claws on the first pair is the same as in *Eresus*, and on the other legs as in the *Attoidæ*.

Fam. II. ATTOIDÆ.

Syn.: 1850. **Attides** C. KOCH, Uebers. d. Arachn.-Syst., 5, p. 42.

This family, perhaps the most sharply defined and most natural within the whole order of *Araneæ*, is without difficulty distinguished from the *Eresoidæ* by the peculiar position and relative size of the eyes. The claws are in all cases only two on each tarsus [1]); they are long and slender, a little sinuated (i. e. with a slight ∽-formed curvature), and spring at a right or slightly acute angle from the upper end of the narrow and high part formed by their base. The tooth-armature is very various, and ordinarily different on the inner and outer claw, the number of teeth on the former being usually far greater than on the latter. The teeth, when there are any, occupy only the outer half of the claw's length; near the base there are no teeth, except now and then on the first pair of legs, the claws of which are often shorter and more uniformly curved than those of the other legs. The 4th pair of legs has usually the claws longest and most copiously provided with teeth. The claw-tuft is formed of hairs that are either flattened and gradually more or less dilated towards the end, or dilated and flattened at the extremity only; in this respect the tufts on the different pairs of legs are often very different; they are sometimes, on the 1st pair, continued as a scopula on the underside of the tarsus. In all the species that have been examined, the female's palpi are destitute of a terminal claw, a circumstance, which in other families, with the exception of the *Scytodoidæ*, only occurs exceptionally. I believe it is only the species of this family, that justify the name "jumping-spiders", given to the whole

1) *Attus phrynoides* WALCK. (Ins. Apt., I, p. 479) is said to have on its extraordinarily long 1st pair of legs *(pedes raptorii)* only *one* toothless claw. This species ought undoubtedly to form a separate genus, to which also *Attus obisioides* DOLESCH. (Bijdr. t. d. Kenn. d. Arachn. v. d. Ind. Arch., p. 433) ought to be referred. This new genus, characterized by the long trochanteres of the fore-legs, may be called *Diolenius* (διωλένιος, with outstretched arms).

sub-order. — All the European species may be referred to one and the same sub-family *(Attinæ)*; among exotic forms perhaps *Lyssomanes* HENTZ [1]), ought to be considered as the type of a separate sub-family, characterized by the eyes being arranged in *four* transversal rows: the lateral eyes of the first row in the ordinary Attoidæ are in fact in *Lyssomanes* removed so high up that they form a separate row about half-way between the first and third pair of eyes. The relative size of the eyes is however exactly the same in *Lyssomanes* as in the Attinæ, i. e. the first pair is considerably larger and the third pair considerably less than the other eyes. (In the *Dinopoidæ*, in which the position of the eyes is the same as in the *Attinæ*, the relative size of the eyes is altogether different: it is in fact the last pair but one, or the eyes of the 2nd row, which in that family are considerably larger than the rest). — Calamistrum and infra-mammillary organ are absent.

There is no family in the whole order of spiders, which, on account of the great similarity between the species, is so difficult to resolve into good genera, as this, while at the same time its extraordinary richness in species renders such a resolution in the highest degree desirable. In the works of the older writers, from LATREILLE and WALCKENAER inclusively, the whole family constitutes but one genus, *Salticus* LATR. or *Attus* WALCK., which by many arachnologists, among whom is BLACKWALL, is still preserved undivided. But already in 1832 HENTZ [2]) detached from *Attus* WALCK. the genus *Synemosyna*, which partly answers to *Leptorchestes* NOB. or *Salticus* C. KOCH (non SUND.), as also *Epiblemum* (ad part. = *Calliethera* C. KOCH). SUNDEVALL [3]), who is followed by WESTRING, the following year divided *Attus* WALCK. into two genera, *Salticus* and *Attus*, which easily admit of distinction. This on the contrary is not the case with most of the Attoid-genera proposed by C. KOCH (in Uebers. d. Arachn.-Syst., Die Arachniden, etc.) between 1835 and 1850, and which have been pretty generally received, in spite of the imperfect manner in which they have been characterized. WHITE in 1841 [4]) formed the genus *Homalattus* and in 1846 [5]) *Dineresus [Deineresus]*, both exotic. OHLERT [6]) has endeavoured to define more accurately those of KOCH's genera, which belong to the Prus-

1) Aran. of the United States, *in* Boston Journ. of Nat. Hist., V, p. 197.
2) On North American Spiders, p. 108.
3) Svenska Spindl. Beskr., *in* Vet. Akad. Handl. f. 1832, p. 199, 201.
4) Descr. of new or little known Arachn., p. 446.
5) Descr. of a new genus of Arachn. etc., p. 179.
6) Aran. d. Prov. Preuss., p. 148—150.

sian Fauna; but his attempts do not appear to me to have fully succeeded, chiefly on account of the insufficient materials he had at his disposal. SIMON in 1864 [1]) combined C. KOCH's many genera so as to form *five*, *Rhanis* C. KOCH (= *Rhene* THOR.: Vid. p. 37), *Attus* WALCK., *Cyrtonota* SIM., *Heliophanus* C. KOCH and *Salticus* (LATR.), of which the last four belong to the Fauna of Europe; the greatest part of KOCH's genera (and sub-genera) SIMON accepted as separate sub-genera or "groups" [2]). SIMON's classification of the Attoidæ here referred to, appears to me very defective, and can hardly be considered as making any advance towards the solution of the difficult problem; the genus *Cyrtonota*, in which he includes KOCH's *Calliethera* together with *Philia*, *Plexippus*, etc. especially is very unnatural. SIMON himself has moreover since abandoned this division and adopted another quite different; he now [3]) divides the European Attoidæ into 10 genera (of which two, *Menemerus* and *Yllenus*, are new) according to characteristics principally derived from the *form of the male's palpi and mandibles.* This division has indeed the advantage of being based upon fixed and easily observable differences of form, but it has also the great defect of applying only to one (and that the rarer) sex; it is impossible to say to which of SIMON's genera a female specimen belongs, as long as the male of the same species is unknown, unless it should happen, that the females of that genus are also distinguished by some common feature; but in such case that feature ought to have been included among the characteristics of the genus. I have already (p. 19, 83) stated my objections to the adoption of genera depending upon characteristics that apply only to one sex, or that are derived from a difference of form in the organs of copulation alone.

What has here been said, sufficiently indicates my opinion, that a natural arrangement of the Attoidæ is as yet a *pium desiderium.* For my own part I have awhile hesitated between two methods of proceeding — either to adopt only *three* genera, *Salticus* (*Pyrophorus* C. KOCH), *Leptorchestes* (*Salticus* C. KOCH) and *Attus;* — or to adopt and endeavour as well as possible to characterize those of the genera formed by C. KOCH, which belong to the European Fauna. These genera are in fact pretty well known as regards their general appearance, and they have also been acknowledged

1) Hist. Nat. d. Araignées, p. 307. — *Dinopis* [*Deinopis*] MAC LEAY, which SIMON also refers to the Attoidæ, is in our opinion the type of a separate family, *Dinopoidæ.* Vid. p. 43.

2) For *Attus Doumercii* WALCK. he proposed *Lagenicola* as a new sub-genus of *Attus* (loc. cit., p. 316).

3) Monogr. d. espèces Europ. de la fam. d. Attides, p. 16.

by several arachnologists. They moreover on the whole form tolerably natural groups, although KOCH has not succeeded in giving any reliable diagnosis of them. I have determined on adopting the second, far more difficult alternative, because I believe the division of the genus *Attus* WALCK. into several smaller genera to be a matter of great practical importance, especially on account of the great number of exotic species that have been described, and which furnish an amount of materials which it will be scarcely possible to manage, unless one can distribute them among smaller generic groups. I am however by no means satisfied with the result of the experiment I have made, and the following arrangement, of the many defects of which I am perfectly conscious, must therefore be looked upon as merely provisional. It may however possibly, even if but negatively, contribute in some measure to the solution of the problem. None but a person having at his disposal far more comprehensive materials for research than I can command, can hope to arrive at any fully satisfactory result.

All C. KOCH'S European genera have been here employed, with the exception of *Icelus* [1]), which is founded on a feature (the back of the mandibles raised to a sharp ridge) belonging only to one sex, the males. Two of his sub-genera, *Ballus* and *Dia* (*Ælurops* NOB.) have been promoted to the rank of genera, the others I have been obliged to pass by. I have also endeavoured to give a place in my scheme to the genera *Menemerus* and *Yllenus* formed by SIMON.

§ Pars cephalica parte thoracica abrupte altior. Quadrangulus oculorum (ex oculis seriei 3^{tiæ} et lateralibus seriei 1^{mæ} formatus) vix vel non longior quam latior. Corpus longum et angustum. Pedes tenues. . . . . . . 1. *Salticus*.

§§ Pars cephalica parte thoracica non altior.

† Quadrangulus oculorum longior quam latior: oculi seriei 3^{tiæ} fere in medio cephalothorace siti. Corpus longum et angustum; pedes tenues. 2. *Leptorchestes*.

†† Quadrangulus oculorum saltem postice latior quam longior.

* Metatarsi et tibiæ omnes aculeis carentes. Cephalothorax duplo fere longior quam latior, humilis, dorso sub-recto. Oculi seriei 1^{mæ} contingentes: medii eorum a margine clypei vix emarginati spatio brevissimo remoti [2]). . . . . 3. *Epiblemum*.

1) The name *Icelus* was already in 1844 by KRØYER given to a genus of fishes.

2) In order to judge rightly of the eyes' distance from the edge of the clypeus and of the form of the latter, it is necessary to remove at least a part of the thick covering of hair which ordinarily conceals the edge: moreover the membrane, which unites the base of the mandibles, and which is sometimes covered with hair, and frequently visible under the edge of the clypeus, must not be reckoned as part of the clypeus.

** Metatarsi pedum saltem anteriorum evidenter aculeati.

A. Oculi seriei 3tiæ non longius a margine cephalothoracis quam inter se remoti.

a. Cephalothorax plerumque duplo fere longior quam latior, minus humilis, dorso evidenter arcuato. Quadrangulus oculorum postice paullo latior. Oculi seriei 1mæ sub-contingentes, a margine clypei profunde emarginati et sub-nudi spatio brevissimo tantum remoti. 4. *Heliophanus.*

b. Cephalothorax non duplo longior quam latior.

α. Cephalothorax humilis valde, dorso sub-plano. Oculi seriei 3tiæ plerumque multo longius inter se quam a margine cephalothoracis remoti.

1. Quadrangulus oculorum postice evidenter latior quam antice; oculi seriei 3tiæ non multo ante medium cephalothoracis siti. Pars cephalica magna, lata. Oculi medii seriei 1mæ a margine clypei vix emarginati satis remoti. . . . . . . 5. *Ballus.*

2. Quadrangulus oculorum postice vix vel non latior quam antice. Oculi seriei 3tiæ longe ante medium cephalothoracis siti; oculi seriei 1mæ disjuncti; medii eorum a margine clypei vix emarginati spatio remoti quod $\frac{1}{3}$ diametri oculi plerumque æquat. Corpus satis longum et depressum . . . . . 6. *Marpessa.*

β. Cephalothorax altus, antice non angustatus, dorso evidenter arcuato. Oculi seriei 3tiæ parum longius inter se quam a margine cephalothoracis remoti. Quadrangulus oculorum postice non latior quam antice. Oculi seriei 1mæ contingentes: medii eorum a margine clypei vix emarginati spatio remoti quod $\frac{1}{4}$ diametri oculi non superat. (Pictura abdominis sæpissime ex colore ipsius cutis, non ex colore pilorum pendet). . . . . 9. *Euophrys.*

B. Oculi seriei 3tiæ longius a margine cephalothoracis quam inter se remoti.

a. Cephalothorax minus altus, dorso leviter tantum arcuato, parte cephalica parum declivi, ita ut oculi seriei 3tiæ vix diāmetro sua altius quam oculi laterales seriei 1mæ sint siti. Oculi seriei 1mæ sub-rectæ inter se proximi, sed non contingentes: medii eorum a margine clypei fortiter emarginati spatio remoti quod dimidiam diametrum oculi æquat. Corpus longius, sub-depressum. 7. *Menemerus.*

b. Cephalothorax altus, immo altissimus, parte cephalica adeo declivi ut oculi seriei 3tiæ multo altius quam oculi laterales seriei 1mæ siti sint.

α. Metatarsi pedum posteriorum circa apicem tantum aculeis armati. Quadrangulus oculorum postice paullo latior quam antice. Oculorum series 1ma paullo recurva: medii eorum a margine clypei evidentius emarginati spatio remoti, quod dimidiam diametrum oculi fere æquat. Corpus longius villosum. 8. *Dendryphantes.*

β. Metatarsi pedum posteriorum non tantum ad apicem aculeati.

I. Oculi medii seriei 1^{mae}, quum desuper inspiciatur cephalothorax, ante frontem eminentes.

1. Mandibulæ facie circa duplo longiores (an etiam in ♀?). Oculi seriei 1^{mae} sub-recurvæ disjuncti; medii eorum a margine clypei, profunde emarginati et sparse tantum pilosi, spatio remoti, quod dimidiam diametrum oculi vix æquat. Pedes longiores. . . . . 10. *Philæus*.

2. Mandibulæ facie non vel paullo tantum altiores. Oculi seriei 1^{mae} rectæ vel sub-recurvæ a margine clypei, pilis densis plerumque tecti, spatio remoti quod dimidiam oculi diametrum plerumque superat. . . . . 11. *Attus*.

II. Frons adeo prominens, ut oculi medii seriei 1^{mae}, quum desuper inspiciatur cephalothorax, a margine frontis occultentur. Series oculorum 1ma recurva; medii eorum a margine clypei dense pilosi spatio remoti, quod dimidiam diametrum oculi superat. Pedes posteriores anterioribus longiores.

1. Tibia pedum 4^{ti} paris evidenter brevior quam metatarsus cum tarso. . . . . 12. *Ælurops*.

2. Tibia pedum 4^{ti} paris æque saltem longa ac metatarsus cum tarso. Ungues præsertim horum pedum longissimi, dentibus longissimis pectinati. . . . . 13. *Yllenus*.

SIMON also takes up *Plexippus* among the European Attoidæ, and gives as the chief features that distinguish it from nearly related genera the following characteristics of ♂: "*patte machoire* (the palpus) *grêle*, *très longue*, *à tarse moins large que la jambe*" [1]). He assigns to it only one European species, *P. Adansonii* SAV. et AUD. I do not know to which genus this to me unknown spider ought properly to be aggregated: SIMON indeed calls his *Plexippus*: "*Plexippus* C. KOCH ex parte"; but he also says of it: "Tel que nous le concevons ce genre n'a aucun rapport avec celui de M. KOCH" [2]), and I therefore do not venture to take up *Plexippus* KOCH among the European genera.

Gen. 1. SALTICUS (LATR.). 1804.

Deriv.: *salticus*, dancing, leaping.

Syn.: 1804. **Salticus** LATR., Nouv. Dict. d'Hist. Nat., XXIV, p. 135 (*ad part.*).
1805. **Attus** WALCK., Tabl. d. Aran., p. 22 ("2e Fam. Les Voltigeuses, *Volatiliæ*" *ad partem*).

1) Monogr. d. espèces Europ. de la fam. d. Attides, p. 6 (16).
2) Ibid., p. 178 (644).

1833. **Salticus** SUND., Sv. Spindl. Beskr., *in* Vet.-Akad. Handl. f. 1832, p. 199.
†1837. **Pyrophorus** C. KOCH, Uebers. d. Arachn.-Syst., 1, p. 29.
1861. **Salticus** WESTR., Aran. Suec., p. 543.
1861. „ BLACKW., Spid. of Gr. Brit., I, p. 47 (*ad partem*).
1864. „ [Salticа]: sub-gen. **Pyrophorus** [**Pyrophora**] SIM., H. N. d. Araignées, p. 336.
1868. **Pyrophorus** SIM., Monogr. d. espèces europ. de la fam. d. Attides, p. 6 (16).
1869. **Pyroderes** ID., ibid., p. 248 (714).

Type: *Salticus formicarius* (DE GEER).

When SUNDEVALL in 1833 (loc. cit.) divided *Salticus* LATR. or *Attus* WALCK. into two genera, *Salticus* and *Attus*, it was for a species of the genus afterwards by C. KOCH called *Pyrophorus*, that he preserved the former, older name, and not for a *Salticus* C. KOCH, which genus was to him unknown. This appears immediately from the description of SUNDEVALL'S *Salticus formicarius*, the ♂ of which has the mandibles "*fere porrectæ, supra planæ*" etc. The very generic diagnosis of *Salticus* SUND. (*"Pars cephalica abrupte altior quam thoracica; ... oculi ... aream quadratam ... delineantes"*) is suitable only to *Pyrophorus*, and not to *Salticus* KOCH, and this last genus cannot therefore be considered as corresponding to *Salticus* SUND. even *ad partem*. As the name *Salticus* came by a mistake only — KOCH erroneously supposed his *Salticus formicarius* to be identical with the species, to which SUNDEVALL had assigned that name — to be applied by KOCH, and after him by OHLERT and others, to an entirely different genus from that so denominated by SUNDEVALL, whereas the real *Salticus* (LATR.) SUND. was by KOCH rechristened *Pyrophorus*, we must of course restore to that genus its original name. *Pyrophorus* is moreover, as SIMON has already remarked, the universally received name given by ILLIGER in 1809 to the so called "American fire-flies", belonging to the Elaterida (Coleopt.). The genus *Salticus* KOCH we call *Leptorchestes*.

The tarsal claws of *Salticus formicarius* are of the usual form, long and slender; on the 4th pair the inner claw has about 8 and the outer about 5 very short, thick, blunt teeth. The hairs in the claw-tuft are dilated spade-wise at the apex.

Gen. 2. LEPTORCHESTES N.

Deriv.: λεπτός, slender; ὀρχηστής, dancer.

Syn.: 1832. **Synemosyna** HENTZ, On North Amer. Spid., p. 108 (*ad partem*).
1836. **Attus** LUC., Attus venator, *in* GUÉR., Mag. de Zool., 6e Année, Cl. VIII, Pl. 15.

1837. **Salticus** C. KOCH, Uebers. d. Arachn.-Syst., 1, p. 29.
1864. „ [**Saltica**]: sub-gen. *id.* SIM., H. N. d. Araignées, p. 335 (*ad max. part.*).
1868. „ SIM., Monogr. d. espèces europ. de la fam. d. Attides, p. 6 (16), 241 (707).

Type: *Leptorchestes formicæformis* LUC.

Concerning *Salticus* (LATR.) SUND. see preceding genus. As type for *Leptorchestes* (*Salticus* C. KOCH) I have taken LUCAS' *Attus formicæformis*, which is identical with *Salticus formicarius* C. KOCH. The right *Salticus formicarius* (DE GEER) and SUND., is the same as *Pyrophorus semirufus* C. KOCH.

The genus *Synemosyna* HENTZ answers properly speaking to *Janus* C. KOCH, but under the former name species have also been included, which belong to *Leptorchestes*, and perhaps even to *Salticus* SUND. *Synemosyna formica* HENTZ [1]), which appears to be typical for the genus, is a *Janus*, and it is therefore this latter name, which must give place to the older appellation *Synemosyna:* the name *Janus* is moreover already appropriated (see p. 36).

In the typical species the claws are small, of quite an ordinary form, sinuated, with about 7 teeth gradually increasing in length on the inner and about 3 on the outer claw. The hairs of the claw-tuft are much dilated at the extremity.

Gen. 3. EPIBLEMUM (HENTZ). 1832.

Deriv.: ἐπίβλημα (ἐπί, on, βάλλω, throw), that which is thrown on or over (in allusion to the animal's swift motions, or the projecting mandibles of ♂).

Syn.: 1832. **Epiblemum** HENTZ, On North Amer. Spid., p. 108 (*ad partem*).
1837. **Calliethera** C. KOCH, Uebers. d. Arachn.-Syst., 1, p. 30 (*ad partem*).
1850. „ ID., ibid., 5, p. 45 (*ad max. part.*).
1861. **Attus** WESTR., Aran. Suec., p. 543 (*ad partem*).
1861. **Salticus** BLACKW., Spid. of Gr. Brit., I, p. 47 (*ad partem*).
1864. **Cyrtonota**: sub-gen. **Calliethera** SIM., H. N. d. Araignées, p. 324, 327 (*ad partem*).
1868. **Calliethera** [**Calliethorus**] ID., Monogr. d. espèces europ. de la fam. d. Attides, p. 6 (16), 180 (646) (*ad max. part.*).

Type: *Epiblemum faustum* HENTZ.

In the above-cited passage, where HENTZ proposes the genus *Epiblemum*, he distinguishes it from *Attus* WALCK. by the mandibles being

1) Aran. of the United States, *in* Boston Journ. of Nat. Hist., V, p. 368, Pl. XXII, fig. 18.

"nearly horizontal, slender, as long as the cephalothorax, tooth as long." Of the two species adduced, *E. faustum* and *E. palmarum*, the first-named is made type of the new genus. But that *E. faustum* is so extremely like our European *Calliethera histrionica* C. Koch and *C. scenica*, that I imagine it to be identical with one or other of them [1]), and they must accordingly resign their generic name *Calliethera* for the older name *Epiblemum*. — *E. palmarum* is perhaps a *Plexippus* C. Koch, and certainly does not belong to the same genus as *E. faustum*.

Simon refers to *Calliethera* also for inst. the species *C. infima* [*-us*] Sim., which its whole appearance, the spines on its legs, etc. indicate in my opinion to belong to *Heliophanus* C. Koch.

The male *Epiblemum*, like the male *Salticus*, is distinguished by its almost horizontal, projecting mandibles. — The eyes of the 3rd row are farther from each other than from the margin of the cephalothorax. The claws are very long and slender, and the teeth on the inner claw very numerous (about 15 in *E. histrionica* on the 4th pair), on the outer claw on the contrary few (in the above named species about 3); the number is however very variable. The hairs of the claw-tufts are gradually somewhat dilated.

Gen. 4. HELIOPHANUS C. Koch. 1833.

Deriv.: ἥλιος, sun; φαίνω, show, shine.

Syn.: 1833. **Heliophanus** C. Koch, *in* Herr.-Schæff., Deutschl. Ins., 119, 1, 2.
1837. „ id., Uebers. d. Arachn.-Syst., 1, p. 29.
1861. **Attus** Westr., Aran. Suec., p. 543 (*ad partem*).
1861. **Salticus** Blackw., Spid. of Gr. Brit., I, p. 47 (*ad partem*).
1864. **Heliophanus** [Heliophana] Sim., H. N. d. Araignées, p. 332 (*saltem ad part.*).
1868. „ Sim., Monogr. d. espèces europ. de la fam. d. Attides, p. 6 (16), 201 (667).
1868. **Calliethera** [Calliethera] id., ibid., p. 6 (16), 180 (646) (*ad partem*).

Type: *Heliophanus cupreus* (Walck.).

The males of this genus, which is easily recognized by its general appearance, are usually distinguished, as Koch has already remarked, by

1) Conf. the description and figure of *E. faustum* in Aran. of the United States (Boston Journ. of Nat. Hist., V, p. 367, Pl. XXII, fig. 17). — Blackwall, who does not consider *E. histrionicum* as specifically different from *E. scenicum*, includes "*Salticus scenicus*" in a catalogue of spiders from *Canada*. (Notice of Spid. capt. by Potter in Canada, p. 34).

a strong tooth on the underside of the pars femoralis of the palpus. In some species the cephalothorax is not double as long as it is broad, but it is easy to distinguish them from other, nearly related genera by the closeness of the anterior central eyes and their inconsiderable distance from the deeply emarginated edge of the almost naked clypeus, together with the peculiar colour (black, abdomen more or less metallic, legs generally yellow or spotted with yellow). The eyes of the 3rd row are usually, but not always, more widely separated from each other than from the margin of the cephalothorax. The claws are of the usual form, nearly similar to those of *Epiblemum*, but the teeth are less numerous on the inner claw. In *H. cupreus* I have counted 6 fine teeth upon the inner and 2 coarse ones on the outer claw (4th pair). OHLERT states the numbers to be 10 and 1. On the 1st and 3rd pair, according to him, the external claw is without teeth. The hairs of the claw-tuft dilate gradually outwards.

Gen. 5. BALLUS (C. KOCH). 1850.

Deriv.: *βάλλω*, throw.

Syn.: 1834. **Salticus** REUSS, Zool. Misc., Arachn., (*ad part.:*) p. 273 (279).
1837. **Euophrys** C. KOCH, Uebers. d. Arachn.-Syst., 1, p. 33 (*ad partem*).
1846. **Marpessa** [**Marpissa**] ID., Die Arachn., XIII, (*ad part.:*) p. 53.
1850. **Attus**: sub-gen. **Ballus** ID., Uebers. d. Arachn.-Syst., 5, p. 68.
1861. „ WESTR., Aran. Suec., p. 543 (*ad partem*).
1861. **Salticus** BLACKW., Spid. of Gr. Brit., I, p. 47 (*ad partem*).
1864. **Attus** [**Atta**]: sub-gen. *id.:* "groupe" **Ballus** [**Balla**], et sub-gen. **Dendryphantes** SIM., H. N. d. Araignées, p. 310 (*ad partem*).
1868. „ ID., Monogr. d. espèces europ. de la fam. d. Attides, p. 6 (16), 14 (24) (*ad partem*).

Type: *Ballus heterophthalmus* (REUSS).

To this genus we refer not only the species proposed as its type, but also *Attus depressus* WALCK. (*Salticus brevipes* HAHN), which C. KOCH refers to *Marpessa*, and *Salticus obscurus* BLACKW., which is probably nothing else than the male of *Ballus depressus*.

The claws are, at least in *B. depressus*, small, slightly sinuated, of ordinary form: on the 4th pair I have counted about 15 teeth on the inner and about 5 on the outer claw; the outer claw of the 1st pair is destitute of teeth, the inner has about 10 teeth. The hairs of the claw-tuft are dilated at the extremity.

Gen. 6. MARPESSA (C. KOCH). 1846.

Deriv.: undoubtedly **Μάρπησσα**, a mythol. prop. name; the word ought therefore to be written ***Marpessa***, not ***Marpissa***.

Syn.: 1837. **Dendryphantes** C. KOCH, Uebers. d. Arachn.-Syst., 1, p. 31 (*ad partem*).
†1846. **Icelus** ID., Die Arachn., XIII, (*saltem ad part.:*) p. 174.
1846. **Marpessa** [**Marpissa**] ID., ibid., p. 56 *et sequ.* } (*ad max. part.*).
1850. „ „ ID., Uebers. d. Arachn.-Syst., 5, p. 47. } (*ad max. part.*).
1861. **Attus** WESTR., Aran. Suec., p. 543 (*ad partem*).
1861. **Salticus** BLACKW., Spid. of Gr. Brit., I, p. 47 (*ad partem*).
1864. **Attus** [**Atta**]: sub-gen. **Dendryphantes** SIM., H. N. d. Araignées, p. 310 (*ad partem*).
1864. **Cyrtonota**: sub-gen. **Phidippus** [**Phidippia**]: "groupe" **Plexippus** [**Plexippa**] ID., ibid., p. 324 (*ad partem*).
1868. **Marpessa** [**Marpissus**] ID., Monogr. d. espèces europ. de la fam. Attides, p. 6 (16), 7 (17).
1868. **Attus** ID., ibid., p. 6 (16), 196 (692) (*ad partem*).
1868. **Menemerus** ID., ibid., p. 6 (16), 196 (692) (*ad partem*).

Type: *Marpessa muscosa* (CLERCK).

The lamina of the male's clava palpalis is not in all the spiders that we assign to this genus "*élargi en palette*," as in the typical species, which feature SIMON however takes as characteristic of the genus. As we have above defined it, it includes among the Attoidæ with which I am acquainted, not only *M. muscosa*, *M. radiata* (GRUBE) and *M. hamata* C. KOCH [1]), but also *Salticus pulchellus* HAHN, *Menemerus falsificus* SIM. and *Attus Lucasii* SIM., which last-mentioned two species, together with several other European Attoidæ, SIMON himself had the kindness to send me. In its general appearance this genus occupies a place between *Epiblemum* and *Menemerus*. *M. pulchella* (HAHN) seems to form a transition to the former genus. Concerning *Icelus* C. KOCH see p. 206. — The claws are somewhat shorter and stronger than in most other *Attoidæ*, at least in *M. muscosa*, in which species I have counted about 15 fine teeth on the inner, and from 3 to 6 on the outer claw.

---

1) *Icelus notabilis* C. KOCH is the male to his ***Marpessa hamata;*** both are stated to be from Naples. I have myself captured them in Rome. SIMON has obligingly sent me both ♂ and ♀ under the name of ***Attus striatus*** WALCK. (***Attus striatus*** (CLERCK) is quite another spider). ***Marpessa hamata*** SIM. is not identical with C. KOCH's spider of that name.

Gen. 7. MENEMERUS (SIMON). 1868.

Deriv.: *μήνη*, moon; *μήρος*, thigh.

Syn.: 1829. **Salticus** HAHN, Monogr. d. Spinn., 5 (*ad part.:*) Tab. 3, fig. *B.*
1868. **Menemerus** SIM., Monogr. d. espèces europ. de la fam. d. Attides, p. 6 (16), 196 (692) (*ad partem*).
1868. **Attus** ID., ibid., p. 6 (16), 14 (24) (*ad partem*).

Type: *Menemerus semi-limbatus* (HAHN).

Of the typical species (= *M. vigoratus* (C. KOCH) SIM.) I have taken several specimens at Naples (whence also HAHN'S specimen came), at Rome and at Nice. As the characteristic feature of the genus *Menemerus*, which distinguishes it from nearly related genera, SIMON states that the pars femoralis of the male's palpus is "*inerme et renflée en massue*". But that character does not apply to all the species, which, according to my definition of the genus, it comprehends, and of the species again, which SIMON reckons to *Menemerus*, I refer e. g. *M. falsificus* SIM. to *Marpessa*. In general appearance *Menemerus* closely resembles *Marpessa:* in cases of doubt however *Menemerus* may be recognized by the distance between the two eyes of the 3rd series being somewhat *less* than that between them and the margin of the cephalothorax, which is *not* the case in *Marpessa*. The claws in this genus are quite of the common form, but little sinuated; in the typical species I have found the outer claw without teeth both on the 1st pair, where the inner claw has about 10, and on the 4th pair, where it has about 15 fine teeth.

Gen. 8. DENDRYPHANTES (C. KOCH). 1837.

Deriv.: *δένδρον*, tree; *ὑφάντης*, weaver.

Syn.: 1837. **Dendryphantes** C. KOCH, Uebers. d. Arachn.-Syst., 1, p. 31 (*ad partem*).
1850. „ ID., ibid., 5, p. 60 (*saltem ad partem*).
1861. **Attus** WESTR., Aran. Suec., p. 543 (*ad partem*).
1864. „ **[Atta]**: sub-gen. **Dendryphantes** SIM., H. N. d. Araignées, p. 310 (*ad partem*).
1867. **Dendryphantes** OHL., Preuss. Spinn., p. 149, 156.
1868. **Attus** SIM., Monogr. d. espèces europ. de la fam. d. Attides, p. 6 (16), 14 (24) (*ad partem*).

Type: *Dendryphantes hastatus* (CLERCK).

In his Monogr. d. espèces Europ. de la fam. d. Attides, p. 6 (16), SIMON mentions as the the characteristic of his *Dendryphantes:* "*digital* (bulbus genitalis) *à découvert sous la jambe"* (pars tibialis), as distinguishing it from *Attus*, *Marpessa*, *Yllenus* and others, which have the *"digital enveloppé en dessus par le tarse"* (lamina bulbi or pars tarsalis). The species of SIMON'S *Dendryphantes* known to me (among which I have however never met with the fullgrown ♂), namely *D. gesticulator* SIM. and *D. dorsatus* C. KOCH [1]), belong to *Attus* according to our definition of that genus. OHLERT had already previously to SIMON defined the genus *Dendryphantes* so that *D. hastatus* must be considered as its type, and this determination, which we adopt, thus has the right of priority in preference to that which SIMON has made for the genus.

The claws are of the ordinary form, little sinuated, with numerous teeth on the inner claw. In *D. hastatus* on the 1st pair of legs I have counted above 20 close-set, very fine comb-teeth, but only 4 coarse and distant teeth on the outer claw. In another specimen the inner claw of the 4th pair had about 18, the outer about 7 teeth. The hairs of the claw-tufts are slightly dilated at the extreme apex.

Gen. 9. EUOPHRYS (C. KOCH). 1835.

Deriv.: *εὖ*, well; *ὀφρύς*, eye-brow.

Syn.: 1834. Euophrys C. KOCH, *in* HERR.-SCHÆFF., Deutschl. Ins., 123, (*ad part.:*) 7, 8.
1837. „ ID., Uebers. d. Arachn.-Syst., 1, p. 33 (*ad partem*).

---

1) In specimens, which I look upon as *young* males of this species, not only is the short tibial joint of the palpus, but also its long tarsal joint enlarged and broader than the preceding joints; the inferior and exterior part of the tibial joint is swelled, but shows no separate bulbus — all just as in the figures of *D. bilineatus* (WALCK.), which SIMON has given loc. cit., Pl. II (VI), fig. 13*a*, and which therefore appear to me to represent the palpus of a *not yet fully developed* male. In ♂ ad., according to SIMON, the tarsal joint is alike in both sexes, small and cylindrical, only a little longer in the male, whose *tibial* joint is on the underside incrassated and hollowed out, and *contains the bulbus genitalis* (?). Such a relation would indeed, as SIMON rightly observes, distinguish these spiders from the other species of the family; it would even separate their from *all* other spiders, for, as far as we know, the bulbus genitalis in all other cases belongs to the *tarsal* and not to the tibial joint. — Also in the younger males of some other Attoidæ, e. g. *Menemerus semilimbatus* or *vigoratus*, the palpus has a form like that in the above described species of *Dendryphantes* SIM.: the tibial joint is very short and only indistinctly separated from the long palpal joint: both these joints are broader than the preceding, and the tibial joint incrassated on the underside. I suspect that the bulbus genitalis is here formed within the *two* last joints of the palpus, though, when freed at the last change of the integument, it adheres to the tarsal joint.

1848. **Attus** ID., Die Arachn., XIV, (*ad part.:*) p. 44—49.
1850. „ ID., Uebers. d. Arachn.-Syst., 5, p. 68 (*excl.* sub-gen. *Ballo*).
1861. „ WESTR., Aran. Suec., p. 543 (*ad partem*).
1861. **Salticus** BLACKW., Spid. of Gr. Brit., I, p. 47 (*ad partem*).
1864. **Attus** [**Atta**]: "groupe" *id.* SIM., H. N. d. Araignées, p. 310.
1868. „ ID., Monogr. d. espèces europ. de la fam. d. Attides, p. 6 (16), 14 (24) (*ad partem*).

Type: *Euophrys frontalis* (WALCK.).

When C. KOCH in 1833, in HERR.-SCHÆFF., Deutschl. Ins., N:o 119, for the first time mentioned an *Attus*, it was *A. terebratus* (CLERCK) that he described under that name. In the same work he, in 1834, described, under the new generic name *Euophrys*, two species, which he called *E. festiva* and *E. frontalis*. Somewhat later (1837), in Uebers. d. Arachn.-Syst., 1, he endeavoured to give the characteristics which distinguish *Attus* and *Euophrys:* he there registered, as belonging to *Attus*, *A. arcuatus* (CLERCK), as also *A. terebratus* (ID.), the position of the eyes and the male's palpi of which he figured, and which species therefore ought to be considered as the type of the genus. To *Euophrys* he refers several species, which are very nearly related to *A. terebratus* and *arcuatus*, but moreover also e. g. *E. petrensis*, which is more nearly related to *E. frontalis*. This latter species is now not mentioned, neither is *E. festiva*. Several years later, in Die Arachniden XIII and XIV (1846, 1848), we find that KOCH has completely altered his view of the genera *Attus* and *Euophrys: A. terebratus* and *A. arcuatus* are now referred to *Euophrys*, whereas *E. frontalis* and *E. petrensis* are aggregated to *Attus*. So also lastly in Uebers. d. Arachn.-Syst., 5 (1850). From what has now been said it is evident, that KOCH at different times has defined the genera *Attus* and *Euophrys* in totally different and irreconcileable ways. As he in 1837, when WALCKENAER'S great genus *Attus* was broken up by him, defined the genus, for which he preserved WALCKENAER'S name, so, that *A. terebratus* was to be the type for *Attus* (WALCK.) KOCH, it is clear, that the name *Euophrys*, if not to be absolutely cashiered, must be applied to some one or more of the species described under that name, which can *not* be referred to the same genus as *A. terebratus*, and preferentially to that species among them, which was *first* described under the name *Euophrys:* accordingly to *E. frontalis*. (The contemporaneously described *E. festiva* = *E. striata* KOCH [non CLERCK] is an *Attus* (WALCK.) NOB.). — *Euophrys* (KOCH) NOB. must accordingly be = sub-gen. *Attus* KOCH 1850.

Like *Attus*, the species of *Euophrys* have a high cephalothorax, but as the back of the cephalothorax is as broad as its base, and the hindermost eyes, are situated near the side-edges of the back, the distance between them is greater or at least not less than that between the eye and the border of the cephalothorax (which does not gradually diminish in breadth towards the front). The eye-area occupies a larger proportion of the cephalothorax, than in *Attus:* in *E. reticulata* (BLACKW.) = *E. frontalis* ♀ (WESTR.) the hindermost eyes are situated actually almost in the middle of the cephalothorax. *E. petrensis* C. KOCH is the only species of this genus known to me, in which the design of the abdomen formed by the distribution of its colours depends on a tolerably thick covering of hair; ordinarily the hair is thin and the markings occasioned by the pigment situated in the skin itself. — The claws are long and very slender, with few or no teeth; in *E. frontalis* I have observed on the inner claw of the 1st pair two very small teeth. The hairs of the claw-tuft are sensibly dilated at the extremity.

Gen. 10. PHILÆUS N.

Deriv.: *Φιλαῖος*, proper name.

Syn.: 1837. **Calliethera** C. KOCH, Uebers. d. Arachn.-Syst., 1, p. 30 (*ad partem*).
†1846. **Philia** ID., Die Arachn., XIII, p. 54, 56.
1850. „ ID., Uebers. d. Arachn.-Syst., 5, p. 45.
1861. **Attus** WESTR., Aran. Suec., p. 543 (*ad partem*).
1864. **Cyrtonota**: sub-gen. **Philia** SIM., H. N. d. Araignées, p. 324, 327 (*saltem ad partem*).
1868. **Attus** ID., Monogr. d. espèces europ. de la fam. d. Attides, p. 6 (16), 14 (24) (*ad partem*).

Type: *Philæus sanguinolentus* (LINN.).

I am not sure that the distinctive features of this genus set forth by me are quite trustworthy, for I have met with the *male* only of one of its species, *Ph. sanguinolentus*. The genus however seems to be perfectly well distinguished from *Attus* and other nearly related forms even by its entire general appearance. — As the name *Philia*, according to AGASSIZ' Nomencl. Zool., was already in 1842 appropriated by SCHIÖDTE to a genus of Hemiptera, I could not preserve it, but have replaced it with the somewhat similar name *Philæus*.

In *Ph. sanguinolentus* the inner claw of the 1st pair of legs has about 20 close-set teeth gradually and slightly increasing in length, and the outer claw about 6 coarse, sparse teeth. The hairs of the claw-tuft are long, slightly dilated at the extremity. On the 4th pair of legs the number of teeth is respectively about 13 and 5.

Gen. 11. ATTUS (WALCK.). 1805.

Deriv.: ἄττω = ἀΐσσω, move with quick, sudden motion.

Syn.: 1805. **Attus** WALCK., Tabl. d. Aran., p. 22 (*ad partem*).
1833. „ C. KOCH, *in* HERR.-SCHÆFF., Deutschl. Ins., 119, 3, 4.
1837. „ ID., Uebers. d. Arachn.-Syst., 1, p. 32.
1837. **Euophrys** ID., ibid., p. 33 (*ad partem*).
1850. „ ID., ibid., 5, p. 60 (*ad max. part.*).
1861. **Attus** WESTR., Aran. Suec., p. 543 (*ad partem*).
1861. **Salticus** BLACKW., Spid. of Gr. Brit., 1, p. 47 (*ad partem*).
1864. **Attus [Atta]**: SIM., H. N. d. Araignées, p. 324 (*ad partem*).
1868. „ „ ID., Monogr. d. espèces europ. de la fam. d. Attides, p. 6 (16), 14 (24) (*ad partem*).
1868. **Dendryphantes** ID., ibid., p. 6 (16), 168 (634) (*saltem ad partem*).

Type: *Attus terebratus* (CLERCK).

When C. KOCH in 1837 (loc. cit.) divided the old genus *Attus* WALCK. or *Salticus* LATR. into a number of smaller genera, he preserved the Walckenaerian name for a generic group that includes *A. terebratus* (CLERCK) and *A. arcuatus* (ID.). Since several species, which KOCH in the same work referred to *Euophrys*, ought also to be reckoned to the same genus, he some years afterwards transferred that appellation to the genus *Attus*, and gave the name of *Attus* to a portion of the species, which he had formerly called *Euophrys*. Such alterations of names no one of course can have the right of making, and we have accordingly restored the generic name *Attus* to the spiders, which KOCH *first* under that name detached from WALCKENAER'S *Attus*. Of *Euophrys* we have already treated p. 216.

The genus *Attus*, as we have above defined it, includes the great majority of European Attoidæ. Perhaps one or more well defined genera might with advantage still be detached from it; I have not however, possibly for want of sufficient material for examination, been able to do so. As I define this genus, it corresponds to KOCH'S *Euophrys* 1850, with the exclusion of the sub-genera *Dia* and *Parthenia*, which I considered might very well be united into one separate genus: *Ælurops*.

The armature of the claws in the genus *Attus* is tolerably various. Generally speaking the teeth of the inner claw are close-set and far more numerous than those of the outer claw; but occasionally, e. g. on the 4th pair of legs in *A. crucifer*, the number is small and about equal on both claws. Sometimes the teeth gradually and uniformly increase in length towards the point of the claw, sometimes they are of almost equal length throughout; their length as compared with their breadth is also very different in different species. In many species the outer claw is toothless, or has but a couple of coarse teeth far apart, while the inner claw is finely and closely pectinated. The hairs of the claw-tufts are usually gradually dilated towards the extremity.

Gen. 12. ÆLUROPS n.

Deriv.: *αἴλουρος*, cat; *ὤψ*, face.

Syn.: 1850. **Euophrys**: sub-gen. †**Dia** et †**Parthenia** C. Koch, Uebers. d. Arachn.-Syst., 5, p. 60 (*saltem ad part.*).
1861. **Attus** Westr., Aran. Suec., p. 453 (*ad partem*).
1864. „ [**Atta**]: sub-gen. *id.:* "groupes" **Dia** et **Parthenia** Sim., H. N. d. Araignées, p. 310, 312, 313 (*saltem ad part.*).
1868. „ Sim., Monogr. d. espèces europ. de la fam. d. Attides, p. 6 (16, 14 (24) (*ad partem*).

Type: *Ælurops* v-*insignitus* (Clerck).

To this genus, besides the typical species, I refer e. g. also *Salticus fasciatus* Hahn, both remarkable for the projecting edge of the forehead, which conceals the central eyes of the first row, when the cephalothorax is looked at perpendicularly from above. As the names *Dia* and *Parthenia* were both already appropriated before Koch applied them to the two subgenera, that we here have united into one genus (vid. 36, 37), I have been obliged to form a new generic name for them. — The species of this and the genus immediately following appear to me to be the most highly developed European forms in the whole family. They leap with extraordinary vigour. Their claws are long and sinuated: in *Æl.* v-*insignitus* ♀ the claws of the 4th pair of legs have, much in front of their middle, about 3 or 4 large, sparse teeth, ♂ about 6. On the 1st pair, the claws of which are much shorter and more uniformly curved, the teeth are still fewer in number, at least in ♂. The claw-tuft is continued as a scopula beneath a part of the tarsus of the 1st pair; and the hairs of it are, nearer the extremity, gradually dilated in the form of tongues.

Gen. 13. YLLENUS (Sim.). 1868.

Deriv.: From some proper name.

Syn.: 1868. **Yllenus** Sim., Monogr. d. espèces europ. de la fam. d. Attides, p. 6 (16), 166 (632).

Type: *Yllenus arenarius* Sim. [1]).

1) For this species Simon cites "Menge, Schrift. d. Naturforsch. Gesellsch. in Danzig, 1866"; but I have not found it described either there or any where else

Of this genus, which, according to SIMON, is distinguished by the lamina of the palpal clava being "*relevé en crête*", I have seen only one individual, a male of the typical species, which SIMON had the goodness to send me. That spider in its entire appearance agrees most accurately with *Ælurops* v-*insignitus*, but it differs not only by the peculiar structure of the palpi, but also by the far greater length of the posterior legs, especially the tibiæ. The claws also are particularly strongly developed: they are very long, even longer than those of *Ælurops*, slender and sinuated, especially on the hindermost legs, where they have in front of the middle a row of about 12 very long, closely set comb-teeth. The claws of the 1st pair, whose tarsi, like those of the 2nd pair, are on the underside clothed with hairs dilated at the apex, are much shorter than those of the posterior pairs of legs, but still long, slightly and uniformly curved, with about as many teeth of the same form as on the following pairs, but here the row of teeth commences nearer to the base of the claw. The claws, especially on the hinder legs, are so large and visible, that both they and their pectination may be observed with a good single lens. The hairs of the claw-tuft are dilated near the extremity in the posterior legs; in the 1st pair the dilatation is more gradual.

Fossil spiders have in the preceding pages not be taken into account, simply because I am not by actual inspection acquainted with any, and I therefore was not in a condition to form from observations of my own an opinion of the relations between them and now existing forms. Some short notices on this subject, with special reference to those extinct *genera*, which (as far as I am aware) up to the present time have been published, may however be of interest to some few arachnologists, and I offer them the more readily, because I have not found, that in any work on the classification of spiders proper attention has been paid to the fossil forms.

These animals, as the usually soft and perishable character of their integuments would lead us to expect, have left but few traces of their exist-

previous to SIMON'S description loc. cit. In KOCH and BERENDT, Die im Bernstein befindl. Crust., Myriapod., Arachn. etc., p. 93, MENGE has, it is true, mentioned a Prussian spider under the name of *Phidippus arenarius*, which *perhaps* is the same as *Yllenus arenarius* SIM., but it is not characterized, and accordingly I could not refer to MENGE as authority for the name.

ence in the fossiliferous deposits, and it is only in *Amber* that we meet with them numerously represented. The oldest known spiders belong to the *Coal formation*, in the strata of which a few specimens have been found in Bohemia [1]) and Silesia [2]), and probably also in England [3]). Only one species belonging to that period is in sufficiently good preservation to be tolerably well characterized, viz. the *Protolycosa anthracophila* described by Römer, which was discovered in a piece of argillaceous slate at Kattowitz in Upper Silesia. It forms the type of the genus

Protolycosa Röm. 1866 [4]). This spider, which is about 5 lines long, is by Römer placed in the vicinity of *Lycosa;* but this appears to me not to be right. The eyes and spinners, if indeed these organs ever existed, have unfortunately perished; nor is it possible to form any clear idea of the appearance of the mandibles, and it is therefore impossible to determine with absolute certainty the systematic position of the animal; nevertheless its general appearance and especially its *extremely coarse and strong legs and palpi* seem to me unequivocally to mark this genus as belonging to the *Territelariæ*, and among these it is that wonderful East Indian genus *Liphistius* Schiödte, that *Protolycosa* most nearly resembles. Not only do these two genera agree in the unusual relative length of the legs — in *Liphistius* the proportion of the different pairs is 4, 2, 3, 1, in *Protolycosa* 4, 2, 3, 1, and thus in both *the 1st pair is the shortest of all;* — but in *Protolycosa* also the dorsal integument of the abdomen is of *a horny substance*, and, according to Römer's figures, *divided into transversal segments*, each furnished with a cross-row of tubercles, just as is the case with *Liphistius* Schiödte [5]). I conceive then that *Protolycosa* ought to be assigned

1) "*Palaranea borassifolia* Frič" (!!), Vid. Feistmantel, K., Die Steinkohlenbecken in der Umgebung von Radnic, p. 66, *in* Archiv f. d. naturwissensch. Landesdurchforschung von Böhmen, Bd I (Prag 1869); Conf. also *Reuss, A. E., Kurze Uebersicht der Geognostischen Verhältnisse Böhmens, p. 59 (Prag 1854), and Römer, F., Protolycosa anthracophila, eine fossile Spinne aus dem Steinkohlengebirge Oberschlesiens, *in* Leonhard and Bronn (Geinitz), Neues Jahrbuch für Mineralogie, Geologie und Palæontologie, Jahrg. 1866, Hft 2, p. 143. (Stuttgard 1866).

2) Römer, loc. cit., p. 136—143, Taf. III, fig. 1—3.

3) Conf. *Lhwyd (Luidius), E., Lithophylacii Britannici Ichnographia etc., Tab. IV (London 1690); *Parkinson, J., The Organic Remains of a former world etc., III, Pl. 17, fig. 3—6 (London 1811); as also a citation from Lhwyd's Epist. III, in Buckland's Geolog. and Miner., I, p. 406 (of Ed. 2).

4) Deriv.: πρῶτος, first, and *Lycosa*.

5) Conf. Schiödte, Om en afvigende Slægt af Spindlernes Orden, p. 6—7.

to the family *Liphistioidæ* NOB. (vid. p. 43), unless it be preferred to create a new family especially for it, a proceeding, which perhaps *the unusually short femoral joints of the palpi* (see RÖMER'S figures), as also *two backward-directed spines in the midst of each side of the abdomen* might justify.

Phalangites MÜNST. 1839 = *Palpipes* ROTH 1851. In the lithographic limestone of Solenhofen in Bavaria, belonging to the *Jurassic formation*, MÜNSTER detected the impression of a previously unknown animal, which, on account of its resemblance to a *Phalangium*, he called *Phalangites priscus* [1]). ROTH [2]), who had at his disposal several specimens, which he divides into two species, thought he could clearly perceive the contour of an abdomen separated from the cephalothorax, and observed two long, jointed and cross-ringed organs, attached to the abdomen and united at the base, which he considered to be *spinners*, and he accordingly aggregated these animals to the Order of Spiders. He named the genus *Palpipes*, and considered that it ought to be referred to the *Mygalides (Territelariæ);* he characterizes it as follows: "Cephalothorax ab abdomine discretus. Palpi maximi, in pedes mutati. Pedum paria longitudine diversa. Tarsi monomeri, ungui valido simplici terminati. Papillæ textoriæ duæ magnæ exsertæ, vel aliud quoddam organum bipartitum, cornutum, articulatum, in medio ventre situm, cornubus antice vergentibus." — The figure given by ROTH of *P. priscus* really gives the impression of a spider with uncommonly long and thin legs and very long, leg-like palpi. Examples of still existing spiders with but one tarsal claw are not wanting (*Sparassus abnormis* BLACKW., *Attus (Diolenius) phrynoides* WALCK.: See above pp. 170 and 203); very long *cross-ringed spinners* occur also in another fossil spider, *Gerdia myura* MENGE, of which we shall speak farther on. Their abnormal position and direction in *Phalangites* may be a consequence of the animal's having been crushed and the relative position of the parts thus changed. In the mean time it is maintained by V. MEYER [3]), that what ROTH looked upon as the coutour of

---

1) MÜNSTER, G., Graf zu, Phalangites priscus, *in* EJUSD. Beiträge zur Petrefakten-kunde, Hft 1, p. 84, Taf. VIII, fig. 3, 4. (Bayreuth 1839).

2) ROTH, J., Ueber fossile Spinnen des lithografischen Schiefers, *in* Gelehrte Anzeigen, Herausgegeben von Mitgliedern d. K. Bayer. Akademie der Wissenschaften, Bd XXXII, p. 164—167. (München 1851).

3) MEYER, HERM. v., Zu Palpipes priscus aus dem lithographischen Schiefer in Bayern, *in* EJUSD. Palæontographica, Beiträge zur Naturgeschichte der Vorwelt, Bd X, Lief. 6, p. 299—304, Taf. L, fig. 1—4 (Cassel 1863). — See also a letter from V. MEYER to BRONN, *in* LEONHARD and BRONN'S Neues Jahrbuch f. Min., Geol. etc., 1861, p. 561. BRONN there surmises that *Phalangites* should be compared with the *Pantopoda (Pycnogonoidea)*.

an abdomen, is the impression of a 5th pair of short and slender legs, and that accordingly the animal does not belong to the *Arachnoidea*, neither to the *Opiliones*, nor to the *Araneæ*, but to the *Crustacea*. This view appears to me to have but little probability, as giving no satisfactory explanation of the organs observed in many specimens, and by ROTH supposed to be *spinners*. To consider them with V. MEYER as *antennæ*, would seem dangerous, as they are always found on or near the abdomen (Conf. ROTH, loc. cit.). That the contour of the abdomen gives the impression of a pair of jointed and converging extremities, might be explained by considering the abdomen itself to have been segmentated. At all events the animals in question are so peculiar, that they not only form a separate family, ***Phalangitoidæ***, but even a group of a higher order, which may be called ***FILIGRADÆ***; if, as I suppose, this group belong to the order of Spiders, it ought, as a separate *sub-order*, characterized especially by *single-jointed tarsi armed with but one coarse claw*, to take a place below both *Scytodoidæ* and *Filistatoidæ*, uniting them with the *Opiliones*.

Numerous representatives of the order of Spiders from the *tertiary* formations are already known. They appear all to belong to the *miocene*, or (the amber spiders) perhaps to a still older period. From the fresh-water formations near Aix in Provence MARCEL DE SERRES [1]) has produced a "*Tegenaria*", as also a "*Phalangium*" said to resemble *Phalangium phaleratum* PANZER, i. e. *Asagena phalerata*. I imagine it to be this last-named species, that is figured in BUCKLAND'S Geology and Mineralogy [2]), and for which the same place of discovery is alleged; it closely resembles a *Theridium*. I propose to call it *Th. Bucklandii*. In the sulphur-impregnated tertiary strata of Radoboj in Croatia several spiders are also said to be found [3]). VON HEYDEN describes the remains of two spiders, discovered in the Browncoal strata of the Siebengebirge on the Rhine, which he calls *Gea Krantzii* [4]) and *Argyroneta antiqua* [5]). The first seems to me to be a species of *Epeira*; the second is certainly no *Argyroneta*, but represents, if the figure can be relied upon, a peculiar genus, which may be called

1) Notes géologiques sur la Provence, *in* Actes de la Société Linnéenne de Bordeaux, T. XIII, p. 34. (Bordeaux 1844).

2) BUCKLAND, W., Geology and Mineralogy considered with reference to Natural Theology (2nd Edit.) II, p. 79, Pl. 46", fig. 12. (London 1837).

3) QUENSTEDT, F. A., Handbuch d. Petrefaktenkunde (2nd Ed.), p. 268. (Tübingen 1867). I do not know whence QUENSTEDT has taken this statement.

4) HEYDEN, C. v., Fossile Insekten aus der Rheinischen Braunkohle (MEYER'S Palæontographica, VIII, Lief. I, p. 2, Taf. II, fig. 11. (1859).

5) Ibid., p. 1, Taf. 2, fig. 12.

Elvina N.[1]). This genus appears to be distinguished by *the palpi being evidently thicker than the legs.* Its nearer relationships it is not possible from v. HEYDEN'S description and figure to determine: probably it may belong to the *Tubitelariæ*, and *possibly* to the *Agalenoidæ (Argyronetinæ).*

In the also miocene fresh-water strata of Œningen (near the Lake of Constance in Switzerland), OSWALD HEER[2]) has met with no less than 28 species of spiders, which it is however difficult to affiliate to any certain genera, as the position of the eyes etc. cannot be distinguished. HEER thinks they may be referred to 10 genera, which, with one exception, are still existing. These spiders are not described, but eleven species have been figured and named: of these one is assigned to *Epeira*, 3 to *Theridium*, 1 to *Argyroneta* [*-necta*], 1 to *Clubiona*, 1 to *Micaria* [*Macaria*] and 3 to "*Thomisus*". But scarcely one of these species appears to be in so good a state of preservation that the identifications can be considered as fully certain. *Theridium maculipes* HEER (loc. cit., p. 356, fig. 219) is more like an *Asagena* than a *Theridium*. *Thomisus œningensis* HEER (fig. 215) would seem to be a *Xysticus*. *Clubiona Eseri* HEER, which is stated to be very like *Cl. lanata* KOCH et BER. (of which more hereafter) is assuredly no *Clubiona*, nor is *Argyroneta longipes* HEER any *Argyroneta*. These two species seem to form each its own separate genus. For one species HEER forms, as we have above stated, a new genus:

Schellenbergia HEER 1865[3]). Of his *S. rotundata* (fig. 211) HEER says, that it is distinguished by "the short palpi with a large, globular terminal joint, short and almost globular abdomen, pressed close to the breast, and provided with transversal impressions. The third pair of legs is the shortest, all the others being of nearly equal length. The thighs are furnished with a longitudinal rib." The animal (a ♂) belongs without doubt to the *Retitelariæ*, and appears to me to stand between *Episinus* and *Ero*.

In the fossil vegetable resin known under the name of *amber*, which is met with in various Brown-coal strata, and is copiously thrown by the waves on the southern coasts of the Baltic, especially the coast of Prussia and the Kurische Haaff, and which also belongs to the tertiary ("oligocene") period, numerous spiders are found, and are, in general, well preserved. The principal work on the subject of these Amber Spiders is that of KOCH and BERENDT: Die im Bernstein befindtlichen Crustaceen, Myriapoden, Arachniden

1) *Elvina*, mythol. proper name.
2) Die Urwelt der Schweitz, p. 355—358. (Zürich 1865).
3) J. R. SCHELLENBERG, a Swiss entomologist.

und Apteren der Vorwelt [1]), which, after the death of the authors, was published by A. MENGE, and provided by him with many important additions and corrections. The number of Spider-species found in Amber appears, according to KOCH'S and MENGE'S works, to amount to about 130; of these nearly 100 are fully described and figured, for the most part in KOCH and BERENDT'S above-named work, two others in a lately published paper of MENGE [2]). Of several of the remaining species MENGE has, partly in KOCH and BERENDT'S work, partly in a separate memoir [3]), given more or less detailed descriptive notices. As we are now about to give a short account of the extinct genera made known by KOCH and MENGE in the above-mentioned works, it will probably be best, in consequence of their somewhat considerable number, to treat each family separately, in the order in which they have been classed in the foregoing pages. It should however be remarked that the characteristics of many of these genera are by the said authors only touched upon in a few words and cursorily, so that it is not always possible to form a sure judgment of their systematic position.

a. *Epeiroidæ.* To this family we refer the following genera:

Græa N. [4]) = *Gea* (KOCH et BER.) 1854. — The fossil species, which KOCH and MENGE reckon to *Gea*, differ from *Epeira* by having the *anterior* central eyes much larger than the posterior, and sitting close together on small protuberances (vid. KOCH and BERENDT, p. 22—24; MENGE, Lebenszeichen, p. 6). This is however by no means the case in the now existing, East Indian species of *Gea*, *G. spinipes* C. KOCH; for in that species the *posterior* central eyes, which are placed uncommonly far backward, are larger than the anterior, according to KOCH himself [5]). The exstinct spiders in question cannot therefore be affiliated to *Gea* C. KOCH 1843, but form an independent genus, for which we propose the name *Græa*, with *G. epeiroidea* (K. et B.) as the type.

Antopia MENGE 1854 [6]). This genus is distinguished by its conically prominent head; the central eyes form a trapezoid, and are larger

1) Also with the title: BERENDT, G. C., Die im Bernstein befindlichen Organischen Reste der Vorwelt, Bd I, Abtheil. II. Berlin 1854. — In Bd I, Abth. I of this work (Berlin 1845) there is a list of the spiders described by C. KOCH in Bd I, Abtheil. II.

2) Ueber einen Scorpion und zwei Spinnen im Bernstein (Schriften der Naturforschenden Gesellschaft in Danzig, Bd II, Hft 2, 1869).

3) Lebenszeichen vorweltlicher, im Bernstein eingeschlossener Thiere. Danzig [1856].

4) Γραῖαι, αἱ, mythol. name (γραῖα, old woman).

5) Die Arachn., X, p. 101, Tab. CCCLII, fig. 823. (1843).

6) ἀντώπιος, looking straight forward.

and placed higher up than the lateral eyes (vid. KOCH et BER., p. 43; Lebenszeich., p. 7). — Type: *A. punctulata* (K. et B.), by KOCH described as a *Mizalia*.

Siga MENGE 1854 [1]). Is said to be nearly related to *Zilla:* "The head is prominent, the posterior central eyes farther apart than the anterior, the palpi of the male provided with an involuted (zusammengerollten) flagellum". — *S. crinita* MENGE (KOCH and BER., p. 27).

Androgeus K. et B. 1854 [2]). The head is triangularly or conically pointed; the eyes are arranged in two longitudinal rows diverging from the front backwards, and thus occupy a triangular area, the point of which is formed by the anterior central eyes; the hindermost eyes are placed far backwards on the back of the cephalothorax, much as in *Hyptiotes* and *Poltys*, which latter genus also in the form of its head resembles *Androgeus*. KOCH united these three genera in his family *Mithraides (Mithracidæ)*. *Androgeus* probably belongs to our *Uloborinæ*, and assuredly not to the *Laterigradæ*, with which MENGE thinks it ought to be classed. — Type: *A. triqueter* K. et B. — Conf. KOCH and BER., p. 27—29; Lebenszeich., p. 9.

Of still existing genera, *Epeira* and *Zilla* are said to be represented; the species of *Zilla* described in KOCH and BER. do not however belong to *Zilla*, as we have fixed the limits of that genus, and probably not even to the Epeiroidæ, but to the *Theridioidæ:* they are said by MENGE to resemble "*Meta tigrina*" (*Linyphia socialis* SUND.) in the position of the eyes and in the legs (vid. KOCH and BER., p. 27).

b. *Theridioidæ*. The following genera appear to belong to this family:

Flegia K. et B. 1854 [3]). Is nearly related to *Episinus*, according to MENGE, but the eyes are placed on a prominent elevation sloping behind. The cephalothorax is rounded, its pars cephalica small; the abdomen ovate, the legs long; the palpi of the male are very long, with a very large clava. The posterior central eyes are larger than the anterior. — Type: *F. longimana* K. et B. — Conf. KOCH and BER., p. 30.

Corynitis MENGE 1854 [4]). Nearly allied with *Flegia;* it is distinguished "by its larger anterior central eyes, and by the male's still longer palpi, the fourth joint of which is slender at the base, incrassated in the

1) *Σιγή*, proper name.

2) *Ἀνδρόγεως*, *Androgeus*, proper name.

3) Deriv. to me unknown. (*Φλεγίας*, *Φλεγύα* and *Φλεγύας* are proper names; *φλεγεός*, burning, shining).

4) *κορύνη*, club.

form of a club at the extremity, with the clava itself almost spherical." — *C. spinosa* MENGE. — Vid. KOCH and BER., p. 30.

Anandrus MENGE 1856 [1]). Of this genus MENGE only says that it is "nearly related to *Linyphia*, but the male's palpi and organs of copulation are very small." (Lebenszeich., p. 7).

Thyelia K. et B. 1854 [2]). In the position of the eyes *Thyelia* approximates to *Clubiona:* they are placed in two parallel or only slightly converging rows; the four posterior eyes, which are placed at about the same distance from each other, form an almost straight line, as do also the four anterior eyes; the central eyes form a trapezoid broader behind. — From most of the figures (as f. inst. that of the typical species) given in KOCH and BER., *Thyelia* appears to belong to the Theridioidæ, but other species, viz. *Th. marginata* (Pl. VI, fig. 45) and *Th. anomala* (Pl. V, fig. 39) more nearly resemble the *Agalenoidæ* (to which family the genus was referred by KOCH), and probably do not belong to *Thyelia* (Conf. MENGE in KOCH et BER., p. 56). — MENGE says loc. cit. that *Thyelia* differs from its relations *Clubiona* and *Amaurobius* by a "narrower head and laterally projecting spines on the else fine-haired legs"; in Lebenszeich., p. 7, he classes it with the Theridioidæ on account of the short spinners and the armature of the legs. — Type: *Thyelia tristis* K. et B. — Conf. KOCH and BER., p. 50—56.

Clya K. et B. 1854 [3]). — Is considered by KOCH to approximate to *Eucharia* (*Steatoda* NOB.) in the form of the body, the legs and the palpi. The head is elevated above the rounded pars thoracica; the abdomen is short, very convex. The eyes of the posterior series are placed on a sharp prominent ridge curved backwards; the central eyes, which are of the same magnitude, form a square; the lateral eyes are nearer together and about half as large as the central eyes. — Type: *C. lugubris* K. et B. — Vid. KOCH and BER., p. 31.

As possibly belonging to the Theridioidæ, MENGE mentions:

Dielacata MENGE 1854 [4]). Nothing more is said of this genus, than that it has only *two* spinners, and two tracheal stigmata before the spinners. — *D. superba* MENGE. (KOCH and BER., p. 94; Lebenszeich., p. 9).

---

1) *α priv.*, and ἀνήρ, man, male (probably with reference to the small sexual organs of ♂).

2) Perhaps from θυηλή, victim.

3) Deriv. unknown.

4) δίς, twice; ἠλακάτη, distaff.

The following still existing genera are stated to have representatives in the Prussian amber: *Ero*, *Theridium*, *Erigone*, *Walckenaera [Micryphantes]*, *Euryopis [-us]* and *Linyphia*.

c. *Scytodoidæ*. MENGE mentions (Lebenszeich., p. 9) a species of *Pholcus*, as also a new genus, perhaps belonging to this family:

Phalangopus MENGE 1854 [1]), of which however it is only said, that it is related to *Pholcus*, with long, slender legs, but with the eyes placed otherwise. — *Ph. subtilis* MENGE. Vid. KOCH and BER., p. 94; Lebenszeich., p. 9.

d. ***Mizalioidæ*** N. The curious genus *Mizalia*, which in KOCH and BERENDT is classed among the *Theridioidæ*, but which MENGE (Lebenszeich., p. 8) refers to his *Clubionida* (= *Drassoidæ* + *Dysderoidæ* NOB.), appears to me to form the type of a quite peculiar family, perhaps most related to the *Urocteoidæ* in the sub-order *Tubitelariæ*. The characters of this family may be seen from those of the only known genus:

Mizalia (K. et B.) 1854 [2]). The cephalothorax is in the form of a broad inverted heart; the pars cephalica, which has the same height as the slightly convex, broad pars thoracica, is drawn out in a kind of snout before the eyes. The eyes are about equal in magnitude and placed in two transversal rows on the superior side of the head: the anterior, shorter row is curved backwards, the posterior row is nearly straight. The legs are rather short and strong (as are also the palpi), their relative length 1, $\overparen{2, 4}$, 3; the abdomen is short, ovate; the superior or posterior spinners are slender, conically pointed, the intermediate spinners cylindrical and *more than double as long as the inferior* (anterior), which are truncated, conical and thicker. — Type: *M. rostrata* K. et B. — Conf. KOCH and BER., p. 42—45.

e. *Hersilioidæ*. Besides an *Hersilia* (of which genus no species is known now to exist in Europe: Conf. p. 115), a new genus belonging to this family has been found in the Prussian amber:

Gerdia MENGE 1869 [3]). This remarkable genus is nearly related to *Hersilia;* but the head is raised into a high vertical boss, and the legs are destitute of the long, third tarsal joint found in *Hersilia* (according to MENGE the tarsi are only two-jointed). The very long three-jointed superior spinners are curved downwards towards their extremity; their long third joint

---

1) *φάλαγξ*, joint; *πούς*, leg (or perhaps *Phalangium* and *πρύς*).
2) Deriv. unknown to me. Probably a proper name.
3) "*Γέρδια*, *textrix*": MENGE.

appears to he thickly annulated. — Type: *G. myura* MENGE. — Conf. MENGE, Ueber einen Scorp. u. zwei Spinn. im Bernstein, p. 8—9.

f. *Agalenoidæ.* The amber fauna contains several species of *Amaurobius (Cœlotes?)*, *Tegenaria* and *Agalena*, but probably not of *Textrix*, to which genus KOCH had referred a couple of species. — Conf. MENGE in KOCH et BER., p. 49, 50.

g. *Drassoidæ.* The following genera I place in this family:

Anatone MENGE 1854 [1]). Of this genus MENGE says (in KOCH and BER., p. 84) that the eyes are placed as in *Philodromus*, but the four anterior eyes are scarcely half as large as the four posterior. In Lebenszeich., p. 8, *Anatone* is said to differ from *Zora* only in having the posterior central and lateral eyes placed nearer together. One species, *A. spinipes* MENGE is stated to stand very close to *Zora spinimana* KOCH. MENGE refers the genus to the *Lycosoidæ.*

Sosibius *[Sosybius]* K. et B. 1854 [2]). Is according to MENGE (Lebenszeich., p. 8) so nearly related to *Clubiona*, as scarcely to be distinguishable from that genus. — The four anterior eyes are placed near the margin of the clypeus in an almost straight line; the posterior central eyes are very small, almost invisible; the anterior central eyes are somewhat smaller than the lateral eyes, which are about equally large. — Type: *S. minor* K. et B. Vid. MENGE, in KOCH and BER., p. 70. — KOCH, who believed that the eyes were arranged in quite another way, united this genus with *Eriodon* and *Selenops* (!) into a family, which he called *Eriodontidæ* (loc. cit., p. 69).

Erithus MENGE 1854 [3]). The lateral and the anterior central eyes, which are all large, flat and close together, are arranged in a single row curved backwards, near the margin of the clypeus; the posterior central eyes are smaller and placed on the superior side of the head. Nothing more is said of this genus. — *E. applanatus* MENGE (KOCH and BER., p. 69).

Heteromma MENGE 1856 [4]). Is said to unite *Clubiona* and *Melanophora* with *Segestria.* Six large eyes are placed quite as in *Segestria;* behind them are two very small eyes (the posterior central eyes), the diameter of which is scarcely equal to $\frac{1}{8}$ of that of the anterior central eyes; to this is to be added the peculiarity, that the abdomen is short-petiolated. MENGE Lebenszeich., p. 8. — *H. intersecta* MENGE.

1) "*ἀνάτονος, sursum tendens*": MENGE (*ἀνά*, up; *τείνω*, stretch).
2) *Σωσίβιος*, proper name; *Sosybius* ought therefore to be written *Sosibius.*
3) *ἔριθος*, labourer; also, female weaver.
4) *ἕτερος*, another, dissimilar; *ὄμμα*, eye.

The following two genera, which are said to be related to *Clubiona* (MENGE, Lebenszeich., p. 9), ought perhaps also to be classed among the Drassoidæ:

Spheconia MENGE 1854 [1]): it is stated to have "a longshafted, fusiform abdomen and long spinners": — *S. brevipes* MENGE; and

Idmonia MENGE 1854 [2]): "the ellipsoïdally arched pars cephalica is separated from the in front heart-shaped pars thoracica; the eyes enclose an ellipsis." — *I. virginea* MENGE. — Vid. KOCH and BER., p. 94.

Of the genera *Clubiona*, *Anyphæna*, *Micaria [Macaria]*, *Drassus*, *Melanophora* and *Gnaphosa [Pythonissa]* several species are described or mentioned in KOCH and BER. and MENGE. Of the genus *Clubiona* however at least one of the species described by KOCH, *C. lanata* (loc. cit., p. 67, Tab. VII, fig. 60) appears to me to belong to a quite different and peculiar genus.

h. *Dysderoidæ.* To this family belongs:

Therea K. et B. 1854 [3]). The pars cephalica is distinctly separated from and higher than the pars thoracica. Six eyes, all close together; the central eyes occupy a trapezoid somewhat broader in front, on each side of which is an obliquely placed lateral eye; the posterior central eyes are a little smaller than the others, which are eqúal in size. The genus is else nearly related to *Dysdera.* — Type: *Therea petiolata* K. et B. — Vid. KOCH and BER., p. 75.

Many other spiders belonging to this family, of the genera *Segestria* and *Dysdera* (10 species of the former genus!), have been found in amber.

i. *Theraphosoidæ.* To this family the following genus no doubt belongs:

Clostes MENGE 1869 [4]). *C. priscus* MENGE, the only known species, resembles, according to MENGE, in the form of its body *Clotho* and *Cteniza*, in the spinners, *Mygale.* The eyes, which are placed on a quadrangular elevation of the head, in two rows, occupy a large, transversal area and are arranged in a manner very unusual in the Theraphosoidæ: the four central eyes form a square, enclosed in a rectangle formed by the four lateral eyes. The superior spinners are very long, three-jointed; the tarsi

---

1) σφήξ, wasp.

2) Ἴδμων, proper name (ἴδμων, skilful). — *Idmonea* LAMOUR. [Polypi] 1821.

3) θήρειος, *ferinus*, savage, brutal.

4) "*κλωστής, qui stamina digitis torquet*": MENGE (κλώθω, spin).

are armed with three claws, as in *Nemesia (Cteniza)*; the superior claws are pectinated. — Conf. MENGE, Ueber einen Scorp. u. zwei Spinn. im Bernstein, p. 6, 7.

k. *Thomisoidæ*. We assign to this family:

Clythia K. et B. 1854 [1]). The eyes are placed in two parallel rows curved backwards; the four anterior eyes are small, equal in size, the posterior four much larger, also equal in size. The legs are rather strong, not much longer than the body, armed with spines on the underside of the tibiæ and metatarsi; the tarsi are thick, with long, pectinated claws. — In its general appearance and the structure of the tarsi these spiders are, according to MENGE (KOCH and BER., p. 46), more intimately related to *Ocypete (Heteropoda, Sparassus)* than to the *Theridioidæ*, to which family he however in Lebenszeich., p. 7, reckons them, as also KOCH had done (KOCH and BER., p. 94). — Type: *C. alma* K. et B., l. c., p. 45.

Athera MENGE 1854 [2]). "Long and slender, the anterior central eyes small and close together, the posterior central eyes more than double as large, far apart; on each side of the last are the large lateral eyes. *A. exilis*." (KOCH and BER., p. 94). — Nothing more is known of the genus, which in Lebenszeich., p. 9, is taken up among the *Thomisoidæ*.

Opisthophylax MENGE 1856 [3]). Of this genus MENGE only says: "Eyes as in *Philodromus*, but the posterior central eyes are very large and looking forwards, and the posterior lateral eyes placed far backwards. *O. exarata*." (Lebenszeich., p. 9).

Syphax K. et B. 1854 [4]). This genus is nearly related to *Xysticus*. The pars cephalica is large and broad; the two anterior central eyes are exceedingly small, the two posterior larger, farther apart; the anterior lateral eyes are very large. The row formed by the 4 anterior eyes is curved slightly backwards or almost straight, the posterior row is curved more strongly backwards. — Type: *S. megacephalus* K. et B. — Conf. KOCH and BER., p. 77.

The now existing genera *Heteropoda* [*Ocypete* KOCH, *Oxypete* MENGE], *Artanes [Artamus]*, *Philodromus* and *Misumena* (? — "*Thomisus*" MENGE) have also representatives among the amber spiders.

---

1) Perhaps = *Clytia*, mythol. proper name. — *Clytia* is a name already many times appropriated. [*Clytia* LAMOUROUX [Polypi] 1812; *Clytia* HÜBN. [Lepidopt.] 1816, etc.].

2) "*ἀ-ϑὴϱ, non fera et venenata*": MENGE. (*ἄϑηϱος*, without chase, game).

3) *ὄπισϑεν*, behind; *φύλαξ*, guard; "*retrospiciens*": MENGE.

4) *Σύφαξ*, proper name.

l. ***Archæoidæ*** *[Archæidæ]* K. et B. This family has been created by KOCH for the remarkable genus

Archæa K. et B. 1854 [1]). The large head is much and, in general, spherically elevated above the pars thoracica, which is narrower behind. The eyes are arranged in two rhomb-like groups, one on each side of the head. The mandibles are strong, often very elongated, with a long claw. The palpi are uncommonly small and slender, especially in the female. The legs are rather long and slender, prop. 1, 2, 4, 3. — Type: *A. paradoxa* K. et B. — KOCH considered this genus not to be related to any other known spiders; MENGE first (in KOCH and BER., p. 22) believed it had most affinity with *Tetragnatha:* the form of the legs as given in KOCH'S figures, reminds one in fact much of that genus and of *Pachygnatha*, which latter genus some species also resemble in their large, diverging mandibles. But in Lebenszeich., p. 9, MENGE refers *Archæa* to the *Laterigradæ* (*Thomisida* MENGE), on account of the form and direction of the fore-legs (which is said to be the same as in the *Laterigradæ*) and of the short and slender posterior legs. The position of the eyes is quite the same as in *Platythomisus* DOLESCH. (vid. sup., p. 170). I therefore place *Archæa* among the Laterigradæ; but this genus may perhaps for the present best be taken as the type of a separate family, distinguished by its ovate cephalothorax with the curiously formed pars cephalica, by the extraordinarily small palpi, and the large mandibles. — Conf. KOCH and BER., p. 19—22.

m. *Lycosoidæ.* By MENGE the following genus is referred to the Lycosoidæ, of which family no more representatives appear to have been found in amber:

Linoptes MENGE 1854 [2]). Nothing more is said of this genus, than that it has a slender body, long legs, long and slender abdomen and eyes resembling those of *Lycosa.* — *L. oculeus* MENGE. Vid. KOCH and BER., p. 94; Lebenszeich., p. 8.

n. *Eresoidæ.* Two species of the genus *Eresus.*

o. *Attoidæ.* To this family several species belong, which are described in KOCH and BER. under the generic names of *Phidippus* and *Leda.* The genus *Leda* appears to be founded on a damaged specimen, and cannot therefore be retained; moreover the name is already appropriated [3]). The species which KOCH reckons to *Phidippus*, do *not*, according to MENGE,

1) ἀρχαῖος, primitive, ancient.
2) λινόπτης, guarder of a net.
3) *Leda* SCHUM. [Moll.] 1817.

belong to that still existing, exotic genus. MENGE creates for them — with the exception of one species, which he assigns to *Euophrys* C. KOCH (*Attus* (WALCK.) NOB.) — a new genus:

Gorgopis MENGE 1854 [1]). This genus, which is said to be nearly related to *Euophrys* C. KOCH (*Attus* NOB.), appears to be characterized principally by having the small eyes of the 2nd series placed at a *very* short distance behind the anterior lateral eyes; they are also somewhat less distant from each other than are the two eyes of the 3rd row. The fourth pair of legs is longer than the others. — Type: *G. frenata* (K. et B.). — Conf. KOCH and BER., p. 93).

Propetes MENGE 1854 [2]). Of this genus MENGE only says that it differs from the genera, into which KOCH has resolved WALCKENAER'S *Attus*, by having the eyes of the 2nd row but slightly smaller than those of the 3rd row, and larger than in now living species. — Type: *P. felinus* MENGE. Vid. KOCH and BER., p. 93.

One species of KOCH'S exstinct *Phidippi*, is, as we have already said, by MENGE referred to *Attus*, or *Euophrys* MENGE; in Lebenszeich. (p. 9) that name is however not mentioned, but instead of it: "*Salticus* 1 sp."

Lastly I ought to cite the genus:

Mastigusa MENGE 1854 [3]), whose affinities are entirely unknown: of the only mentioned species, *M. acuminata* MENGE, it is stated that the male has on its palpi "flagella which are curved backwards in form of a ram's horn (widderhorn-ähnlich nach hinten gebogenen Geisseln) and are almost as long as the body." Vid. KOCH and BER., p. 94.

Three more genera *Onca*, *Epeiridion* and *Ocia* are mentioned by MENGE (KOCH and BER., p. 8 and 24; Lebenszeichn., p. 8), but they are not at all characterized. The two former are said to belong to the *Epeiroidæ*, the last named is taken up among the *Thomisoidæ*.

I beg here to express my grateful acknowledgement to those Arachnologists who, since the printing of this treatise was commenced, have assisted me by the communication of valuable information or specimens of interesting species. In addition to the gentlemen named in pp. 2 and 3,

1) γοργῶπις, fierce-looking.
2) προπετής, rash, hasty.
3) μαστιγόω, whip, scourge.

I beg with thankfulness to mention Mr E. SIMON, Prof. A. MENGE, Prof. E. GRUBE, Prof. R. LEUCKART, Prof. J. G. SCHIÖDTE, and more especially the Rev. O. P. CAMBRIDGE, through whose kindness I have had the opportunity of examining a large number of English Spiders.

ADDENDA.

Pag. I—XXIV:

1869. BARTA, E., Verzeichniss der Spinnen des nördlichen Böhmens. (Archiv für die Naturwissenschaftliche Landesdurchforschung von Böhmen, Bd I).

BERENDT, Vid. KOCH and BERENDT.

1865. BERGSØE, V., Iagttagelser om den Italienske Tarantel og Bidrag til Tarantismens Historie i Middelalderen og nyere Tid. (Naturhist. Tidskrift, 3 Række, Bd III).

1869. CAMBRIDGE, O. P-., Part I. of Catalogue of a collection of Ceylon Araneidea lately received from Mr J. Nietner, with descriptions of new species and characters of a new genus. (The Linnean Society's Journ., Zool., Vol. X).

1869. —ID.— Descriptions and sketches of two new species of Araneidea, with characters of a new genus. (*ibid.*).

1868. HENTZ, M., and SCUDDER, S. H., Supplement to the descriptions and figures of the Araneides of the United States by Nicholas Marcellus Hentz. Edited by Samuel H. Scudder (Proceed. of the Boston Society of Natural History, Vol. XI).

1854. KOCH, C. L., and BERENDT, G. C., Die im Bernstein befindlichen Crustaceen, Myriapoden, Arachniden und Apteren der Vorwelt. Berlin 1854. [*With additions by* A. MENGE]. — *Also with the title:* Die im Bernstein befindlichen organischen Reste der Vorwelt, gesammelt, in Verbindung mit Mehreren bearbeitet und herausgegeben von G. C. BERENDT. Bd I. Abtheil. II: Die im Bernstein befintlichen Crustaceen, Myriapoden, Arachniden und Apteren der Vorwelt.

MENGE in KOCH and BER., Vid. KOCH and BERENDT.

SCHIÖDTE, J. C., Specimen Faunæ subterraneæ, Vid. (p. XX) ID., Bidrag til den underjordiske Fauna.

SCUDDER, Vid. HENTZ and SCUDDER.

1835. WESTWOOD, J. O., [*Gastracanthus:*] (Transact. of the Entom. Soc. of London, Vol. I. Proceed.).

Pag. 54, lin. 25:

(The *Zilla montana* of WESTRING we propose to call *Z. Strœmii*, in memory of the Norwegian Zoologist H. STRØM).

Pag. 65 (after *Uloborinæ*):

In a paper recently published (Descr. and sketches of two new spec. of Aran. etc.), CAMBRIDGE has given descriptions and figures of a highly remarkable genus from Ceylon, *Miagrammopes* CAMBR., which as he, no doubt rightly, thinks, is most nearly related to *Uloborus* and *Hyptiotes* (*Mithras*). What in the first place gives this genus a peculiar interest, is the circumstance of its having *only four eyes*, placed in a transversal row across the pars cephalica; so that now a veritable four-eyed spider is at last discovered! (Conf. p. 28, note 1). But *Miagrammopes* is still more remarkable by the *absence of a separate sternal plate*, the legs being simply articulated to the lower side of the cephalothorax, which forms the sternal surface. This unique character would perhaps warrant the formation of a special family within the sub-order Orbitelariæ for the genus in question; but in every other point of systematical importance it appears to me to agree with the *Uloborinæ*. — Two species, *M. Thwaitesii* and *M. Ferdinandi*, are described and figured.

Pag. 81 (in the *Syn.* of *Linyphia*):

1845. **Meta** C. KOCH (*ad part.:*) Die Arachn., XII, p. 130.

Pag. 85 (in the *Syn.* of *Erigone*):

1830. **Linyphia** SUND., Sv. Spindl. Beskr., *in* Vet.-Akad. Handl. f. 1829, p. 211 (*ad partem*).
1834. **Theridium** REUSS, Zool. Misc., Arachn., (*ad part.:*) p. 222 (228).

Pag. 86 (in the *Syn.* of *Walckenaera*):

1830. **Linyphia** SUND., Sv. Spindl. Beskr., *in* Vet.-Akad. Handl. f. 1829, p. 211 (*ad partem*).
1831. **Theridium** HAHN, Die Arachn. I (*ad part.:*) saltem p. 91, 92, Tab. XXII, fig. 69, 70. — Monogr. Aran., 6, Tab. IV, fig. *C*.

INDEX.

The names printed in *Italics* are *Synonyms*.

ERRATA.

Pag. v,	*line*	17 *from bottom:*	*for* descripzione	*read*	descrizione
» VII	»	4 »	» segestriformes	»	segestriformis
» XIV	»	15 »	» 1848	»	1849
» 3	»	16 *from top*	» a	»	as
» 11	»	18 »	» *Spodrus*	»	*Sphodrus*
» 14	»	3 *from bottom*	» *Epesinus*	»	*Episinus*
» »	»	15 »	» Ar	»	As
» »	»	18 »	» number	»	a number
» 15	»	17, 18 »	» applicable	»	appropriate
» 27	»	18 »	» och	»	and
» 28	»	4, 5 *from top*	» this whole system of	»	the very basis of this
» 30	»	13 *from bottom*	» and the	»	and
» 32	»	18 »	» non	»	not
» 41	»	2, 4 »	» TELLKAMP	»	TELLKAMPF
» 42	»	5, 8 *from top*	» (13)	»	(15)
» »	»	7 »	» (10)...(14)	»	(12)...(16)
» 44	»	6 »	» *Œcobiidiæ*	»	*Œcobiidæ*
» 46	»	8 *from bottom*	» *Marpissa*	»	*Marpessa*
» 47	»	16 *from top*	» previded	»	provided
» 50	»	15 »	» *notacanthas*	»	*notacantha*
» 53	»	8 *from bottom*	» examinor	»	examino
» 57	»	2 *from top*	» *fusca*	»	*Merianæ*
» 59	»	12 »	» be	»	to be
» 67	»	17 *from bottom*	» 1835	»	1834
» 75	»	9 »	» multo minus	»	spatio multo minore
» 77	»	19 »	» tenues	»	tenues;
» 78	»	1 »	» 1, 2, 3, 4.	»	1, 2, 4, 3.
» 83	»	10, 11 *from top*	» *L. pratensis* BLACKW.	»	*L. hortensis* SUND. (*L. pratensis* BLACKW.).
» 93	»	14 *from bottom*	» Steatoda [Steatodum]	»	Eucharia [Eucharium]
» 98	»	10 »	» disproportionably	»	disproportionately
» 119	»	17 »	» family	»	the family
» 121	»	4 *from top*	» 9.	»	10.
» 154	»	4 *from bottom*	» (BLACKW.)	»	BLACKW.
» 178	»	9 *from top*	» Philodrominæ	»	Philodrominæ (excl. *Micrommata* LATR.)
» 210	»	4 »	» LUC.	»	(LUC.)

LITERATURE[1]).

1842—46. AGASSIZ L., Nomenclator zoologicus, continens nomina systematica generum animalium tam viventium quam fossilium, secundum ordinem alphabeticum disposita, adjectis auctoribus, libris, in quibus reperiuntur, anno editionis, etymologia et familiis, ad quas pertinent, in singulis classibus. Soloduri 1842—1846. — Arachnidæ. Recognovit GUIL. F. ERICHSON.

† 1736. ALBIN, E., A natural history of spiders and other curious insects. London 1736.

* 1840. [AMARY, A.], Statistica fisica ed economica dell' isola di Capri. (Esercitazioni dell' Academia degli aspiranti naturalisti, Vol. II, Part. 1. Napoli 1840).

AUDOUIN, V., *in* Dict. class., *Vid.* Dictionnaire classique d'Hist. Nat.

—ID.— (SAVIGNY *and*), *Vid.* Description de l'Égypte.

1833. —ID.— Observations sur la structure du nid de l'araignée pionnière. (Annales de la Société Entomologique de France, T. II). — [*Also separate with the title:* Observations sur le nid d'une araignée, construit en terre, et remarquable par une grande perfection de travail].

1867. AUSSERER, A, Die Arachniden Tirols nach ihrer horizontalen und verticalen Verbreitung. I. (Verhandlungen der zoologisch-botanischen Gesellschaft in Wien, Bd XVII, 1867).

[1867. —ID.—, Beobachtungen über Lebensweise, Fortpflanzung und Entwickelung der Spinnen. (Zeitschrift des Ferdinandeums, 3 Folge, Hft. XIII)].

1839. BARKER-WEBB, PH., *and* BERTHELOT, S., Histoire Naturelle des Iles Canaries. 3 Tomes. Paris 1835—49. [Tom. II. 2. Entomologie: Arachnides, Myriapodes et Thysanoures, par H. LUCAS].

* 1789. BECHSTEIN, J. M., Ueber den wahren Ursprung des fliegenden Sommers. (LICHTENBERG *and* VOIGT, Magaz. für das Neueste aus der Physik und Naturgeschichte, Bd. VI).

BERTHELOT, S., *Vid.* BARKER-WEBB.

BERTHOLD, A. A., *Vid.* LATREILLE, Nat. Fam. d. Thierreichs.

---

1) Vid. pag. 19. — The complete titles of most of the periodicals here referred to, may be seen in e. g. CARUS and ENGELMANN, Bibliotheca Zoologica. 2 Voll., Leipzig 1861.

1832. BLACKWALL, J., Description of a species of Arachnida hitherto uncharacterized, belonging to the Araneidæ (London and Edinburg Philosophical Magazine, New [3th] Ser., Vol. I.)

[1833. —ID.— Notice of several recent discoveries in the structure and economy of spiders. (Transactions of the Linnean Society, Vol. XVI)].

1833. —ID.— Characters of some undescribed genera and species of Araneidæ. (Lond. and Edinb. Phil. Mag., N. [3th] Ser., Vol. III).

* 1834. —ID.— Researches in Zoology. London 1834.

1834, 36. —ID.— Characters of some undescribed species of Araneidæ. (Lond. and Edinb. Phil. Mag., N. [3th] Ser., Voll. V, VIII).

1837. —ID.— Characters of a new genus and some undescribed species of Araneidæ. (*ibid.*, Vol. X).

[1839. —ID.— On the number and structure of the mammulæ employed by spiders in the process of spinning. (Transact. of the Linn. Soc., Vol. XVIII, P. II)].

1841. —ID.— The difference in the number of eyes with which spiders are provided, proposed as the basis of their distribution into tribes; with descriptions of newly discovered species, and the characters of a new family and three new genera of spiders. (*ibid.*, Vol. XVIII, Part. IV; (1841); *Extr. in* Proceedings of the Linn. Soc., Vol. I, N:o 8 (1840)).

1843. —ID.— A catalogue of spiders either not previously recorded or little known as indigenous to Great Britain, with remarks on their habits and economy. (*ibid.*, Vol. XIX, Part. II (1843); *Extr. in* Proceed. of the Linn. Soc., Vol. I, *and in* Annals and Magazine of Natural History, Vol. X).

1844—54. —ID— Descriptions of some newly discovered species of Araneidea. (Ann. and Mag. of Nat. Hist., Voll. XIII (1844), XVIII (1846); 2 Ser., Voll. X (1852), XI (1853), XIII (1854).

1846. —ID.— Notice of spiders captured by Professor Potter in Canada, with descriptions of such species as appear to be new to science. (*ibid.*, Vol. XVII).

1850. —ID.— Descriptions of some newly discovered species and characters of a new genus of Araneidea. (*ibid.*, 2 Ser. Vol. VI).

1851—52. —ID.— A catalogue of British spiders, including remarks on their structure, functions, œconomy and systematic arrangement. (*ibid.*, 2 Ser., Voll. VII, VIII (1851), IX, X (1852)).

1853, 54, 57. —ID.— Supplement to a catalogue of British spiders, including remarks on their structure, functions, œconomy and systematic arrangement. (*ibid.*, 2 Ser., Voll. XI (1853), XIV (1854), XX (1857)).

1855. —ID— Decriptions of two newly discovered species of Araneidea. (*ibid.*, 2 Ser., Vol. XVI).

1856. —ID.— Descriptions of three newly discovered species of Araneidea. (*ibid.*, 2 Ser., Vol. XVII).

1857. —ID.— Descriptions of the male of Lycosa tarentuloides Maderiana Walck. and of three newly discovered species of the genus Lycosa. (*ibid.*, 2 Ser. Vol. XX).

1858. BLACKWALL, J., Descriptions of six newly discovered species and characters of a new genus of Araneidea (*ibid.*, 3 Ser., Vol. I).

1858. —ID.— Characters of a new genus and descriptions of three recently discovered species of Araneidea. (*ibid.*, 3 Ser,, Vol. II).

1859. —ID.— Descriptions of six recently discovered species, and characters of a new genus of Araneidea. (*ibid.*, 3 Ser., Vol. III).

1859. —ID.— Descriptions of newly discovered spiders captured by James Yate Johnson Esq., in the island of Madeira. (*ibid.*, 3 Ser., Vol. IV).

1861, 64. —ID.— A history of the spiders of Great Britain and Ireland. 2 Parts. London 1861, 1864.

1861. —ID.— Descriptions of several recently discovered spiders. (Ann. and Mag. of Nat. Hist., 3 Ser., Vol. VIII).

1862. —ID.— Descriptions of newly discovered spiders from the island of Madeira. (*ibid.*, 3 Ser., Vol. IX).

1862—63. —ID.— Descriptions of newly discovered spiders captured in Rio Janeiro by John Gray and Hamlet Clark. (*ibid.*, 3 Ser., Voll. X, XI).

1863. —ID.— Notice of a Drassus and Linyphia new to science, and a Neriene hitherto unrecorded as British. (*ibid.*, 3 Ser., Vol. XII).

1864. —ID— Notice of the capture of Mithras paradoxus in England (*ibid.*, 3 Ser., Vol. XIII).

1864. —ID.— Descriptions of seven new species of East-Indian spiders received from the Rev. O. Cambridge. (*ibid.*, 3 Ser., Vol. XIV).

1864. —ID.— Notice of spiders indigenous to the Salvages received from the Barão do Castello de Paiva. (*ibid.*).

1865. —ID.— Descriptions of recently discovered spiders collected in the Cape de Verde Islands by John Gray Esq. (*ibid.*, 3 Ser., Vol. XVI).

1865. —ID.— Descriptions of recently discovered species, and characters of a new genus of Araneidea from the East of Central Africa. (*ibid.*, 3 Ser., Vol. XVI).

1867. —ID.— A list of spiders captured in the south-east region of Equatorial Africa; with descriptions of such species as appear to be new to arachnologists. (*ibid.*, 3 Ser., Vol. XVIII).

1867. —ID.— Notes on spiders, with descriptions of several species supposed to be new to arachnologists (*ibid.*, 3 Ser., Vol. XX).

1868. —ID.— Notice of several species of spiders supposed to be new or little known to arachnologists. (*ibid.*, 4 Ser., Vol. II).

1857. BÖCKH (BÖKH), G., Ueber die Spinnen der Umgebung Presburgs (Verhandlungen des Vereins für Naturkunde zu Presburg, Jahrg. II, 1857. Hft. 2).

1861. —ID— Vorläufige Uebersicht der während der Reise der K. K. Fregatte Novara von den Herren Naturforschern gesammelten Spinnen. (Verhandl. d. zool.-bot. Gesellsch. in Wien, Bd XI, 1861).

1862. —ID.— (FRAUENFELDT, G. v., *and*) Ueber unterirdisch lebenden Spinnen und Fische. (*ibid.*, Bd XII, 1862. Sitzungsbericht).

BRANDT *and* RATZEBURG, Medizin. Zool., *Vid.* RATZEBURG.

1827. BRÉBISSON, L. A. DE, Catalogue des Arachnides, des Myriapodes et des Insectes-Aptères que l'on trouve dans le département du Calvados; d'après la méthode de M. Latreille. (Mémoires de la Société Linnéenne de Normandie, 1826—1827).

Bruchstücke zu einer Fauna der Berberei, *Vid.:* WAGNER, Reisen etc.

BRULLÉ, A., Insectes de Morée, *Vid.* Expédition scient. de Morée.

[1810. BULLMAN, J. C., Ueber die Natur und Entstehung des fliegenden Sommers. (Neue Schriften der Naturforschenden Gesellschaft zu Halle, Hft. 5)].

* 1859. CAMBRIDGE, O. PICKARD-, Remarks on Arachnida taken chiefly in Dorsetshire and Hampshire, with list of 134 species. (Zoologist, 1859).

* 1860. —ID.— Supplement to a note on the Arachnida of Dorset and Hants. in Zool. 6493. (*ibid.* 1860).

* 1860. —ID.— A list of Southport spiders, with some remarks on uniformity of use and meaning of words in natural history; with list of 80 species of spiders. (*ibid.*).

1860. —ID.— Descriptions of two British spiders new to science (Ann. and Mag. of Nat. Hist., 3 Ser., Vol. V).

* 1861. —ID.— Notes on spiders captured in 1860; with list of species containing 56 additions to former lists of British spiders. (Zoologist, 1861).

1861. —ID.— Descriptions of ten new species of spiders recently discovered in England. (Ann. and Mag. of Nat. Hist., 3 Ser., Vol. VII).

1862. —ID.— List of new and rare spiders captured in 1861; being a supplement to the lists in Zool. 6493, 6862, 7553. (Zoologist, 1862).

1862. —ID.— Descriptions of ten new species of British spiders. (*ibid.*).

* 1862. —ID.— Sketch of an arachnological tour in Scotland in 1861; with a list of Scotch spiders. (*ibid.*).

1863. —ID.— Descriptions of twenty four new species of spiders lately discovered in Dorsetshire and Hampshire; together with a list of rare, and some other hitherto unrecorded British spiders. (*ibid.*, 1863).

1868. —ID.— Descriptions of a new genus and six new species of spiders. (The Linnean Society's Journal, Zool., Vol. X).

1869. —ID.— Descriptions and sketches of some new species of Araneidea, with characters of a new genus. (Ann. and Mag. of Nat. Hist., 4 Ser., Vol. III).

1867. CANESTRINI, GIOV., Intorno agli Aracnidi dell' ordine Araneina osservati nel Veneto e nel Trentino. (Commentario della Fauna, Flora e Gea nel Veneto e nel Trentino, Fasc. 2).

1868. —ID.— Nuovi Aracnidi Italiani. (Annuario della Società dei Naturalisti in Modena, Anno III).

* 1868. —ID.— Nuove specie Italiane di animali: II. Nuovi Aracnidi. (Comm. d. Fauna, Flora e Gea, Fasc. 4).

* 1868. CANESTRINI, GIOV., Enumerazione degli Aracnidi dell ordine Araneina osservati nel Veneto. (*ibid.*).

1869. —ID.— *and* PAVESI, P., Araneidi Italiani. (Atti della Società Italiana di Scienze Naturali, Vol. XI, Fasc. III. 1868). *Also separate:* Modena 1869.

1842. CANTOR, TH., General features of Chusan, with remarks on the Flora and Fauna of that island. (Ann. and Mag. of Nat. Hist., Vol. IX).

1789. CEDERHIELM, J., Faunæ Ingricæ Prodromus, exhibens methodicam descriptionem insectorum agri Petropolensis, præmissa mammalium, avium, amphibiorum et piscium enumeratione. Lipsiæ 1789.

CIRILLO, *Vid.* CYRILLUS.

1855. CLARK, H., Notice and description of a new species of spider. (Ann. and Mag. of Nat. Hist., 2 Ser., Vol. XVI).

1757. CLERCK, C., Svenska spindlar, uti sina hufvud-slägter indelte samt under några och sextio särskildte arter beskrifne och med illuminerade figurer upplyste. Aranei suecici, descriptionibus et figuris æneis illustrati, ad genera subalterna redacti, speciebus ultra LX determinati. Stockholmiæ 1757.

* 1843. CONTARINI, N., Cataloghi degli uccelli e degli insetti delle provincie di Padova e di Venezia. Bassano 1843.

—ID.—, *Vid.* Venezia e le sue lagune.

* 1804. COQUEBERT DE MONBRET, A. J., Illustratio iconographica insectorum. Dec. I, Fasc. 3. Paris 1799, 1802, 1804. [Fasc. III].

1841. COSTA, A., Sur les travaux entomologiques de l'académie des aspirants naturalistes de Naples. (Ann. de la Soc. Ent. de France., T. X, Bulletins).

1835. COSTA, O. G. *and* A., Fauna del Regno di Napoli. Napoli 1829—66. Aracnidi [*by* O. G. COSTA; *also under the title:* Monografia degli Aracnidi del Regno di Napoli. — *This part of the work has not been continued*].

1835. COSTA, O. G., Cenni zoologici, ossia descripzione sommaria delle specie nuove di animali discoperti in diverse contrade del Regno nell' anno 1834. (*Also under the title:* Annuario zoologico 1834). Napoli 1834.

* 1843. Cremona e la sua provincia. Cremona 1843.

1817. CUVIER, G., Le Règne Animal, distribué d'après son organisation, pour servir de base à l'histoire naturelle des animaux et d'introduction à l'anatomie comparée. 3 Voll. [Tome III, Contenant les Crustacées, les Arachnides et les Insectes, par M. LATREILLE. Paris 1817].

1829. *Idem liber:* Nouv. [2:de] Éd. 5 Voll. [Tome IV: Crustacés, Arachnides et partie des Insectes, par P. A. LATREILLE. Paris 1829. — *Voll. IV and V also with the title:* Les Crustacées, les Arachnides et les Insectes distribués en familles naturelles, par P. A. LATREILLE. 2 Voll.].

1836. *Idem liber:* [3:me] Édition accompagnée de Planches gravées . . . . ., par une réunion de disciples de Cuvier. [Tome VII:] Les Arachnides. Avec un Atlas, par M. A. DUGÈS et M. MILNE-EDWARDS. Paris.

* 1787. CYRILLUS [CIRILLO], D., Entomologiæ Neapolitanæ specimen I. Neapoli 1787.

1826. DALMAN, J. W., Årsberättelse om nyare zoologiska arbeten och upptäckter, till Kongl. Vetenskaps-Academien afgifven d. 31 Mars 1826. Stockholm 1826.

[1840. DARWIN, CH., Journal of researches into the geology and natural history of the various countries visited by H. M. S. Beagle under the command of Captain Fitz Roy, R. N. from 1832 to 1836. London 1840].

1778. DE GEER, CH., Mémoires pour servir à l'histoire des Insectes. (7 Voll. en 8 Tomes). Stockholm 1772—1778. [Tome VII].

—ID.— Genera et species Insectorum, *Vid.:* RETZIUS, A. J.

*182..[1]) Description de l'Égypte (10 Voll. de texte, 10 Voll. de planches in folio. Paris 1809—13). Histoire naturelle. Zoologie (2 Voll. de texte, 2 Voll. de planches): Arachnides. Planches par J. C. DE SAVIGNY. [Tome II].

1827. *Idem liber:* 2:de Édition, (24 Tom. en 26 Voll. de texte in 8:o, 12 Voll. de planches in folio). Histoire Naturelle. Zoologie: Arachnides. Planches par J. C. DE SAVIGNY, Texte (Explication sommaire des Planches des Arachnides de l'Égypte et de la Syrie publiées par Jules César de Savigny etc.) par V. AUDOUIN. [T. XXII].

1822—31. Dictionnaire classique d'histoire naturelle, par M. M. Audouin, Bourdon, Brongniart, De Candolle, Daudebard de Férussac, Deshayes, Desmoulins, Drapiez, Dumas, Edwards, Flourens, Geoffroy de Saint-Hilaire, Guérin, Guillemin, A. de Jussieu, Kunth, De Lafosse, Lamouroux, Latreille, Lucas, Prévost, Richard et BORY DE SAINT-VINCENT. Ouvrage dirigé par ce dernier collaborateur et dans lequel on a ajouté, pour le porter au niveau de la science, un grand nombre de mots qui n'avaient pu faire partie de la plupart des Dictionnaires antérieurs. 17 Voll. Accompagnés d'un atlas de 160 Planches. Paris 1822—31. [*Articles on spiders by* V. AUDOUIN].

1802—1804. Dictionnaire, Nouveau, d'histoire naturelle appliquée aux arts, principalement à l'agriculture et à l'économie rurale et domestique: par une société de naturalistes et d'agriculteurs. 24 Voll. Paris 1802—1804. [*Art.* Araignée *and* Mygale, *and* Tableau méthodique des Insectes (*in* Vol. XXIV) *by* LATREILLE].

*1816—19. *Idem liber:* 2:de Édition. 36 Voll. Paris 1816—19. [*Art. on spiders by* LATREILLE].

*1833—39. Dictionnaire pittoresque d'histoire naturelle et des phenomènes de la nature, par une société de naturalistes sous la direction de E. GUÉRIN-MÉNEVILLE. 9 Voll. Paris 1833—36.

*1840—48. Dictionnaire universel d'histoire naturelle, résumant et complétant tous les faits présentés par les encyclopédies, les anciens dictionnaires scientifiques, les oeuvres complètes de Buffon, de Lacépède, de Cuvier, et par les meilleurs traités speciaux sur les diverses branches des sciences naturelles etc. — Ouvrage dirigé par CH. D'ORBIGNY. [*Articles on spiders by* H. LUCAS]. 13 Voll. et un Atlas de 288 Planches. Paris 1840—1848. [*A new Edition is in the press*].

---

1) Not earlier than 1825.

* 1843. DIEFFENBACH, E., Travels in New Zealand, with contributions to the Geology, Botany etc. 2 Voll. London 1843. [*Arachn. by* WHITE].

1853. DOBLIKA, K., Beitrag zur Monographie des Spinnengeschlechtes Dysdera. (Verhandl. d. zool.-bot. Gesellsch. in Wien, Bd III, 1853).

1852. DOLESCHALL, L., Systematiches Verzeichniss der im Kaiserthum Oesterreich vorkommenden Spinnen. (Sitzungsberichte d. Mathem.-Naturwissensch. Classe d. Kais. Akademie d. Wissenschaften zu Wien, Bd IX, 1852. [*Also separate*].

1857. —ID.— Bijdrage tot de Kennis der Arachniden van den Indischen Archipel. (Natuurkundig Tijdschrift voor Nederlandsch Indie, Deel XIII (3 Ser., D. III)).

1859. —ID.— Tweede Bijdrage tot de Kennis der Arachniden van den Indischen Archipel. (Acta Societatis Scientiarum Indo-Neerlandiæ, Vol. V).

1794. DORTHES, Observations on the structure and œconomy of some curious species of Aranea. (Transact. of the Linn. Soc., Vol. II).

1864. DOUMERC, A., Descriptions de deux Aranéides des genres Thomise et Épeire du Sénégal. (Ann. de la Soc. Entom. de France, 4 Sér., T. IV).

1820. DUFOUR, LÉON, Description de six Arachnides nouvelles. (Annales générales des Sciences physiques, Vol. IV).

1820. —ID.— Observations sur quelques Arachnides quadripulmonaires. (*ibid.*, Vol. V).

1820. —ID.— Description de cinq Arachnides nouvelles. (*ibid.*).

1820. —ID.— Observations générales sur les Arachnides et description de quelques especes nouvelles ou peu connues. (*ibid.*, Vol. VI).

1824. —ID.— Descriptions et figures de quelques Arachnides. (Annales des Sciences Naturelles, Tome II).

1831. —ID.— Descriptions et figures de quelques Aranéides nouvelles ou mal connues; et procédé pour conserver à sec ces invertébrés dans les collections. (*ibid.*, T. XXII).

1835. —ID.— Observations sur la Tarentule (Lycosa tarantula), avec la figure de cette aranéide. (*ibid.*, 2 Sér., Zool., T. III).

1835. —ID.— Description et figure d'une nouvelle espèce d'Épeire. (*ibid.*).

1836. —ID.— Observations sur la Filistata bicolor. (Ann. de la Soc. Entom. de France, T. V).

1852. —ID.— Sur la Micrommata spongitarsis. (*ibid.*, 2 Sér., T. X, Bull.).

1855. —ID.— Description de deux nouvelles espèces d'Aranéides. I. Épeira thomisoides, nouvelle espèce. II. Sur une nouvelle espèce de Théridion, et note sur le Theridion dispar. (*ibid.*, 3 Sér., T. III).

1358. —ID.— Sur le Drassus segestriformes. (*ibid.*, 3 Sér., T. VI).

1361. —ID.— Notices Entomologiques. I. Sur l'Épeira sericea et le Pompilus croceicornis, avec quelques considérations sur leur habitat géographique. (*ibid.*, 4 Sér., T. I).

1834. DUGÈS, A., Recherches sur l'ordre des Acariens en général et la famille des Trombidiés en particulier. (Ann. d. Sciences Nat., 2 Sér., Zool., Vol. I).

[1835. —ID.— Sur les organes de la respiration dans les Aranéides Ségestria et Dysdera. (Ann. de la Soc. Ent. de France, T. IV, Bull.)].

1836. —ID.— Observations sur les Aranéides. (Ann. d. Sciences Nat., 2 Sér., Vol. VI).

—ID.— [*and* MILNE-EDWARDS], Arachnides du Règne Animal de Cuvier, Atlas, *Vid.* CUVIER, G., Le Règne Animal, [3:me] Éd.

1815(?). Edinburg Cyclopedia, The, conducted by D. BREWSTER. Vol. VII. [*Article* Crustaceology *by* W. E. LEACH].

1830. EICHWALD, E., Zoologia specialis, quam expositis animalibus tum vivis, tum fossilibus potissimum Rossiæ in universum, et Poloniæ in specie, in usum lectionum publicarum in universitate cæsarea Vilnensi habendarum edidit. 2 Voll. Vilna 1830. [T. II].

1841. —ID.— Fauna Caspio-Caucasia, nonnullis observationibus novis illustravit. Petropoli 1841. (Nouveaux Mémoires de la Société Impériale des Naturalistes de Moscou, T. VII).

1868. EISEN, G., *and* STUXBERG, A., Bidrag till kännedomen om Gotska Sandön. (Öfversigt af Kongl. Vetenskaps-Akademiens Förhandlingar, Årgång. XXV, 1868. [*Arachn. determ. by* T. THORELL].

*18... Encyclopædia Britannica. 4th, 5th and 6th Editions. Supplement [*Article* Annulosa *by* W. E. LEACH].

1789, 1811. Encyclopédie méthodique ou par ordre de matières; par une société de gens de lettres, de savans et d'artistes. Histoire Naturelle (10 Tomes, Paris 1782—1832). Insectes [*Article* Araignée *in* T. IV (1789), *and art.* Mygale *in* T. VIII (1811) *by* A. G. OLIVIER].

ERICHSON, GUIL. F., *Vid* AGASSIZ, Nomencl. Zool.

1868. Eugenies, Kongliga Svenska fregatten, Resa omkring jörden, under befäl af C. A. Virgin. Åren 1851—1853. Vetenskapliga iakttagelser, på H. Maj:t Konung Oscar den förstes befallning utgifna af K. Svenska Vetenskaps-Akademien. Hft. 12. Zoologi. Arachnider, 1. [*by* T. THORELL.]

1832. Expédition scientifique de Morée. Section des Sciences physiques. Tom. III. 1:re Partie: Zoologie. 2:e Section. Les Animaux articulés, par M. BRULLÉ. Les Crustacés par M. GUÉRIN. Paris 1832.

1845—49. Exploration scientifique de l'Algérie pendant les années 1840, 1841, 1842. Sciences physiques. Zoologie I: Histoire Naturelle des Animaux articulés par H. LUCAS. Première Partie: Les Crustacés, Arachnides, Myriapodes et Hexapodes. 1 Vol. Paris 1849.

1775. FABRICIUS, J. C., Systema Entomologiæ, sistens insectorum classes, ordines, genera, species, adjectis synonymis, locis, descriptionibus, observationibus. Flensburgi et Lipsiæ 1775.

1777. FABRICIUS, J. C., Genera Insectorum eorumque characteres naturales, secundum numerum, figuram, situm et proportionem omnium partium oris; adjecta mantissa specierum nuper detectarum. Chilonii.

1779. —ID.— Reise nach Norwegen, mit Bemerkungen aus der Naturgeschichte und Œkonomie. Hamburg 1779.

1781. —ID.— Species Insectorum, exhibentes eorum differentias specificas, synonyma auctorum, loca natalia, metamorphosin, adjectis observationibus, descriptionibus. 2 Tomi. Hamburgi et Kilonii 1781. [Tom. I].

1787. —ID.— Mantissa Insectorum, sistens eorum species nuper detectas, adjectis characteribus genericis, differentiis specificis, emendationibus, observationibus. 2 Tomi. Hafniæ 1787. [Tom. I].

1793. —ID.— Entomologia Systematica emendata et aucta. Secundum classes, ordines, genera, species; adjectis synonymis, locis, observationibus, descriptionibus. 4 Tomi. Hafniæ 1792—1794. [Tom. II].

1798. —ID.— Supplementum Entomologiæ Systematicæ. Hafniæ 1798.

182..—30. Faune Française, ou histoire naturelle générale et particulière des animaux qui se trouvent en France, constamment ou passagèrement, à la surface du sol, dans les eaux qui le baignent et dans le littoral des mers qui le bornent; par M.M. P. Vieillot, A. G. Desmarest, H. Ducrotay de Blainville, Audinet-Serville, Lepelletier de Saint-Fargeau et C. A. Walckenaer. 29 Livr. de texte, 29 Cahiers de planches. Paris 1820—30. [*Arachn.* (*Aranéides*) *by* C. A. WALCKENAER. — Pag. 1—96 (Livr. 11—12): 182.. [1]); p. 97—176 (Livr. 26) 1830, p. 177—240 (Livr. 29): 1830. — *The work has never been completed*].

1775. FORSKÅL, P. Descriptiones animalium, avium, amphibiorum, piscium, insectorum, vermium, quæ in itinere orientali observavit. Post mortem auctoris edidit C. NIEBUHR. Adjuncta est materia medica Kahirina atque tabula Maris Rubri geographica. Hauniæ 1775.

1776. —ID.— Icones rerum naturalium quas in itinere orientali depingi curavit. Post mortem auctoris ad Regis mandatum æri incisas edidit C. NIEBUHR. Hauniæ 1776.

* 1785. FOURCROY, A. F. DE, Entomologia Parisiensis. 2 Voll. Paris 1785.

FRAUENFELD, G. v., *Vid.* BÖCKH (FRAUENFELD *and*).

† 1720—38. FRISCH, J. L., Beschreibung von allerley Insecten in Teutsch-Land, nebst nützlichen Anmerckungen und nöthigen Abbildungen von diesem kriechenden und fliegenden inländischen Gewürme etc. Th. 1—13 [*in* 1 Vol.]. Berlin 1720—38.

1775. FUESSLIN, J. C., Verzeichniss der ihm bekannten Schweitzerischen Inseckten, mit einer ausgemahlten Kupfertafel: nebst der Ankündigung eines neuen Insecten Werkes. Zürich und Winterthur 1775.

1) According to AUDOUIN, Résumé d'Entomologie, I, p. 244 (Encyclopédie Portative) 1829, this Number was published in 1826; but the genus *Philodromus* formed in it is cited already in LATREILLE's Fam. Nat. du Règne Anim., printed in 1825.

1847?—49. Gay, C., Historia fisica y politica de Chile, segun documentos adquiridos en esta Republica duranti doze años de residencia en ella, y publicata bajo los auspicios del supremo gobierno. Zoologia. T. III, IV: Aracnidos [*by* H. Nicolet]. Paris 1849. [T. III].

† 1762. Geoffroy, E. L., Histoire abrégée des Insectes qui se trouvent aux environs de Paris. 2 Tomes. Paris 1762.

1799—1800. —id.— [*Idem liber:*] Histoire abrégée des Insectes, dans laquelle ces animaux sont rangés suivant un ordre méthodique; Nouvelle Édition, revue, corrigée et augmentée d'un supplément considérable. 2 Tomes. Paris An VII de la République Française.

1863. Giebel, C. G., Drei und zwanzig neue und einige bekannte Spinnen der Hallischen Sammlung. (Zeitschrift für die gesammten Naturwissenschaften, Bd XXI, 1863).

1867. —id.— Zur Schweitzerischen Spinnenfauna. (*ibid.*, Bd XXX.)

Gmelin, *Vid.* Linnæus, Systema Naturæ, Ed. XIII.

Gœze, *Vid.:* Lister, Naturgeschichte d. Spinnen.

1842. Graells, M. de la Paz, Notice sur divers faits qui confirment la propriété venimeuse du Latrodectus malmignatus Walckenaer. (Ann. de la Soc. Entom. de France, T. XI.)

[1842. Grube, A. E., Einige Resultate aus Untersuchungen über die Anatomie der Araneiden. (Müller's Archiv für Anatomie und Physiologie, Jahrg. 1842)].

1859. —id.— Verzeichniss der Arachnoiden Liv-, Kur- und Ehstlands. (Archiv für die Naturkunde Liv-, Ehst- und Kurlands, Ser. 2, Bd I. [*Also separate:* Dorpat 1859.]

1861. —id.— Beschreibungen neuer, von den Herren L. v. Schrenck, Maack, C. v. Ditmar u. a. im Amurlande und Ostsibirien gesammelter Araneiden. (Bulletin de l'Académie des Sciences de S:t Petersbourg, T. IV; Mélanges biologiques tirés du Bull., T. IV, 1).

1862. Günther, A., On an apparently undescribed spider from Cochin China. (Ann. and Mag. of Nat. Hist., 3 Ser., Vol. X).

Guérin (-Méneville), F. E., Dict. pittor. d'hist. nat., *Vid.* Dictionnaire pittoresque d'hist. nat.

* 1837(?) —id.— Iconographie du Règne Animal de G. Cuvier, ou représentation d'après nature de l'une des espèces les plus remarquables et souvent non encore figurées, de chaque genre d'animaux, pouvant servir d'atlas à tous les traités de Zoologie. 7 Voll. Paris 1829—44 [*Arachn.*]

1820—36. Hahn, C. W., Monographia Aranearum. Monographie der Spinnen. 8 Hefte. Nürnberg 1820—36 [Hft. 1—3: 1820—22 [1]); 4: 182..; 5: 1829; 6: 1831; 7: 1833; 8: 1836].

1831—48. —id.— *and* Koch, C. L., Die Arachniden. Getreu nach der Natur abgebildet und beschrieben. 16 Voll. Nürnberg 1831—1848 [*Voll. I—II by*

1) According to Ersch, Handbuch der Deutschen Litteratur.

HAHN, *Voll. III—XVI by* KOCH. — Vol. I: 1831; II: 1834; III: 1836; IV: 1838; V—VII: 1839; VIII: 1841; IX: 1842; X: 1843; XI—XII: 1845; XIII: 1846; XIV—XVI: 1848].

1834. HAMMERSCHMIDT, Neue Spinnen. (OKEN's Isis, 1834).

1857. HASSELT, A. W. M. VAN, [*On some rare Dutch spiders, in*] Verslag van de dertiende algemeene Vergadering der Nederlandsche Entomologische Vereeniging. (Tijdschrift voor Entomologie, Deel I, 2).

1858. —ID.— Over huid- en kleurverwisseling van Dolomedes fimbriatus (Hahn), in verband met zijne soortbepaling en die van andere spinnen uit dit geslacht. (*ibid.*, D. I, 6).

1860. —ID.— Studien over de z.g. Curaçaosche Oranje-Spin, eene nog weinig bekende Latrodectus-soort. (*ibid.*, D. III).

1865. —ID.— [*Notice of* VINSON, Aranéides des Iles de la Réun., Maurice et Madagascar, *in*] Verslag van de twintigste algemeene Vergadering der Nederlandsche Entomologische Vereeniging gehouden te Amersfoort, den 30:sten Julij 1864. (*ibid.*, D. VIII).

1869. —ID.— [*On the occurrence of Atypus Sulzeri and Pholcus opilionoides in Holland, in*] Verslag van de drie- en twintigste algemeene Vergadering der Nederlandsche Entomologische Vereeniging. (*ibid.*, D. XII).

HEINEKEN, C., *and* LOWE, *Vid.* LOWE.

1832. HENTZ, N. M., On North-American spiders. (SILLIMAN's American Journal of Science and Arts, Vol. XXI).

1841. —ID.— Description of an American spider, constituting a new sub-genus of the tribe Inæquitelæ of Latreille. (*ibid.*, Vol. XLI).

1842—50. —ID.— Descriptions and figures of the Araneides of the United States. (Boston Journal of Natural History, Vol. IV, 1 (1842) [*Extr. in the* Proceedings of the Boston Society of Natural History, I, 1844]; Vol. IV, 2 (1842); Vol. IV, 3 (1843); Vol. V, 2 (1845); Vol. V, 3 (1846); Vol. V, 4 (1847); Vol. VI, 1, 2 (1850)).

HERKLOTZ, J. A., Bouwstoffe v. eene Fauna v. Nederland, *Vid.* SIX, Lijst van Spinnen.

HERRICH-SCHÆFFER, G. A. W., Deutschl. Ins., *Vid.* PANZER, Faunæ Ins. Germ. Initia.

1867. KEMPELEN, L. v., Bemerkungen über Spinnen im Allgemeinen und eine Untersuchung von Drassus lapidicola insbesondere. (Verhandl. d. k. k. zool.-bot. Gesellsch. in Wien, Bd XVII, 1867).

1867. —ID.— Thysa pythonissæformis. Eine neue Gattung and Art. (*ibid.*).

1849. KESSLER, K., Beitrag zur Naturgeschichte und Anatomie der Gattung Lycosa. (Bulletin de la Société Impériale des Naturalistes de Moscou, XXII, 1849, II).

1862. KEYSERLING, E. v., Beschreibung einer neuen Spinne aus den Höhlen von Lesina. (Verhandl. d. zool.-bot. Gesellsch. in Wien, Bd XII, 1862).

1863. —ID.— Beschreibungen neuer Spinnen. (*ibid.*, Bd XIII, 1863).

1864. KEYSERLING, E. v., Beschreibungen neuer und wenig bekannter Arten aus der Familie Orbitelæ Latr. oder Epeïridæ Sund. (Sitzungsberichte der Isis zu Dresden, 1863). *Also separate:* Dresden 1864.

1865. —ID.— Beiträge zur Kenntniss der Orbitelæ Latr. (Verhandl. d. zool.-bot. Gesellsch. in Wien, Bd XV, 1865).

KOCH, C. L., *in* HERR.-SCHÆFF., Deutschl. Insekten, *Vid.* PANZER, Faunæ Ins. Germ. initia.

1835—41. —ID.— Deutschlands Crustaceen, Myriapoden und Arachniden. Ein Beitrag zur deutschen Fauna. Herausgegeben von HERRICH-SCHÆFFER. 40 Hfte.

—ID.— Die Arachniden, *Vid.* HAHN *and* KOCH, Die Arachniden.

1837, 50. —ID.— Uebersicht des Arachniden-Systems. 5 Hfte. Nürnberg 1837—1850. [Hft 1: 1837; Hft 5: 1850].

—ID.— Arachn. d. Regentsch. Algier, *Vid.* WAGNER, Reisen.

1847. —ID.— System der Myriapoden, mit den Verzeichnissen und Berichtigungen zu Deutschlands Crustaceen, Myriapoden und Arachniden, Hft. 1—40. Regensburg 1847. — *Also with the title:* Kritische Revision der Insecten-Fauna Deutschlands von D:r Panzer und D:r Herrich-Schæffer, Hft. 1—190. III Bändchen, enthaltend die Verzeichnissen und Berichtigungen zu Deutschlands Crustaceen, Myriapoden und Arachniden, und ein System der Myriapoden, von C. L. KOCH.

1855. KOCH, L., Zur Charakteristik des Artenunterschiedes bei den Spinnen im Allgemeinen und insbesondere der Gattung Amaurobius. (Korrespondenz-Blatt des zoologisch-mineralogischen Vereines in Regensburg, 9 Jahrg., 1855).

—ID.— Die Thiere Andalusiens, Arachn., *Vid.* ROSENHAUER.

1862. —ID.— Zur Arachniden-gattung Tetragnatha Walck. (Korresp.-Blatt d. zool.-miner. Vereins in Regensburg, 16 Jahrg., 1862).

1864. —ID.— Die europäischen Arten der Arachnidengattung Cheiracanthium. (Abhandlungen der Naturhistorischen Gesellschaft zu Nürnberg vom Jahre 1864).

1865. —ID.— Beschreibungen neuer Arachniden und Myriopoden. (Verhandl. d. zool.-bot. Gesellsch. in Wien, Bd XV, 1865).

1866, 67. —ID.— Die Arachniden-Familie der Drassiden. Nürnberg 1866—18.. [Hft. 1—6 (p. 1—304): 1866; Hft 7 (p. 305—352): 1867. [*The work is still in progress*].

1867. —ID.— Beschreibungen neuer Arachniden und Myriapoden. (Verhandl. d. zool.-bot. Gesellsch. in Wien, Bd XVII, 1867).

1867. —ID.— Zur Arachniden- und Myriapoden-Fauna Süd-Europas. (*ibid.*)

1868. —ID.— Die Arachnidengattungen Amaurobius, Cælotes und Cybæus. (Abhandl. d. Naturhist. Gesellsch. in Nürnberg, 1868). *Also separate:* Nürnberg 1868.

1857. Kolenati, F. A., Meletemata Entomologica. Fasc. VII. Einige Arachniden der Caucasischen Länder u. s. w. (Bulletin de la Soc. Imp. des Naturalistes de Moscou, T. XXX, Année 1857, N:o 2).

Kongl. svenska fregatten Eugenies resa, *Vid.* Eugenies resa.

Kritische Revision d. Ins.-Fauna Deutschl., *Vid.* Koch, C. L., System der Myriapoden.

1837. Krynicki, J., Arachnographiæ Rossicæ decas prima. (Bull. de la Soc. Imp. des Natur. de Moscou, Année 1837. N:o V).

1818. Lamarck, J. B. de, Histoire naturelle des Animaux sans Vertèbres, présentant les caractères généraux et particuliers de ces animaux, leurs distribution, leurs classes, leurs familles, leurs genres, et la citation des principales espèces qui s'y rapportent; précédée d'une introduction offrant la détermination des caractères essentiels de l'animal, sa distinction du végétal et des autres corps naturels, enfin l'exposition des principes fondamentaux de la zoologie. 7 Voll. Paris 1815—22. [Tome V].

1838. Lambotte, H., Notice sur le Théridion malmignatte. (Bulletins de l'Académie Royale des Sciences et Belles-Lettres de Bruxelles, Année 1837, T. IV).

1798. Latreille, P. A., Extrait d'un mémoire sur la famille des Araignées mineuses. (Bulletins des Sciences, par la Société Philomatique, T. I, ii, N:o 22).

1798. —id.— Description d'une nouvelle espèce d'araignée. (*ibid.*).

1802. —id.— Histoire naturelle des Fourmis, et recueil de mémoires et d'observations sur les abeilles, les araignées, les faucheurs et d'autres insectes. Paris An X—1802. (p. 332—353: Mémoire sur une nouvelle distribution méthodique des Araignées; lu la Société Philomatique).

1804. —id.— Histoire Naturelle, générale et particulière des Crustacés et des Insectes. Ouvrage faisant suite aux oeuvres de Leclerc de Buffon, et partie du cours complet d'histoire naturelle rédigé par C. S. Sonnini. 14 Voll. Paris An X—XIII. [Vol. VII: An XII].

—id.— *in* Nouv. Dict. d'hist. nat. (Tabl. méthod. d. Ins.), *Vid.* Dictionnaire, nouv., d'hist. nat.

1806, 9. —id.— Genera Crustaceorum et Insectorum, secundum ordinem naturalem in familias disposita, iconibus exemplisque plurimis explicata. 4 Voll. Parisiis et Argentorati 1806, 1807, 1809. [T. I, IV].

1810. —id.— Considérations générales sur l'ordre naturel des animaux composant les classes des Crustacées, des Arachnides et des Insectes. Paris 1810.

—id.— Arachnides du Règne Animal de Cuvier, *Vid.* Cuvier, Le Règne Animal, Nouv. Éd.

1824. —id.— Note sur un nouveau genre d'Aranéides (Ann. d. Sciences Nat., T. III).

1825. —id.— Familles naturelles du Règne Animal, exposées succinctement et dans un ordre analytique, avec l'indication de leurs genres. Paris 1825.

1827. LATREILLE, P. A., Natürliche Familien des Thierreichs. Aus dem Französischen. Mit Anmerkungen und Zusätzen von A. A. BERTHOLD. Weimar 1827.

—ID.— Arachnides du Règne Animal de Cuvier (Les Crustacées, les Arachnides et les Insectes distrib. en fam. nat.), *Vid.* CUVIER, Le Règne Animal, Nouv. (2e) Éd.

1831. —ID.— Cours d'Entomologie, ou de l'histoire naturelle des Crustacés, des Arachnides, des Myriapodes et des Insectes, à l'usage des élèves de l'école du Muséum d'histoire naturelle. Première année: Discours d'ouverture du cours. Tableau de l'histoire de l'entomologie. Généralités de la classe des Crustacés et de celle des Arachnides, des Myriapodes et des Insectes. Exposition méthodique des ordres, des familles et des genres des trois premières classes. (Avec un Atlas composé de 24 Planches). Paris 1831.

1832. —ID.— Vues générales sur les Aranéides à quatre pneumobranches ou quadripulmonaires, suivies d'une notice de quelques éspèces de Mygales inédites et de l'habitation de celle qu'on nomme *Nidulans.* (Nouvelles Annales du Muséum d'histoire naturelle, T. I).

1770. LAXMANN, E., Novæ Insectorum species. (Novi Commentarii Academiæ Scientiarum Imperialis Petropolitanæ. T. XIV, pro anno 1759, Pars prior).

1815. LEACH, W. E., Zoological Miscellany; being descriptions of new and interesting animals. 3 Voll. London 1814—1817. [Vol. II].

—ID.— Edinb. Cyclop., *Art.* Crustaceology, *Vid.* Edinburg Cyclopedia.

† 1774. LEPECHIN, J., Tagebuch der Reise durch verschiedene Provinzen des Russischen Reiches in den Jahren 1768 und 1769. Aus dem Russischen übersetz von C. H. HAASE. 3 Thle. Altenburg 1774, 1775, 1783. [Vol. I. — *The Vol. I of the Russian original is printed in* 1771].

[1848. LEUCKART, R., Ueber den Bau und die Bedeutung der sogenannten Lungen bei den Arachniden (Zeitschrift für wissenschaftliche Zoologie, Bd I)].

[1867. LINCECUM, G., The Tarentula. (The American Naturalist, I, N:o 8)].

† 1735. LINNÆUS [v. LINNÉ], Systema Naturæ sive regna tria naturæ systematice proposita, per classes, ordines, genera et species. Lugduni Batavorum 1735.

1758. —ID.— *idem liber*, Ed. X, reformata. 2 Tomi. Holmiæ 1758. [T. I].

1767. —ID.— *idem liber*, Ed. XII, reformata. 3 Tomi. Holmiæ 1766—1768. [T. I, Pars II].

1789? —ID.— *idem liber*, Ed. XIII, aucta, reformata. Cura J. F. GMELIN. 3 Tomi *in* 10 Voll. Lipsiæ 1788—1793. [T. I, Pars V].

† 1742. —ID.— Animalia per Sueciam observata. (Acta Literaria et Scientiarum Sueciæ. Vol. IV, continens annos 1735—1739). [Ann. 1736].

† 1745. —ID.— Öländska och Gothländska Resa, på Riksens högloflige Ständers befallning förrättad åhr 1741. Stockholm och Upsala 1745.

† 1746. LINNÆUS [v. LINNÉ], C., Fauna Suecica sistens animalia Sueciæ regni: quadrupeda, aves, amphibia, pisces, insecta, vermes, distributa per classes et ordines, genera et species. Cum differentiis specierum, synonymis auctorum, nominibus incolarum, locis habitationum, descriptionibus insectorum. Stockholmiæ 1746.

1761. —ID.— *idem liber*, Ed. II, auctior. Stockholmiæ 1761.

* 1776. —ID.— Vollständiges Natursystem, nach der 12:ten lateinischen Ausgabe und nach Anleitung des Holländ. Houttuyns'schen Werks, mit einer ausführlichen Erklärung ausgefertiget von P. L. S. MÜLLER. 6 Thle in 9 Bdn. Nürnberg 1773—1776. [Supplement- und Register-Band].

—ID.—, Entomologia, cur. et aug. C. DE VILLERS, *Vid.* VILLERS, Linnæi Entom.

† 1678. LISTER, M., Historiæ animalium Angliæ tres tractatus. Unus de araneis. Alter de cochleis tum terrestribus tum fluviatilibus. Tertius de cochleis marinis. Quibus adjectus est quartus de lapidibus ejusdem insulæ ad cochlearum quandam imaginem figuratis. Londini 1678.

† 1778. —ID.— Naturgeschichte der Spinnen überhaupt und der Engelländischen Spinnen insonderheit, aus dem Lateinischen übersetzt, und mit Anmerkungen vermehrt von F. H. W. MARTINI, nach dessen Handschrift aber zum Druck befördert und mit neuen Zusätzen vermehrt von J. A. E. GŒZE. Quedlinburg und Blankenburg 1778.

1831. LOWE, R. T., Descriptions of two species of Araneidæ, natives of Madeira. In a letter to the editor. (Zoological Journal, Vol. V).

1836. LUCAS, H., Observations sur les Aranéides du genre Hersilia, et description de deux espèces nouvelles appartenant à ce genre. (GUÉRIN, Magazin de Zoologie, 6:e Année).

1836. —ID.— Quelques observations sur le genre Atypus et description d'une espèce nouvelle appartenant à ce genre. (Ann. de la Soc. Ent. de France, T. V).

—ID.—, *in* Dict. univ. D'ORBIGNY, *Vid.* Dictionnaire universelle d'hist. nat.

* 1840. —ID.— Histoire naturelle des Crustacés, des Arachnides et des Insectes Thysanoures, faisant suite au Buffon Duménil. Paris 1840.

1843. —ID.— Note sur le Latrodectus malmignathus. (Ann. de la Soc. Ent. de France, 2 Sér. T. I, Bull.).

—ID.— Arachn. des Iles Canaries, *Vid.* BARKER-WEBB.

1844. —ID.— Note monographique sur les Aranéides composant le genre Tegenaria. (Ann. de la Soc. Ent. de France, 2 Sér., T. II).

1845. —ID.— Clotho Durandi trouvé a Nîmes. (*ibid.*, 2 Sér., T. III, Bull.).

—ID.— Arachn. de l'Algérie, *Vid.* Exploration scientifique de l'Algérie.

[1847. —ID.— Sur une ponte d'une Scytodes thoracica Latr. (Ann. de la Soc. Ent. de France, 2 Sér., T. V, Bull.)].

1847. —ID.— Epeira diadema. Sur une variété remarquable de cette Aranéide. (*ibid.*, Bull.).

1847. LUCAS, H., Episinus truncatus. Note sur cette Aranéide. (*ibid.*, Bull.).

1847. —ID.— Latrodectus martius trouvé en France. (*ibid.*, Bull.).

1847. —ID.— Scytodes thoracica. Note au sujet de cet insecte. (*ibid.*, Bull.).

1849. —ID.— Description et figure d'une nouvelle espèce d'Aranéide, appartenant au genre Theridion. (*ibid.*, 2 Sér., T. VII).

1849. —ID.— Salticus formicæformis n. sp. (*ibid.*, Bull.; GUÉRIN-MÉNEVILLE, Revue et Magazin de zoologie, 2 Sér., T. II, 1850).

1849. —ID.— Eresus cinnabarinus trouvé à Passy. (Ann. de la Soc. Ent. de France, 2 Sér., T. VII, Bull.).

1850. —ID.— Histoire naturelle des Crustacés, des Arachnides et des Myriapodes. Précédée de l'histoire naturelle des Annélides par M. le Comte de CASTELNAU. Paris 1850.

1851. —ID.— Observations géographiques sur la Filistata bicolor. (Ann. de la Soc. Ent. de France, 2 Sér., T. IX, Bull.).

1851. —ID.— Observations sur le jeune âge de la Segestria perfida. (*ibid.*, Bull.).

1853. —ID.— Note sur une variété remarquable de l'Epeira scalaris. (*ibid.* 3 Sér., T. I, Bull.).

1853. —ID.— Essai sur les animaux articulés qui habitent l'île de Crète. Revue et Mag. de zool., 2 Sér., T. V (1853), VI (1854). — *Also separate.*

1855. —ID.— Note sur une nouvelle espèce d'Aranéide qui habite l'Espagne méridionale. (Ann. de la Soc. Ent. de France, 3 Sér., T. III).

* 1857. —ID.— Note sur la rétractilité ou la non-rétractilité des ongles dans les tarses des Aranéides du genre Mygale. (Comptes rendus hebdomadaires des séances de l'Académie des Sciences, T. XLV; l'Institut, T. XXVI, N:o 1253).

1859. —ID.— De la manière de vivre, de l'habitat et de la synonymie chronologique de l'Oletera picea. (Ann. de la Soc. Ent. de France, 3 Sér., T. VII, Bull.).

1859. —ID.— Sur les mandibules de la Segestria florentina. (*ibid.*, Bull.).

1860. —ID.— Quelques remarques sur la manière de vivre de la Segestria florentina, Aranéide de la tribu des Quadripulmonaires (*ibid.*, T. VIII).

1863. —ID— Note sur la rétractilité ou la non-rétractilité des ongles des palpes dans les Aranéides du genre Mygale. (*ibid.*, 4 Sér., T. III).

1863. —ID.— Note sur une variété de la Segestria florentina. (*ibid.*).

1864. —ID.— Atypus piceus de Sulzer rencontré aux environs de Fontainebleau. (*ibid.*, 4 Sér., T. IV, Bull.).

1864. —ID.— Note sur des espèces d'Aranéides sous les noms d'Eresus albomarginatus, pulchellus et siculus. (*ibid.*, Bull.).

1865. —ID.— Observations sur le genre Eriodon, Aranéide de la tribu des Théraphoses, précédées de quelques remarques sur les coupes génériques qui composent actuellement cette tribu. (*ibid.*, 4 Sér., T. V).

1834. M...., C., An illustration of the structure of some of the organs of a spider deemed the type of a new genus, and proposed to be called Trichopus libratus. (LOUDON's Magazine of Natural History, Vol. VII).

1839. MAC LEAY, W. S., On some new forms of Arachnida. (Ann. of Nat. Hist., Vol. II).

* 1824. MAIRONI DA PONTE, G., I tre regni della natura nella provincia Bergamasca. (Atti della Società Italiana di Scienze, T. XIX, Fasc. 2° di Fisica). — *Also separate:* Modena 1824.

* 1800. MARMOCCHI, F., Memoria sopra il ragno rosso dell' agro Volterrano. (Atti dell' Academia dei Fisiocritici di Siena, T. VIII).

MARTINI, *Vid.*, LISTER, Naturgesch. d. Spinnen.

1793. MARTYN, TH., Aranei, or a natural history of spiders, including the principal parts of the well-known work on English spiders by Eleazar Albin, as also the whole of the celebrated publication on Swedish spiders by Charles Clerck; revised, enlarged and designed a new. 2 Voll. London 1793.

1861. MEADE, R. H., Description of a new species of spider lately discovered in England. (Ann. and Mag. of Nat. Hist., 3 Ser., Vol. VII).

[1843. MENGE, A., Ueber die Lebensweise der Arachniden. (Neueste Schriften der Naturforschenden Gesellschaft in Danzig, Bd. IV, Hft. 1)].

1850. —ID.— Verzeichniss der Danziger Spinnen. (*ibid.*, Bd IV, Hft. 3).

1866, 68, 69. —ID.— Preussische Spinnen. [I Abtheilung, p. 1—152:] (Schriften d. Nat.-forsch. Gesellsch. in Danzig, Neue Folge, Bd I, Hft. 3, 4); II Abtheil. [p. 153—218:] (*ibid.*, Bd II, Hft. 1); III Abtheil. [p. 219—264:] (*ibid.*, Bd II, Hft. 2). — *Also separate:* Danzig 1866, 1868, 69. — *The work is still in progress.*

1849. MENZEL, A., Kurzer Abriss einer Naturgeschichte der Spinnen. Ein Festgeschenk für die Jugend. Zürich 1849.

1790. MEYER, F. A. A., Ueber einige Spinnen der Göttingischen Gegend. Nebst Anzeige eines vollständigen Cursus über die Thiergeschichte. Göttingen 1790.

1849. MOTSCHOULSKY, V. DE, Note sur deux araignées venimeuses de la Russie méridionale qu'on croit être le Tchim des Kalmouks. (Bull. de la Soc. Imp. d. Natur. de Moscou, T. XXII, Année 1849. N:o 1).

1764. MÜLLER, O. F., Fauna Insectorum Fridrichsdalina, sive methodica descriptio insectorum agri Fridrichsdalensis, cum characteribus genericis et specificis, nominibus trivialibus, locis natalibus, iconibus allegatis, novisque pluribus speciebus additis. Hafniæ et Lipsiæ 1764.

1776. —ID.— Zoologiæ Danicæ Prodromus, seu animalium Daniæ et Norvegiæ indigenarum characteres, nomina et synonyma imprimis popularium. Havniæ 1776.

MÜLLER, P. L. S., Linnæi Vollst. Nat.-Syst., *Vid.* LINNÆUS, Vollständiges Natur-System.

Museum Leskeanum. *Vid.* ZSCHACH, Mus. Lesk.

* 1845. Napoli e le sue vicinanze. Vol. I. Napoli 1845.

NICOLET, H., Aracn. de Chile, *Vid.* GAY, Hist. fisica y polit. de Chile.

1863. NORDMANN, A. v., Erstes Verzeichniss der in Finnland und Lappland gefundenen Spinnen, Araneæ. Vorgetragen in der finnischen Wissenschafs-Societät d. 1 Dec. 1862. (Bidrag till Finlands naturkännedom, etnografi och statistik, Bd VIII).

Nouv. Dict. d'Hist. Nat., *Vid.* Dictionnaire, Nouveau, d'Hist. Nat.

1851. OHLERT, E., Beiträge zur Diagnose und Revision der Preussischen Spinnengattungen. (Programm der höheren Burgschule zu Königsberg i. Pr., April 1851. Königsberg 1851).

1854. —ID.— Beiträge zu einer auf die Klauenbildung gegründeten Diagnose und Anordnung der Preussischen Spinnen. (Verhandl. d. zool.-bot. Gesellsch. in Wien, Bd IV, 1854).

1865. —ID.— Arachnologische Studien. (Programm der Real-Schule auf der Burg zu Königsberg i. Pr., Sept. 1865. Königsberg 1865).

1867. —ID.— Die Araneiden oder echten Spinnen der Provinz Preussen. Leipzig 1867.

† 1772. OLAFSEN, E., Eggert Olafsens og Biarne Povelsens Reise igiennem Island, foranstaltet af Videnskabernes Sælskab i Kiøbenhavn. 2 Voll. Sorøe 1772.

OLIVIER, A. G., Encycl. Méth., *Vid.* Encyclopédie méthodique.

[1862. PACKARD JR, A. S., Entomological Report. (Second annual report upon the natural history and geology of the State of Maine. [Augusta] 1862)].

1771, 73. PALLAS, P. S., Reise durch verschiedene Provinzen des Russischen Reichs 3 Voll. St. Petersburg 1771, 1773, 1776 [Vol. I, II].

1772. —ID.— Spicilegia zoologica. Tomus I, continens quadrupedium, avium, amphibiorum, piscium, insectorum, molluscorum aliorumque marinorum Fasciculos decem. Berolini 1767—1774. [Fasc. 9, *with the title:* Spicilegia zoologica, quibus novæ imprimis et obscuræ animalium species iconibus, descriptionibus atque commentariis illustrantur].

1777. —ID.— Naturgeschichte merkwürdiger Thiere. Durch den Verfasser verteutscht. Bd I, 1—10 Sammlung. Berlin und Stralsund 1769—1778 [9:te Samml.].

1793—1844. PANZER, C. F. W., Faunæ insectorum Germaniæ initia. Deutschlands Insecten. Hft. 1—110. Regensburg 1793—1809. [Hft 1—12: 1793; 13—24: 1794; 25—36: 1796; 37—48: 1797; 49—60: 1798; 61—72: 1799; 73—84: 1801; 85—96: 1804; 97—106: 1806; 107—110: 1809 [1]]. Fortgesetzt von G. W. HERRICH-SCHÆFFER. [*Arachn. by* C. L. KOCH]. Hft. 111—190. Regensburg 1829—1844. [Hft. 111—118: 1829—1832; 119—121: 1833; 122—127: 1834; 128—133: 1835; 134—158: 1836—1838; 159—190: 1839—1844 — *probably* [2]].

1) According to SCHINER, Fauna Austriaca. Die Fliegen, I, p. XXVI.

2) Compare the "Leipziger Messen-Cataloge" for the years mentioned. — I have only had opportunity to consult a part of this work.

1804. PANZER, G. F. W., D. Jacobi Christiani Schæfferi Iconum insectorum circa Ratisbonam indigenorum enumeratio systematica. Systematische Nomenclatur über weiland Herrn Dr. Jacob Christian Schäffers natürlich ausgemahlte Abbildungen Regensburgischer Insekten. Erlangen 1804.

* 1864. PAVESI, P., Notizie naturali e chimico-agronomiche sulla provincia di Pavia. (Aracnidi).

—ID.—, *Vid.* CANESTRINI *and* PAVESI.

1833. PERTY, M., Delectus animalium articulatorum quæ in itinere per Brasiliam ann. 1817—1820 . . . peracta collegerunt J. B. de Spix et de Martius. Digessit, descripsit, pingenda curavit Max. Perty. 3 Fasc. Monachi 1830—1834 [Fasc. 3].

1786. PETAGNA, V., Specimen insectorum Ulterioris Calabriæ. Neapoli 1786.

* 1792. —ID.— Institutiones entomologicæ. 2 Voll. Neapoli 1792.

† 1702—11. PETIVER, J., Gazophylacii naturæ et artis Decades X. London 1702—1711.

PICKARD-CAMBRIDGE, *Vid.* CAMBRIDGE.

* 1761. PODA, N., Insecta Musei Græcensis. Græcii 1761.

1781. POIRET, Sur quelques insectes de Barbarie. Suite. (Observations sur la physique, sur l'histoire naturelle et sur les arts [Journal de Physique], T. XXXI).

PONTE, DA, *Vid.* MAIRONI DA PONTE.

1765. PONTOPPIDAN, E., Kurzgefasste Nachrichten, die Naturhistorie in Dännemark betreffend. Aus dem Dänischen übersetzt. Kopenhagen und Hamburg 1765.

1866. PRACH, H., Monographie der Thomisiden (Krabbenspinnen) der Gegend von Prag, mit einem Anhange, das Verzeichniss der bisher in der Umgebung unserer Hauptstadt aufgefundenen Araneinen enthaltend. (Verhandl. d. zool.-bot. Gesellsch. in Wien, Bd XVI, 1866).

QUOY et GAIMARD, *Vid.* Voyage de la corvette l'Astrolabe.

1821. RAFINESQUE, C. S., Description d'une araignée qui constitue un genre nouveau. (Ann. géner. d. Sciences Phys., T. VIII).

1839. RAIKEM, A., Recherches, observations et expériences sur le Theridion marmignatte de Volterra et sur les effets de sa morsure. (Ann. d. Sciences Nat., 2 Sér., T. XI).

1833. RATZEBURG, J. T. C., (BRANDT, J. F., *and*) Medizinische Zoologie, oder getreue Darstellung und Beschreibung der Thiere, die in der Arzneimittellehre in Betracht kommen, in systematischer Folge herausgegeben. 2 Bd. Berlin 1827—34. [Bd II].

1787. RAZOUMOWSKI, G. DE, Lettre de M. le Comte de Razoumowski à M. Reynier sur une Araignée. (Observ. s. la Phys. [Journ. de Phys.], T. XXXI).

* 1789. —ID.— Histoire naturelle du Jorat et de ses environs, et celle des trois lacs de Neufchâtel, Morat et Bienne; précédé d'un essai sur le climat, les productions, le commerce et les animaux de la partie du pays de Vaud ou de la Suisse romande qui entre dans le plan de cet ouvrage. 2 Voll. Lausanne 1789.

1783. RETZIUS, A. J., Caroli de Geer genera et species insectorum e generosissimi auctoris scriptis extraxit, digessit, latine quod partem reddidit, et terminologiam insectorum Linnæanam addidit. Lipsiæ 1783.

1834. REUSS, A., Zoologische Miscellen. Arachniden. (Museum Senckenbergianum, Bd I).

1857. RINK, H., Grønland, geografisk og statistisk beskrevet (Tillæg N:o 3: Udsigt over Grønlands Land-, Ferskvands- og Strandbreds-Arthropoder, ved J. C. SCHIØDTE). Kiøbenhavn 1857.

1827. RISSO, A., Histoire naturelle des principales productions de l'Europe méridionale. 5 Voll. Paris et Strassbourg 1826—1827 [T. V].

1862. ROBERTSON, J., A new British Mygale. (Brighton Herald 1862; Ann. and Mag. of Nat. Hist., 3 Ser., Vol. X).

1789. RŒMER, J. J., Genera insectorum Linnæi et Fabricii iconibus illustrata. Vitoduri Helvetorum 1789.

† 1761. RŒSEL V. ROSENHOF, A. J., Monatlich herausgegebene Insectenbelustigungen. 4 Voll. Nürnberg 1746—1761. [Vol. IV].

1856. ROSENHAUER, W. G., Die Thiere Andalusiens nach dem Resultate einer Reise zusammengestellt, nebst den Beschreibungen von 249 neuen oder bis jetzt noch unbeschriebenen Gattungen und Arten. Erlangen 1856. [*Arachn. by* L. KOCH].

* 1778. ROSSI, P., Osservazioni insettologiche. (Memorie di Matematica e Fisica della Società Italiana, Vol. IV).

1790. —ID.— Fauna Etrusca, sistens insecta quæ in provinciis Florentina et Pisana præsertim collegit. 2 Voll. Liburni 1790. [T. II].

* 1794. —ID.— Mantissa insectorum, exhibens species nuper in Etruria collectas, adjectis Faunæ Etruscæ illustrationibus et additionibus. 2 Voll. Pisis 1792, 1794. [T. II].

1846. ROSSI, F. W., Neue Arten von Arachniden des K. K. Museums, beschrieben und mit Bemerkungen über verwandte Formen begleitet. (HAIDINGER, Naturwissenschaftliche Abhandlungen, Bd I. Wien 1846).

SAVIGNY, *Vid.* Description de l'Égypte.

† 1767—69. SCHÆFFER, J. C., Icones insectorum circa Ratisbonam indigenorum, coloribus naturam referentibus expressæ. Natürlich ausgemahlte Abbildungen Regensburgischer Insecten. 3 Voll. Regensburg 1767—1769.

* 1804. *Idem liber:* Nova Editio, methodo systematico aucta a G. W. F. PANZER. 4 Voll. Erlangæ 1804. [*Conf.* PANZER, Schæfferi Ic. ins. circa Ratisb. indig. enum. system.].

1847. SCHIÖDTE, J. C., Forelöbig Beretning om Undersögelser om den underjordiske Fauna i Hulerne i Krain og Istrien. (Oversigt af det Kongelige Danske Videnskabernes Selskabs Forhandlinger 1847).

1849. —ID.— Bidrag til den underjordiske Fauna. (Det Kongelige Danske Videnskabernes Selskabs Skrifter, 5 Række, Naturvid. o. mathem. Afd., 1851, Bd. II). — *Also separate.*

1849. —ID.— Om en afvigende Slægt af Spindlernes Orden. (KRØYER, Naturhistorisk Tidskrift, Ny [2] Række, Bd II).

SCHIÖDTE, J. C, Udsigt over Grønlands Land-, Ferskvands- og Strandbreds-Arthropoder, *Vid.* RINK, Grønland.

1865. —ID.— Om Slægten Stalita. (Naturhist. Tidskr., 3 Række, Bd III).

1781. SCHRANCK, F. V. PAULA, Enumeratio insectorum Austriæ indigenorum. Augustæ Vindelicorum 1781.

* 1795. —ID.— Naturhistorische und ökonomische Briefe über das Donaumoor. Mannheim 1795.

1803. —ID.— Fauna Boica. Durchgedachte Geschichte der in Baiern einheimischen und zahmen Thiere. 3 Vol. Vol. I: Nürnberg 1798; Vol. II: Ingolstadt 1801, 1802; Vol. III: Landshut 1803 [Vol. III. 1].

1763. SCOPOLI, J. A., Entomologia Carniolica exhibens insecta Carnioliæ indigena et distributa in genera, species, varietates. Methodo Linnæana. Vindobonæ 1763.

1772. —ID.— Annus I—V Historico-naturalis. 5 Voll. (Observationes zoologicæ *in* Ann. V). Lipsiæ 1769, 1770, 1772.

1849. SEIDEL, Ueber die Schlesischen Arten aus den Familien der Epeirides und Theridides. (Uebersicht der Arbeiten und Veränderungen der Schlesischen Gesellschaft für vaterländische Kultur im Jahre 1848).

[1848. SIEBOLD, C. TH. v., Lehrbuch der vergleichenden Anatomie der wirbellosen Thiere. (*Also with the title:* Lehrbuch der vergleichenden Anatomie von v. SIEBOLD und STANNIUS. Erster Theil: Wirbellose Thiere, von C. TH. v. SIEBOLD). Berlin 1848].

1861. SIEMASCKKO, J. M. v., Verzeichniss der in der Umgegend von St. Petersburg vorkommenden Arachniden. (Horæ Societatis entomologicæ Rossicæ, Fasc. I).

1861. SILL, V., Beitrag zur Kenntniss der Crustaceen, Arachniden und Myriapoden Siebenbürgens. (Verhandlungen und Mittheilungen des Siebenbürgischen Vereins für Naturwissenschaften zu Hermannstadt, Jahrg. XII, N:o 1).

1861. —ID.— Zweiter Beitrag zur Kenntniss der Crustaceen und Arachniden Siebenbürgens. (*ibid.*, N:o 11, 12).

1862. —ID.— Dritter Beitrag zur Kenntniss der Crustaceen und Arachniden Siebenbürgens. (*ibid.*, Jahrg. XIII, 1862).

1864. SIMON, E., Histoire Naturelle des Araignées (Aranéides). Paris 1864.

1866. —ID.— Monographie des espèces européennes du genre Pholcus. (Ann. de la Soc. Ent. de France, 4 Sér., T. VI).

1866. —ID.— Sur quelques araignées d'Espagne. (*ibid.*).

1867. —ID.— Sur trois araignées nouvelles. (Revue et Mag. de Zool., Année 1867).

1868. —ID.— Sur quelques Aranéides du midi de la France. (*ibid.*, 2 Sér. T. XX, Année 1868).

1868, 69. —ID.— Monographie des espèces européennes de la famille des Attides (Attidæ Sundev. — Saltigradæ Latr.). (Ann. de la Soc. Ent. de France, 4 Sér., T. VIII).

1858. SIX, G. A., Lijst van spinnen in de Provinzie Utrecht gevonden en gedetermineerd door G. A. Six. (HERKLOTZ, Bouwstoffe voor eene Fauna van Nederland, Deel II. Leiden 1858).

1858. —ID.— Opmerkingen omtrent de kleurverandering van Epeira Herii Hahn en eenige andere spinnen. (Tijdschrift v. Entomologie, D. I).

1858. —ID.— [Aanvulling van de lijst van inlandsche spinnen, *in*] Verslag van de veertiende algemeene Vergadering der Nederlandsche Entomologische Vereeniging. (*ibid.*, D. II).

1863. —ID.— Nieuwe bijdrage tot de kennis der inlandsche spinnen. (*ibid.*, D. VI).

Statistica fis. ed econ. dell' isola di Capri, *Vid.* AMARY.

1866. STAVELEY, E. F., British Spiders: an introduction to the study of the Araneidæ of Great Britain and Ireland. London 1866.

1810. STRACK, C. F. L., Einige selbstgemachte Beobachtungen über den Sommerflug und die Spinne, die ihn hervorbringt. (Neue Schriften d. Naturforsch. Gesellsch. zu Halle, Heft 5).

1765, 68. STRØM, H., Beskrivelser over Norske Insekter. Første Stykke. (Det Trondhiemske Selskabs Skrifter, Deel III, 1765). Andet Stykke. (*ibid.*, D. IV, 1768),

* 1776. SULZER, J. H., Abgekürzte Geschichte Schweitzerischer und ausländischer Insekten, nach Linnéischer Form, in 32 Tafeln und erläuternden Vignetten. 2 Voll. Winterthur 1776. — *Idem liber: Supplement*, *Vid.* RŒMER.

1823. SUNDEVALL, C. J., Specimen academicum genera Araneidum Sueciæ exhibens. Lundæ 1823.

1830, 32, 33. —ID.— Svenska Spindlarnes Beskrifning. (Kongl. Vetenskaps-Akademiens Handlingar för 1829, 1831, 1832).

1833. —ID.— Conspectus Arachnidum. Dissertatio academica. Londini Gothorum 1833.

1841. —ID.— Årsberättelser om nyare zoologiska arbeten och upptäckter, till Kongl. Vetenskaps-Akademien afgifne för åren 1837—1840. Stockholm 1841.

1844. TELLKAMPF, TH., Beschreibung einiger neuen in der Mammuth-Höhle in Kentucky aufgefundenen Gattungen von Gliederthieren. (Archiv für Naturgeschichte, 10 Jahrg., 1844, Bd I).

1834. TEMPLETON, R., On the spiders of the genus Dysdera, Latr., with the description of a new allied genus. In a letter to the Editor. (Zoological Journal, Vol. V: 1832—1834).

1855. THORELL, T., Om hanen af Scytodes thoracicus Latr. (Öfversigt af Kongl. Vetenskaps-Akademiens Förhandlingar, Årg. XI, 1854).

1855. —ID.— Recensio critica aranearum suecicarum, quas descripserunt Clerckius, Linnæus, De Geerus. P. I. Dissert. Academica. Upsaliæ 1855.

1856. —ID.— Recensio critica aranearum suecicarum, quas descripserunt Clerckius, Linnæus, De Geerus. (Nova Acta Regiæ Societatis Scientiarum Upsaliensis, Ser. 3, Vol. II, Pars Prior. 1856). *Also separate:* Upsaliæ 1856.

1858. THORELL, T., Om Clercks original-spindelsamling. (Öfvers. af K. Vet.-Akad:s Förhandl., Årgång XV, 1858).

1858. —ID.— Till kännedomen om slägtena Mithras och Uloborus. (*ibid.*).

1858. —ID.— Om Epeira marmorea och pyramidata. (*ibid.*).

1860. —ID.— Nya exotiska Epeirider. (*ibid.*, Årg. XVI, 1859).

1868. —ID.— Om Aranea lobata Pallas (A. sericea Oliv.). (*ibid.*, Årg. XXIV, 1867).

—ID.— Eugenies Resa, Arachnider (Araneæ. Species novæ minusve cognitæ), *Vid.* Eugenies Resa omkr. jorden.

—ID.— [*Arachn. from Gotska Sandön*] *Vid.* EISEN and STUXBERG.

* 1794. TOTI, L., Memoria fisico-medica sopra il falangio o ragno venefico dell' agro Volterrano. (Atti dell' Academia dei Fisiocritici di Siena, T. VII).

* 1847. Venezia e le sue lagune. 2 Voll. Venezia 1847. [*Arachn. by* CONTARINI, *in* Vol. II].

VIGORS, N. A., *Vid.* TEMPLETON, On the spid. of the gen. Dysdera.

1789. VILLERS, C. DE, Caroli Linnæi Entomologia, Faunæ Suecicæ descriptionibus aucta; D. D. Scopoli, Geoffroy, De Géer, Fabricii, Schranck, etc. speciebus vel in Systemate non enumeratis vel nuperrime detectis, vel speciebus Galliæ australis locupletata, generum specierumque rariorum iconibus ornata, curante et augente Carolo de Villers. 4 Voll. Lugduni 1789. — *With a Number of Plates under the title:* Nomenclator iconum entomologiæ Linnæanæ ,curante et augente Car. de Villers.

1863. VINSON, A., Aranéides des îles de la Réunion, Maurice et Madagascar. Paris 1863.

* 1826—34. Voyage de la corvette l'Astrolabe, éxécuté par ordre du Roi pendant les années 1826—1829 sous le commandement de M. Dumont d'Urville. Zoologie, par QUOY et GAIMARD. 4 Voll. Paris 1826—1834.

1841. WAGNER, M., Reisen in der Regentschaft Algier, in den Jahren 1836, 1837 und 1838. Mit einem naturhistorischen Anhang und einem Kupferatlas. 3 Voll. Leipzig 1841. (Arachniden und Myriapoden der Regentschaft Algier bearbeitet vom Forstrath KOCH in Regensburg, *in* Vol. III). — *Vol. III also with the title:* Bruchstücke zu einer Fauna der Berberei, mit besonderer Rücksicht auf die geographische Verbreitung der Thiere am Becken des Mittelmeeres, nach der von Moritz Wagner in der Regentschaft Algier gesammelten Materialien, von J. F. Brandt, M. Erdl, W. F. Erichson, C. L. Koch, H. Nathusius, E. A. Rossmässler, H. Schlegel, A. Wagner und R. Wagner. Leipzig 1841.

1802. WALCKENAER, C. A. (DE), Faune Parisienne, Insectes. Ou histoire abrégée des insectes des environs de Paris, classés d'après le systeme de Fabricius; précédée d'un discours sur les insectes en général, pour servir d'introduction à l'étude de l'entomologie. 2 Voll. Paris An XI.—1802.

1805. —ID.— Tableau des Aranéides. Paris 1805.

1806—08. —ID.— Histoire Naturelle des Aranéides. 5 Livr. Paris et Strasbourg 1806—1808. [*The work has not been continued*].

—ID.— Faune Franç., Arachn. (Aranéides de France), *Vid.* Faune Française.

1833. —ID.— Mémoire sur une nouvelle classification des Aranéides. (Ann. de la Soc. Ent. de France, T. II).

1834. WALCKENAER, C. A. DE, Synonymies de l'Aranea domestica de Lyonnet. (*ibid.*, T. III, Bull.).

1835. —ID.— Mémoire sur une nouvelle espèce de Mygale, sur les Théraphoses et sur les divers genres dont se compose cette tribu d'Aranéides. (*ibid.* T. IV).

1837, 41, 47. —ID.— (*and* GERVAIS, P.), Histoire naturelle des Insectes. Aptères. 4 Voll. 1837—1847. [T. I, II, IV].

1847. —ID.— Sur une nouvelle famille du genre Tétragnatha dans les Araignées. (Ann. de la Soc. Ent. de France, 2 Sér., T. V, Bull.).

1851. WALKER, F., List of spiders captured by F. Walker. (Ann. and Mag. of Nat. Hist., Ser. 2, Vol. VII).

* 1855. —ID.— List of spiders found at Piercefield near Chepstow. (Zoologist, 1855).

[1832. WATT, M., Observations on the Aranea geometrica, obtextrix, domestica and other spiders, and particularly on the power they possess of fixing their threads horizontally or at any degree of inclination, to two perpendicular bodies at a considerable distance from each other, so as to suspend the circular part of their web in an open space: Also some remarks on the food of spiders etc. (Memoirs of the Wernerian Natural History Society for the years 1826—1831. Vol. VI)].

1843. WESTRING, N., Om stridulationsorganet hos Asagena serratipes. (Naturhist. Tidskrift, Bd IV, 1842—1843).

[1844, 47. —ID.— Bidrag till historien om insecternas stridulationsorganer. (*ibid.*, 2 Række, Bd I, 1844—45; Bd II, Hft. 4, 1847)].

1851. —ID.— Förteckning öfver till närvarande tid kända, i Sverige förekommande spindelarter, utgörande ett antal af 253, deraf 132 äro nya för den svenska Faunan. (Götheborgs Kongl. Vetenskaps- och Vitterhets-Samhälles Handlingar, Ny Tidsföljd, Hft. 2).

[1858. —ID.— Anvisning att ändamålsenligt insamla och conservera Arachnider, förnämligast med afseende å spindlarne. (*ibid.*, Ny Tidsföljd, Hft. 4)].

[1858. —ID.— Beskrifning på stridulationsorganer hos slägtena Pachycoris Burm. och Scutellera Lamarck, af insektsordningen Hemiptera, jemte öfversigt af alla de hittills bekanta olika sätten för sådane ljuds framalstrande bland andra insekter (*ibid.*)].

1861. —ID.— Araneæ Suecicæ descriptæ. (*ibid.*, Ny Tidsföljd, Hft. 7). ***Also separate:*** Gothoburgi 1861.

1834. WESTWOOD, J. O., Insectorum Arachnoidumque novorum Decades duo. (Zool. Journ., Vol. V, 1832—1834).

1841. WHITE, A., Descriptions of new or little known Arachnida. (Ann. of Nat. Hist., Vol. VII).

1846. —ID.— Description of a new genus of Arachnida, with notes on two other species of spiders. (*ibid.*, Vol. XVIII).

1849. —ID.— Descriptions of apparently new species of Aptera from New Zealand. (Proceedings of the Zoological Society, Part XVII, 1849).

WIDER, *Vid.:* REUSS, Zool. Misc.

† 1788. ZCHACH, J. J., Museum Nath. Gtfr. Leskeanum. Pars entomologica, ad systema entomologiæ Cl. Fabricii ordinata. Lipsiæ 1788.